D1708086

A new generation of large, ground-based telescopes is just coming into operation. They will take astronomical research well into the next century. These extremely powerful telescopes demand specially designed instruments and observing techniques. The VII Canary Islands Winter School of Astrophysics gathered together leading experts from around the world to review this technology. Based on the meeting, this timely volume presents eight specially written chapters covering all aspects of telescope instrumentation.

This book provides an essential reference for all astronomers who will be the users of these large telescopes. It reviews both the challenges involved in designing successful instrumentation and the questions in astronomy they must address. We are taken from the fundamentals of astronomical imaging, low- and high-resolution spectroscopy, and polarimetry up to the state-of-the-art technology in adaptive optics and laser guide stars, interferometry, image pattern recognition, and optical, near and mid-infrared arrays.

This timely volume provides an excellent introduction for graduate students and an invaluable reference for researchers using the latest generation of large astronomical telescopes.

CAMBRIDGE CONTEMPORARY ASTROPHYSICS

Instrumentation for Large Telescopes

CAMBRIDGE CONTEMPORARY ASTROPHYSICS

Instrumentation for Large Telescopes

VII Canary Islands Winter School of Astrophysics

Edited by

J. M. RODRÍGUEZ ESPINOSA
Instituto de Astrofísica de Canarias

A. HERRERO
Instituto de Astrofísica de Canarias

F. SÁNCHEZ
Instituto de Astrofísica de Canarias

CAMBRIDGE
UNIVERSITY PRESS

PUBLISHED BY THE PRESS SYNDICATE OF THE UNIVERSITY OF CAMBRIDGE
The Pitt Building, Trumpington Street, Cambridge CB2 1RP, United Kingdom

CAMBRIDGE UNIVERSITY PRESS
The Edinburgh Building, Cambridge CB2 2RU, United Kingdom
40 West 20th Street, New York, NY 10011–4211, USA
10 Stamford Road, Oakleigh, Melbourne 3166, Australia

First published 1997

Printed in the United Kingdom at the University Press, Cambridge

A catalogue record for this book is available from the British Library

ISBN 0 521 58291 1 hardback

Contents

Techniques for High Angular Resolution Astronomical Imaging

J. M. Beckers

Detectors and Data Analysis Techniques for Wide Field Optical Imaging

M. J. Irwin

Modern Methods of Image Reconstruction

R. C. Puetter

Spectroscopic Techniques for Large Optical/IR Telescopes

K. Taylor

High Resolution Spectroscopy

D. F. Gray

Near Infrared Instrumentation for Large Telescopes

I. S. McLean

Mid-IR Astronomy with Large Telescopes

B. Jones

Polarimetry with large telescopes

S. di Serego Alighieri

Contents

Participants

Aguilar Sánchez, Y.	Sternwarte der Universitt Bonn (Germany)
Alba Villegas, S.	Universidad de Valencia (Spain)
Alvárez Moreno, J.	Universidad Autónoma de Madrid (Spain)
Alvárez Timn	Universidad Autónoma de Madrid (Spain)
Avila-Foucat, R.	Universit de Nice (France)
Bagschik, Klaus	Sternwarte der Universitt Bonn (Germany)
Barbosa Escudero, F.	INAOE (Mexico)
Beckers, Jacques M.	National Solar Observatory/NOAO (U.S.)
Bello Figueroa, C. D.	Universidad de Zaragoza (Spain)
Brescia, Massimo	Osservatorio Astronomico di Capodimonte (Italy)
Carrasco, Esperanza	INAOE (Mexico)
Castellanos, Marcelo	Universidad Autónoma de Madrid (Spain)
Colom-Canales, Cecilia	UNAM (Mexico)
Damm, Peter Christian	Institute of Physics and Astronomy (Denmark)
De Diego Onsurbe, J.A.	UNAM (Mexico)
Del Burgo Díaz, C.	Instituto de Astrofísica de Canarias (Spain)
di Serego Alighieri, S.	Osservatorio Astrofisico Arcetri (Italy)
Eibe, M. Teresa	Armagh Observatory (Ireland)
Eiroa, Carlos	Universidad Autónoma de Madrid (Spain)
Franqueira Pérez, Mercedes	Universidad Complutense de Madrid (Spain)
Gach, Jean-Luc	Observatorie de Marseille (France)
Galadi Enriquez, David	Universitat de Barcelona (Spain)
Galdemard, Philippe	CEA-Service d'Astrophysique (France)
García Lorenzo, M Begoa	Instituto de Astrofísica de Canarias (Spain)
García Navas, Javier	Vilspa IUE Observatory (Spain)
García Vargas, María L.	Vilspa IUE Observatory (Spain)
Gil de Paz, Armando	Universidad Complutense de Madrid (Spain)
González. Pérez, Jos N.	Instituto de Astrofísica de Canarias (Spain)
Gómez, Mercedes	Observatorio Astronómico Córdoba (Argentina)
Gómez-Cambronero, Pedro	Universidad Autónoma de Madrid (Spain)
Gorosabel, Javier	Laeff-Inta (Spain)
Gray, David F.	University of Western Ontario (Canada)
Herranz de la Revilla, Miguel	Instituto de Astrofísica de Andalucía (Spain)
Herrero Davó, A.	Instituto de Astrofísica de Canarias (Spain)
Hervé, Lamy	Universit de Liège (Belgium)
Irwin, M. J.	Institue of Astronomy (U.K.)
Holmberg, Johan	Lund Observatory (Sweden)
Horch, Elliott	Yale University (U.S.)
Iglesias Paramo, Jorge	Instituto de Astrofísica de Canarias (Spain)
Jaunsen, Andreas	Nordic Optical Telescope (Spain)
Jevremovi, Darko	Astronomical Observatory (Yugoslavia)
Jones, Barbara	University of California San Diego (U.S.)
Kalmin, Stanislav Y.	Crimean Astrophysical Observatory (Ucrania)
Keller, Luke	University of Texas at Austin (U.S.)
Kelly, Douglas M.	University of Wyoming (U.S.)
Kimberly, Ann Ennico	Institute of Astronomy Cambridge (U.K.)
Kohley, Ralf	Sternwarte der Universität Bonn (Germany)
Licandro, Javier	Universidad de la República (Uruguay)

Liu, Michael C.	University of California (U.S.)
Luhman, Michael L.	Naval Research Laboratory (U.S.)
Maragoudaki, Fotini	University of Athens (Greece)
Marchetti, Enrico	Center for Space Studies & Activities (Italy)
Marechal, Pierre	Observatorie des Mid-Pyrnes (France)
Martín, Eduardo	Instituto de Astrofísica de Canarias (Spain)
Martín, Susana	Instituo de Astrofísica de Andalucía (Spain)
Mclean, Ian S.	University of California Los Angeles (U.S.)
McMurtry, Craig W.	University of Wyoming (U.S.)
Miralles, Joan-Marc	Observatorie des Mid-Pyrnes (France)
Monteverde Hdez. M. Ilusión	Instituto de Astrofísica de Canarias (Spain)
Mahoney, Neil	University of Durham (U.K.)
Olivares Martín, Jose Ignacio	Instituo de Astrofísica de Andalucía (Spain)
Ortiz García, Elena	Universidad Autónoma de Madrid (Spain)
Otazu Porter, Xavier	Universitat de Barcelona (Spain)
Pérez, Marilo	Laeff-Inta (Spain)
Piché, Francois	Institute of Astronomy Cambridge (U.K.)
Prades Valls, Andres	Universitat de Barcelona (Spain)
Puetter, Richard C.	University of California San Diego (U.S.)
Riccardi, Armando	Osservatorio Astrofisico di Arcetri (Italia)
Richter, Matthew J.	University of Texas at Austin (U.S.)
Rodríguez Espinosa, José M.	Instituto de Astrofísica de Canarias (Spain)
Rodríguez Ramos, José M.	Instituto de Astrofísica de Canarias (Spain)
Rosenberg González, Alfred	Osservatorio Astronomico di Padova (Italy)
Sánchez P., Leonardo J.	Université de Nice (France)
Sánchez-Blanco, Ernesto	Universidad Autónoma de Madrid (Spain)
Sarro Baro, Luis M.	Laeff-Inta (Spain)
Schipani, Pietro	Osservatorio Astronomico di Capodimonte (Italy)
Selam, Selim O.	Ankara University Observatory (Turkey)
Shcherbakov, Victor	Crimean Astrophysical Observatory (Ucrania)
Solomos, Nikolaos	Hellenic Naval Academy (Greece)
Taylor, Keith	Anglo-Australian Telescope (Australia)
Vallejo Chavarino, C.	Universidad Complutense de Madrid (Spain)
Vasilyev, Sergey V.	Kharlov State University (Ucrania)
Vega Beltrán, Juan C.	Osservatorio Astronomico di Padova (Italy)
Villar Martín, Montserrat	European Southern Observatory (Germany)
Vitali, Fabrizio	Osservatorio Astronomico di Roma (Italy)
Woerle, Klaus	M.P.I. für Extraterrestrische Physik (Germany)
Zerbi, Filippo M.	Università di Pavia (Italy)

Preface

Astronomy is entering a new observational era with the advent of several Large Telescopes, 8 to 10 metre in size, which will shape the kind of Astrophysics that will be done in the next century. Scientific focal plane instruments have always been recognized as key factors enabling astronomers to obtain the maximum performance out of the telescope in which they are installed. Building instruments is therefore not only a state of the art endeavour, but the ultimate way of reaching the observational limits of the new generation of telescopes. Instruments also define the type of science that the new telescopes will be capable of addressing in an optimal way. It is clear therefore that whatever instruments are built in the comming years they will influence the kind of science that is done well into the 21st century.

The goal of the 1995 Canary Islands Winter School of Astrophysics was to bring together advanced graduate students, recent postdocs and interested scientists and engineers, with a group of prominent specialists in the field of astronomical instrumentation, to make a comprehensive review of the driving science and techniques behind the instrumentation being developed for large ground based telescopes. This book is unique indeed in that it combines the scientific ideas behind the instruments, something at times not appreciated by engineers, with the techniques required to design and build scientific instruments, something that few astronomers grasp during their education.

One of the most important features of the new generation of large telescopes is the strong push for image quality. Therefore the book goes through the principles of high resolution imaging, including Adaptive Optics (AO) techniques and laser guide stars, and optical interferometry. In particular, the changes that the use of AO imposes on the design of focal plane instruments for large telescopes is reviewed.

Large Telescopes are not particularly suited to wide field astronomy, however deep cosmological surveys demand the large collecting area of the new generation of Large Telescopes together with large format CCD devices, to carry out deep studies of as wide as possible well defined areas of the sky. Thus the book makes a review of wide field techniques and the constraints that these impose on the design of both telescope and imaging instruments. The problems of faint object detections are also studied in detail. There is furthermore a chapter devoted to Image Restoration techniques that can be applied to the high quality data expected from large Telescopes. Indeed Image Restoration techniques are becoming important tools on their own in the hands of astrophysicists.

Spectroscopy, as one of the techniques that will benefit most from the large collecting power of the new Telescopes, is also covered comprehensively. Scientific motives are examined as are various approaches to building optical spectrometers. Starting with the classical grating spectrometer, other more sophisticated techniques are also reviewed, like, for instance, Fabry-Perot and Fourier Transform spectrometers, as well as multi-object spectrographs. An interesting discussion on doing away with slits is also included.

Huge improvements are expected from the use of large Telescopes in the thermal IR, where the increase in collecting area together with the fact that these telescopes will be diffraction or near diffraction limited in the thermal IR, will produce sensitivity increases, for background limited observations, that are proportional to the square of the mirror diameter. Infrared instrumentation is discussed in detail, both for imaging and spectroscopy, and both for the near and thermal infrared. Examples of actual instruments are given, including instruments working at or soon to be installed at the Keck telescopes.

Polarimetry, being always photon starved, is another of the observing techniques that will greatly benefit from the large collecting power of the new generation of large Telescopes. The scientific case for adding spectropolarimetric capabilities to Large telescopes

is also reviewed, with real examples of exciting science which can be done with spectropolarimetric techniques, and the description of the polarimetric capabilities of some
sucessful instruments currently used, or planned, at major telescopes.

The book also discusses both visible and infrared array detectors. It contains, in
fact, what may be the most up to date recollection of existing or planned array devices.
Large Telescopes are expensive, thus observing efficiency is an important key driver, and
large format arrays are important ingredients in designing new cost-effective focal plane
instrumentation.

Throughout the book, many references to existing or almost completed instruments
are made. In fact the authors are in many cases active astronomers that are playing an
important role in the design and construction of focal plane instruments for existing or
planned Large Telescopes.

José M. Rodríguez Espinosa
Instituto de Astrofísica de Canarias, Tenerife
January 1997

Acknowledgements

It is a pleasure to acknowledge the participation in the school of the eight Professors whose skill in presenting the material in a clear, concise, and very often amusing way, and the 80 students whose spirited discussions and presentations all contributed to making the school a very enjoyable event. I am indebted to Ana M. Pérez for her help in putting together the manuscript. Terry Mahoney has checked the English for some chapters. Begoña López spotted some mayor typos after a careful and quick reading of the manuscript. Ramón Castro has played an important role in improving the quality of many figures which he has kindly prepared for the book. His dedication is appreciated.

Techniques For High Angular Resolution Astronomical Imaging

By J. M. BECKERS[1]

[1]National Solar Observatory/NOAO
Tucson, AZ 85726 USA

1. Introduction

Astronomical telescopes are devices which collect as much radiation from astronomical (stellar) objects and put it in an as sharp (small) an image as possible. Both collecting area and angular resolution play a role. The relative merit of these two functions has changed over the years in optical astronomy, with the angular resolution initially dominating and then, as the atmospheric seeing limit was reached, the collecting area becoming the most important factor. Therefore it is the habit these days to express the quality of a telescope by its (collecting) diameter rather than by its angular resolution. With the introduction of techniques which overcome the limits set by atmospheric seeing, the emphasis is changing back to angular resolution. This time, however, it is set by the diffraction limit of the telescope so that both angular resolution and collecting power of a telescope will be determined by its diameter. Both telescope functions will therefore go hand-in-hand.

Although image selection and various speckle image reconstruction techniques have been successful in giving diffraction limited images (see, e.g., the paper by Oskar von der Lühe in the First Canary Island Winter School, 1989), the most powerful and promising technique for all astronomical applications is the one using adaptive optics. That is because, for an unresolved image, it puts most of the collected photons in an as small an image as possible which benefits both in discriminating against the sky background, in doing high spectral and spatial resolution spectroscopy and in doing interferometric imaging with telescope arrays. For resolved objects adaptive optics allows imaging without the complications of image reconstruction techniques applied to short exposure, noisy images. It therefore extends diffraction limited imaging to much fainter and complex objects.

The first lectures in this series therefore concentrates on a description of the principles of astronomical adaptive optics and on the present status of its applications. They are summarized in Section 3 of these notes. Although very powerful in giving high resolution images, the angular resolution of images obtained with individual telescopes is ultimately limited by telescope diffraction to e.g. 0.02 arcsec with the 10 meter Keck telescope at 1 μm wavelength. Higher angular resolution requires larger telescopes or interferometric arrays of telescopes. As was the case in radio astronomy a few decades ago, optical astronomy is reaching the size limits of single aperture telescopes, and, as in radio astronomy, the development of interferometric arrays of optical telescopes is increasingly gaining momentum as a means for obtaining very high angular resolution observations. In Section 4 I will focus on the status of the development of interferometric arrays of optical telescopes. Before discussing adaptive optics and interferometric imaging, it is necessary to shortly introduce some of the basics of atmospheric optics.

2. Some Basics About Atmospheric Wavefront Distortions

2.1. *Introduction*

I will summarize only the wavefront distortions resulting from the refractive properties of the Earth atmosphere to the extent that they are relevant to the implementation of adaptive optics and interferometric imaging. I refer to other excellent reviews for a more detailed description (Roddier 1981, 1987, 1989; Woolf, 1982).

It is common to rely on the work by Tatarski (1961) for the propagation of waves in an atmosphere with fully developed turbulence characterized by the eddy decay from larger to smaller elements in which the largest element L_u (the so-called "upper scale of turbulence") is the scale at which the original turbulence is generated (Kolmogorov, 1941). In addition there is a lower scale of turbulence L^l, set by molecular friction, at which the eddy turbulence is converted into heat. It is very small and is commonly ignored. It is convenient to describe the behavior of properties of such a turbulent field statistically in the form of its structure function $D(\rho)$. For the temperature the (three dimensional) structure function for Kolmogorov turbulence equals :

$$D_T(\rho) \equiv <\mid T(\mathbf{r}+\rho) - T(\mathbf{r}) \mid^2>_\mathbf{r} K^2 \tag{2.1}$$

which is the variance in temperature between two points a distance ρ apart. For Kolmogorov turbulence $D_T(\rho)$ equals :

$$D_T(\mid \rho \mid) = C_T^2 \mid \rho \mid^{2/3} K^2 \tag{2.2}$$

where C_T^2 is commonly referred to as the structure constant of temperature variations. Temperature variations in the atmosphere result in density variations and hence in variations in the refractive index n. So similar to the structure function and structure constant of the temperature there is a structure function and constant for the refractive index :

$$D_n(\mid \rho \mid) = C_n{}^2 \mid \rho \mid^{2/3} \tag{2.3}$$

where

$$C_n = 7.8 \times 10^{-5}(P/T^2)C_T \tag{2.4}$$

with P in millibars and T in degrees Kelvin.

There is little dispute about the validity of the Kolmogorov turbulence structure at small spatial scales, less than L_u. There is, however, substantial disagreement about the size of L_u. Measurements show large variations of C_T^2 on the scale of a few meters with height and anisotropies on the scale of a meter. Such variations would be consistent with L_u values of the order of a few meters. Some observations of the two-dimensional structure function $D_\phi(\mathrm{x})$ of the stellar wavefront phases incident on the telescope show a behavior consistent with a much larger L_u (up to kilometers) but other observations are consistent with L_u values near 5 to 15 meters. Recent observations with the 10 meter Keck telescope show atmospheric image motions which are substantially less than would be expected for very large L_u, indicating L_u values of a few tens of meters. It is likely that L_u is generally not a unique quantity anyway, and that turbulent energy is fed into the atmosphere at many different scales set both by surface heating and high shears in atmospheric wind profiles, and by the telescope environment.

Although the details of the atmospheric physics per se is not of particular interest to adaptive optics and interferometers, the resulting effects on the wavefront are. Especially the structure function of the wavefront both for the total atmosphere and for different atmospheric layers, as well as its temporal variation, are of particular interest. For large telescopes and their adaptive optics it is normally accepted to be close to that predicted by Tatarski/Kolmogorov (although the Keck results dictate caution). For optical interferometers with baselines which are generally much larger than the large telescope diameters the Tatarski/Kolmogorov approximation is also often used. Its use should, however, be seriously questioned.

2.2. *Spatial Wavefront Structure*

The stellar wavefront incident on the telescope has spatial variations both in phase and amplitude (both combined in the "complex amplitude"). Of these the phase variations are the most important in image formation and seeing.

2.2.1. *Phase Variations*

The phase structure function at the entrance of the telescope for Tatarski/Kolmogorov turbulence equals :

$$D_\phi(\mathbf{x}) = <\mid \phi(\mathbf{y} + \mathbf{x}) - \phi(\mathbf{y}) \mid^2>_\mathbf{y} = 6.88 r_o^{-5/3} x^{5/3} \text{rad}^2 \qquad (2.5)$$

where the wavelength (λ) and zenith distance (ζ) dependent coherence length r_o (the so-called "Fried parameter") equals :

$$r_o(\lambda, \zeta) = 0.185 \lambda^{6/5} \cos^{3/5}\zeta (\int C_n^2 dh)^{-3/5} \qquad (2.6)$$

When not otherwise indicated the Fried parameter r_o in this review (and generally elsewhere) refers to $r_o(0.5\mu, 0°)$. The seeing dominated image size d (FWHM) in a telescope relates to r_o as d $\approx \lambda/r_o$ for $r_o <$ telescope diameter D and otherwise to λ/D.

2.2.2. *Amplitude Variations*

Amplitude/Intensity variations (also called scintillation) across the telescope aperture contribute much less to image quality degradation than phase variations and are therefore generally ignored in the planning and evaluation of adaptive optics systems. The Roddiers showed that scintillation does contribute to the quality of image restoration at the $\approx 15\%$ level at visible wavelengths (0.5 μm), decreasing rapidly towards longer wavelengths (3% at 2.2 μm). Except for that paper I am not aware of any other publication dealing with the effect of amplitude variations on adaptive optics.

2.2.3. *Modal Representation of the Wavefront*

In describing the phase variations in the wavefront for a circular aperture like a telescope it is often useful to express the phase variations in terms of the set of orthogonal Zernike polynomials $Z_j(n, m)$ in which n is the degree of a radial polynomial and m the azimuthal frequency of a sinusoidal/cosinusoidal wave. Noll gives normalized versions for $Z_j(n, m)$ in which the normalization is done in such a way that the RMS value of each polynomial over the circle equals 1. Table 1 lists the low order terms of Z_j together with their meaning and the mean square residual amplitude Δ_j in the phase variations at the telescope entrance caused by Kolmogorov turbulence after removal of the first j terms. For large j one has approximately:

$$\Delta_j \approx 0.2944 j^{-0.866} (D/r_o)^{5/3} \text{rad}^2 \qquad (2.7)$$

From Table 1 one derives for the RMS phase variation ϕ_{RMS} across a circular aperture without any correction:

$$\phi_{\text{RMS}} = 0.162 (D/r_o)^{5/6} \text{waves} \qquad (2.8)$$

and after tilt correction in both directions only

$$\phi_{\text{RMS}} = 0.053 (D/r_o)^{5/6} \text{waves} \qquad (2.9)$$

Z_j	n	m	Expression	Description	Δ_j/S
Z_1	0	0	1	constant	1.030
Z_2	1	1	$2r \ \sin\varphi$	tilt	0.582
Z_3	1	1	$2r \ \cos\varphi$	tilt	0.134
Z_4	2	1	$\sqrt{3}(2r^2 - 1)$	defocus	0.111
Z_5	2	2	$\sqrt{6}r^2 \ \sin2\varphi$	astigmatism	0.0880
Z_6	2	2	$\sqrt{6}r^2 \ \cos2\varphi$	astigmatism	0.0648
Z_7	3	1	$\sqrt{8}(3r^3 - 2r) \ \sin\varphi$	coma	0.0587
Z_8	3	1	$\sqrt{8}(3r^3 - 2r) \ \cos\varphi$	coma	0.0525
Z_9	3	3	$\sqrt{8} \ r^3 \ \sin3\varphi$	trifoil	0.0463
Z_{10}	3	3	$\sqrt{8} \ r^3 \ \cos3\varphi$	trifoil	0.0401
Z_{11}	4	0	$\sqrt{5}(6r^4 - 6r^2 + 1)$	spherical	0.0377

TABLE 1. Modified Zernike polynomials & their Mean Square Residual Amplitude Δ_j (in rad^2) for Kolmogorov Turbulence, after Removal of the first j Zernike Polynomials. r = distance from center circle, φ = azimuth angle; $S = (D/r_o)^{5/3}$

Most of the phase variations in the wavefront can therefore be removed by simple rapid guiding of the telescope or by other wavefront tip-tilt methods.

For small wavefront disturbances, the fractional decrease (1-SR) in the central intensity of an unresolved star image from a perfect diffraction limited image (SR being commonly called the "Strehl Ratio") equals approximately:

$$1 - SR \approx \Delta \approx 1 - e^{-\Delta}. \tag{2.10}$$

These are referred to as the Maréchal and the extended Maréchal approximations respectively (Δ in rad^2). For SR = 80 % this implies D = 0.4 r_o without any correction at all and D = 1.4 r_o with tip-tilt correction alone. For D/r_o = 60 (e.g., an 8 meter telescope with 0.75 arcsec seeing) it will take according to equation 2.7 the adaptive optics correction of the first 3640 Zernike terms to reach a Strehl Ratio of 80 % (ignoring the effects of amplitude variations).

2.3. Height Variation of Atmospheric Optical Disturbances

2.3.1. C_n^2 Variations

Figure 1 reproduces the variation of the average value of C_n^2 with height as given by the so- called Hufnagel/Valley model and extrapolated to low altitudes for day and night conditions as done by Roddier. The actual $C_n^2(h)$ profile varies from site to site and from time to time. In addition the curve in Figure 1 does not show the very rapid fluctuations in C_n^2 with height observed in the balloon flights referred to earlier. Figure 1 should therefore only be used as an approximation. One often distinguishes three layers in the $C_n^2(h)$ profile : (i) the "Surface Layer" near the telescope (between about 1 to 20 meters) subject to, e.g., wind-surface interactions and manmade seeing, (ii) the "Planetary Boundary Layer" up to $\approx$ 1000 meters subject to the diurnal solar heating cycle, and (iii) the "Free Atmosphere" above this. The increase of C_n^2 at h $\approx$ 10 km is related to the high wind shear regime in the tropopause. Above it the refractive index variations rapidly decrease, with an effective upper limit to atmospheric seeing occurring at h $\approx$ 25 km.

2.3.2. The Isoplanatic Angle

The isoplanatic angle θ_o is commonly defined as the radius of a circle in the sky over which the atmospheric wavefront disturbances, and their resulting instantaneous (speckle) point-spread-functions, can be considered identical. A good approximation for θ_o is

$$\theta_o = 0.314 r_o / H, \qquad (2.11)$$

where H is the average distance of the seeing layer, or

$$H = sec\zeta \left(\frac{\int C_n^2 h^{5/3} dh}{\int C_n^2 dh} \right)^{3/5}. \qquad (2.12)$$

At this distance the Strehl Ratio has decreased by an amount depending on D/r_o. For $D/r_o = 5, 10, 20, 50, 100$ and ∞ the Strehl Ratio decreases to respectively 71, 62, 56, 49, 47 and 36 %.

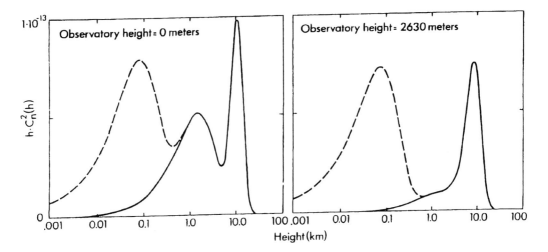

FIGURE 1. Average C_{n^2} profile with Local Height h_L (in km). Left: for a Sea Level site. Right: for a 2630 meter High Mountain Site.

Sometimes the term isoplanatic patch is used in a broader sense to refer, for example, to the distance over which image motions are practically identical (as compared to their seeing dominated width). In that case it is useful to talk about the "isoplanatic patch for image motion" θ_{motion} which, as for the wavefront (θ_o), equals $\approx 0.3\ D/H$. For $r_o = 13.3$cm (0.75 arcsec seeing at 0.5μm), D = 8 meters and H = 5000 meters, $(\theta_o$ equals 1.7 arcsec but θ_{motion} equals 100 arcsec. θ_{motion} has also been referred to as the "isokinetic patch" size. For diffraction limited images the isoplanatic patch size for image motions is much smaller.

2.4. *Temporal Variation of the Wavefront*

The temporal variations of the wavefront are predominantly determined by the wind velocities at the different heights in the atmosphere since the turbulent elements responsible for the seeing live longer than the time it takes for them to move across their diameter. These wind velocities amount to typically $V_{wind} = 10$m/sec, reaching frequently 30 m/sec and higher at the ≈ 12 km tropopause layer. Since the wind directions and velocities vary with height, the temporal behavior of the wavefront is complex and hard to characterize. Typical time scales are

$$\tau_o \approx 0.314 r_o / V_{wind} \qquad (2.13)$$

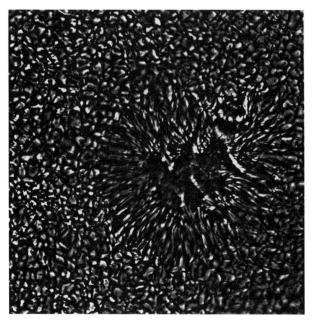

FIGURE 2. Image of a Sunspot obtained with the National Solar Observatory 76 cm Vacuum Tower Telescope at Sacramento Peak at 490 nm by Means of Image Selection Techniques. The Image is Diffraction-Limited at 0.15 arcsec.

or 0.004 sec for the wavefront changes and $\tau_{motion} \approx 0.314\,D/V_{wind} = 0.25$ sec for seeing image motion. For wind velocities V(h) varying with height the average wind velocity equals:

$$V_{wind} = \left(\frac{\int C_n^2 v^{5/3} dh}{\int C_n^2 dh} \right)^{3/5}. \qquad (2.14)$$

The quantity $f_o = 1/\tau$ is closely related to the so-called Greenwood Frequency f_G which is often used in the specification of adaptive optics control systems. It is generally taken to equal:

$$f_G = 0.43 V_{wind}/r_o = 0.135 f_o. \qquad (2.15)$$

2.5. *Image Selection Methods*

The description of the wavefront disturbances given above is a statistical description. At anyone time the actual wavefront entering the telescope may be substantially better or worse than that given by the statistical description. It often pays to observe only during those moments when the wavefront is significantly better than average even though not the photons collected during the other times are thrown away. This "image selection at moments of good seeing" is especially powerful for solar observations for which the number of photons in broad band observations is plentiful. Figure 2 shows a solar image so obtained at the National Solar Observatory 76 cm Vacuum Tower Telescope. But also for many nighttime observations image selection combined with shift-and-add summation of short exposure images can often give high resolution images. Table 2 shows the gain in angular resolution obtained from doing so for the best 10, 1 and 0.1 percentile of images for different telescope diameters D.

D/r_{circ}	10 percentile	1 percentile	0.1 percentile
3.0	2.7	2.9	3.0
4.0	3.1	3.4	3.6
5.0	3.1	3.6	4.0
7.0	2.8	3.4	4.1
10.0	2.4	3.0	3.8
15.0	2.1	2.6	3.2
20.0	1.9	2.3	3.0
50.0	1.5	1.7	2.1
100.0	1.4	1.6	1.8

TABLE 2. Gain in Angular Resolution Obtained with Image Selection (from Hecquet and Coupinot, 1985)

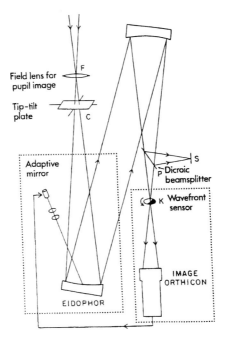

FIGURE 3. The Adaptive Optics Concept as Originally Proposed by Babcock. F Images the Entrance Pupil onto the Adaptive Mirror (in this case the "Eidophor"). The Shape of the Adaptive Mirror is Servo Controlled by a Wavefront Sensor (in this case a rotating knife edge device K) which Follows a Beamsplitter P which Sends the Other Part of the Light to the Astronomer S. The Tip-tilt plate C Removes the Overall Image Motion.

3. Adaptive Optics

3.1. *Principles of Adaptive Optics*

The concept to use adaptive optics for compensating atmospheric seeing originates with Horace W. Babcock in 1953. Although his paper deals with its application to astronomical imaging at visible wavelengths, it appears to be the first description of a much broader discipline which was to find its application to military, laser power, medical (ophthalmology), and probably other applications. Figure 3 is adapted from Babcock's paper.

Adaptive optics removes the wavefront distortions introduced by the earth atmosphere

Name	Location	D(cm)	Adapt. E.	WFS	N_{act}	LGS	Comments
COME-ON+	La Silla	360	Piston	H-S	52		in operation
SOR	Albuquerque	150	Piston	H-S	241	yes	in operation
ACE	Mt. Wilson	152	Piston	H-S	69	yes	in operation
Un. Hawaii	Mauna Kea	360	Bimorph	CS	13		in operation
WCE	Yerkes	100	Piston	H-S	69		experimental
Beijing	Beijing	220	Piston ?	H-S ?	21		experimental
LLNL	Lick	300	Piston	H-S	69	yes	experimental
ChAOS	APO	350	Piston	H-S	97	(yes)	experimental

TABLE 3. Summary of Active Astronomical Adaptive Optics Systems. D = Telescope Diameter; WFS = Wavefront Sensor Type; N_{act} = Number of Actuators; LGS = Laser Guide Star; H-S = Hartmann-Shack Sensor; CS = Curvature Sensor.

by means of an optical component which is introduced in the light beam and which can introduce a controllable wavefront distortion which both spatially and temporally counteracts that of the atmosphere. This optical component is generally, but not always, a mirror whose surface can be distorted. To control it the wavefront distortions have to be known. They are measured by means of a wavefront sensor using either the object under study for its measurement or a nearby stellar or laser generated object (also referred to as natural or laser guide stars). In case the wavefront is measured with the required accuracy and spatial and temporal resolution, and in case the adaptive mirror control is perfect, the atmospheric effects are removed and the telescope will give a diffraction limited image.

Apparently not knowing about Babcock's paper, V.P. Linnik published in 1957 independently a description of the concept of adaptive optics. His publication was the first one, however, to suggest the use of artificial guide stars for doing the wavefront sensing.

After a few initially unsuccessful attempts to implement the adaptive optics concept in astronomy, recent advances in military and civilian technology have led in the last decade to a large number of astronomical adaptive optics system developments some of which are now starting to pay off in important astronomical results. Table 3 summarizes these current efforts to the extent that they are in existence at a telescope. A more complete listing, including systems under development and being planned, can be found in Beckers (1993) and Thompson (1995).

3.2. *Wavefront Sensing*

Since it is impossible to compare directly the wavefront incident on the atmosphere interferometrically with that reaching the telescope, one has to resort to measurements which assess the (spatially) differential wavefront distortions within the telescope pupil. Mostly measurements of the wavefront gradients (tilts) are used. Other ways of assessing the wavefront are wavefront curvature analysis, and neural nets are also being used. The most important quality criterion for wavefront sensors for astronomical applications is of course their sensitivity for faint sources. Astronomical wavefront sensors using stellar signals therefore have to work broadband, in white light. Since the wavefront tilts and curvatures are almost achromatic, one often measures the wavefront at a different wavelength than the one being used for astronomical observations (so-called polychromatic adaptive optics systems). I summarize below methods for wavefront sensing which have been or are being used.

3.2.1. *Foucoult/Knife-Edge Wavefront Sensor*

Babcock proposed the use of the common knife-edge test to measure wavefront tilts (Fig 3). The knife edge is placed in the stellar image. To a good approximation the intensity distribution in the following pupil image then represents the wavefront gradients/tilts in a direction at right angles to the knife edge. By splitting the stellar image into two and using two knife edges in orthogonal directions the full wavefront tilt is measured. The same can be done by rapid rotation of the knife edge around the stellar image.

3.2.2. *Shearing Interferometer Wavefront Sensor*

By laterally shifting (or shearing) the wavefront and mixing it with itself, interference patterns are obtained which correspond to the wavefront tilt in the shear direction. Since the distance of the fringes are proportional to the wavelength used, gratings are commonly used to obtain an amount of shearing which is also proportional to the wavelength, thus resulting in the desired broadband, white light signal. As is the case with the knife-edge sensor, it is necessary to make two orthogonal measurements to assess the full wavefront tilt.

3.2.3. *Hartmann-Shack Wavefront Sensor*

Most commonly used in astronomy is the Hartmann-Shack sensor. It is an improved version of the classical Hartmann test proposed by Shack & Platt. It replaces the Hartmann screen in the pupil by an array of small lenslets in an image of the pupil. The lenslet array forms an array of images whose positions are measured to give the full vectorial wavefront tilt in the areas of the pupil covered by each lenslet. The advantage of the lenslet modification is of course the enhanced sensitivity since almost all the light collected is used for wavefront measurement. In addition it is not necessary to divide the light into two, as is the case for the knife-edge and shearing interferometer devices so that all photons per subaperture can be used to measure the wavefront tilt.

3.2.4. *Wavefront Curvature Sensor*

An out-of-focus image of a star shows intensity patterns which are due to both amplitude variations (scintillation) and curvature variations in the incident wavefront. The amplitude variations are identical in out-of-focus images taken on opposite sides of the focus, whereas the curvature effects are opposite in sign. This allows the measurement of the wavefront curvature by subtraction of the two images. Roddier is using this method in controlling his adaptive optics system on Mauna Kea.

From Table 3 it is clear that the Hartmann-Shack wavefront sensor is preferred by most modern adaptive optics systems. Curvature sensing. and with it curvature control using bimorph mirrors, is primarily being pursued by Roddier's team at the University of Hawaii. It has resulted in some of the best astronomical results to-date. In addition to the techniques listed above, others have been suggested and tried. They include neural network and phase diversity wavefront estimates.

3.3. *Limiting Sensitivities*

The limiting sensitivity for astronomical observations is critical to the performance of adaptive optics systems. It is essential that the best wavefront sensing techniques are used and that the efficiency of the wavefront sensing system is optimized by maximum photon detection. In Hartmann-Shack sensors the dimensions of the lenslets is generally taken to correspond approximately to r_0. For estimating the required sensitivity for wavefront sensing, a position accuracy of 10% RMS of the Hartmann-Shack sensor image

Band	λ (μm)	r_0 (cm)	τ_0 (sec)	V_{lim}	θ_0 (arcsec)	% Sky Coverage
U	0.365	9.0	0.009	7.4	1.2	2 x 10-5
B	0.44	11.4	0.011	8.2	1.5	6 x 10-5
V	0.55	14.9	0.015	9.0	1.9	3 x 10-4
R	0.70	20.0	0.020	10.0	2.6	1 x 10-3
I	0.90	27.0	0.027	11.0	3.5	0.01
J	1.25	40	0.040	12.2	5.1	0.05
H	1.62	55	0.055	13.3	7.0	0.22
K	2.2	79	0.079	14.4	10	1.32
L	3.4	133	0.133	16.2	17	14.5
M	5.0	210	0.21	17.7	27	71
N	10	500	0.50	20.4	64	100

TABLE 4. Limiting V Magnitude for Polychromatic Wavefront Sensing & Sky Coverage at Average Galactic Latitude for different spectral bands. Conditions are: 0.75 arcsec seeing at 0.5 μm; $\tau_{det} = 0.3\tau_0 = 0.3\, r_0/V_{wind}$; $V_{wind} = 10$ m/sec; H = 5000 meters; photon detection efficiency = 20%; spectral bandwidth 300 nm; SNR = 100 per Hartmann-Shack image; detector noise = 5 e-

sizes (λ/r_0) appears reasonable. For photoelectron noise limited detectors this means the detection of ≈ 100 photon events per detection time τ_{det}. For detector or sky noise limited applications (e.g. infrared detectors) it means the detection of $100 \times$ the detector quantum noise per detection time τ_{det}. Very low read-out noise CCD arrays have recently become the detector of choice.

Table 4 lists the limiting magnitudes for wavefront sensing for such a visible light wavefront sensor used in the polychromatic wavefront sensing mode for the different photometric bands as well as the resulting sky coverage taking into account the star numbers. Using natural stars for wavefront sensing results in very little sky coverage at visible wavelengths. The sky coverage increases rapidly towards the infrared which is, together with the decreased complexity and cost, the reason that most current efforts focus on the 1 to 5 μm wavelength region. Sky coverage increases rapidly with improved seeing (approximately proportional to r_0^5 for low sky coverage). However, only with the introduction of Laser Guide Stars (see section 3.7) will it be possible to reach almost full sky coverage at all wavelengths.

3.4. *Adaptive Optical Elements*

It is up to the adaptive optical component to correct the wavefront. These optical components are generally mirrors. They have to be designed so that their shape can be adjusted to match the instantaneous wavefront distortion well. They are characterized by their size, by the number of adjustable subareas (or elements), by the number of actuators, by the wavefront influence function of each actuator, by the speed at which they can be adjusted, and by their stroke. Since most of the wavefront distortion is in the form of wavefront tilt across the aperture, the function of the adaptive optical component is often divided among two components, one a tip-tilt mirror covering the full aperture, the other one the adaptive mirror which corrects the higher order wavefront distortions. Adaptive elements that are being used include:

3.4.1. *Segmented Tip-Tilt Mirrors*

Since most wavefront sensors measure the wavefront tilt over a number of subareas of the telescope aperture, it is advantageous for the control of the wavefront to correct these tilts with an array of mirrors which match those subareas and which individually directly correct the measured tilts. In each segment both piston, and tip and tilt are normally adjusted. The piston is measured separately generally using interferometry on an artificial source to force quasi mirror surface continuity.

3.4.2. *Continuous Faceplate Piston Mirrors*

Continuous faceplate adaptive mirrors automatically maintain continuity and therefore can work with a reduced number of actuators. Actuators are generally of the push-pull type using mostly piezoelectric or electrostrictive materials.

3.4.3. *Curvature Control Mirrors*

The University of Hawaii system uses this type of adaptive mirror for its curvature sensing/control adaptive optics system. The curvature actuation is done by bimorph techniques using oppositely polarized piezoelectric materials. They have the same advantage as the segmented mirror as far as the capability of using direct wavefront signal - actuator command control goes, but, in this case, without the need to control an additional variable (the piston).

A number of other ways exist in which the optical path across an aperture can be varied. There is the original suggestion for an adaptive mirror using an electron beam scanner on a film of oil as actuator (the Eidophor). Membrane mirrors use deformation of a thin membrane by electrostatic forces. Liquid crystal technology is also being explored as a way to correct the wavefront.

3.5. *System Considerations*

In almost all adaptive optics systems a computer model of the system performance accompanies the experimental efforts. It includes a number of factors which contribute to the determination of the performance of a system at its "design wavelength" (r_E = element/segment/ lenslet size = r_o). The expressions given here, using natural guide stars, are approximate.

Finite Spatial Resolution (or *"wavefront fitting error"*) The finite spatial resolution of both wavefront sensor and adaptive mirror lead to a residual wavefront variance Δ of:

$$\Delta_{\text{spatial}} \approx 0.34 (r_E/r_o)^{5/3} \text{rad}^2 \tag{3.16}$$

Finite Temporal Resolution (or *"time delay error"* or *"servo error"*)

The assessment of the effect of finite temporal resolution on the performance depends on the servo characteristics of the system. Following Greenwood one has:

$$\Delta_{\text{temporal}} \approx (f_G/f_{\text{servo}})^{5/3} \text{rad}^2 \tag{3.17}$$

where f_{servo} is the closed-loop servo bandwidth at -3dB of the adaptive optics control system.

Photon Noise As discussed in section 3.3 one would like for wavefront sensing to have a Signal-to-Noise ratio (SNR) of at least 10 per subaperture in the Hartmann-Shack tilt measurements, or $N_{\text{pe}} = 100$ photon events in a detector which is photon noise limited. In general one has

$$\Delta_{\text{photon}} \approx 4 SNR^{-2} \text{rad}^2 \approx 4/N_{\text{pe}} \text{rad}^2. \tag{3.18}$$

3.6. *Restrictions of Adaptive Optics Using Natural Guide Stars*

Adaptive optics (using natural guide stars) have some major limitations. Among these are: (i) the *sky coverage limitation* already shown in Table 4, (ii) the limitation to relatively *bright objects*, (iii) *the variation of the point-spread-function* with position across the field-of-view (the isoplanatic patch and beyond) which complicates relative photometry and additional image restoration/super-resolution techniques, and (iv) *light losses* and *increased infrared emissivity* due to the additional optical elements which are necessary before the image is formed. Recent developments in the implementation of laser guide stars and associated techniques promise to remove most of these limitations.

3.7. *Removing These Restrictions*

3.7.1. *Full Sky Coverage by Means of Laser Guide Stars*

By far the largest limitation to the application of adaptive optics to astronomy is the very limited sky coverage when using natural guide stars for wavefront sensing. Similar limitations existed for many military applications of adaptive optics. The solution uses artificial, laser guide stars (LGSs) for wavefront sensing which are created by laser light scattering in high layers of the atmosphere. LGSs using Rayleigh scattering in lower atmospheric layers and scattering off the 90 km high mesospheric neutral sodium layer have been achieved. The latter is of most interest for astronomical applications since the resulting laser guide star is located well outside the atmospheric seeing layers and the furthest away. I will therefore describe only the sodium laser guide star concept in the following.

Concept Description

The sketch in Figure 4 describes the concept of the laser guide stars. A laser tuned to the sodium D_2 lines is pointed at the ≈ 11.5 km thick layer of enhanced neutral sodium located at an altitude of approximately 90 km. The optical thickness of this layer at the sodium line center equals about 0.05. When illuminated by the laser sodium, atoms are radiatively excited from the $^2S_{1/2}$ ground layer to the $^2P_{3/2}$ layer from which they depart either by a spontaneous emission (in $\approx 10^{-8}$ sec) back to the ground level emitting photons in all directions, or by stimulated emission to the same level emitting photons in the same direction as the incoming photons. Some of the spontaneously emitted photons return to earth and reach the telescope to be viewed as the laser guide star. They are used for the sensing of the wavefront. Since the scattering occurs over a range of heights the phases of the returning photons are random. Since Hartmann-Shack- and curvature-sensors are geometrical optics devices this does not affect the wavefront tilt or curvature sensing. Increasing the energy in the sodium laser beam will increase the intensity of the laser guide star up to the point that the intensity becomes high enough to cause the stimulated emission to dominate the spontaneous photon emission. At that point the laser guide star brightness ceases to increase and "saturation" occurs. For a 50 cm, or a ≈ 1 arcsec apparent, diameter laser guide star this occurs at a laser power of about 5 kW in the pulse.

Status of LGS Experimentation

Laser guide stars as bright as $m_V = 6$ will soon be available allowing their use for visible wavelength adaptive optics. Laser guide star aided adaptive optics systems have been successfully demonstrated on astronomical objects using Rayleigh scattering laser guide stars. Sodium laser guide stars have been used for wavefront sensing and are reported to have recently been used for controlling full adaptive optics systems at the Lick Observatory.

Focus Anisoplanatism and its Removal

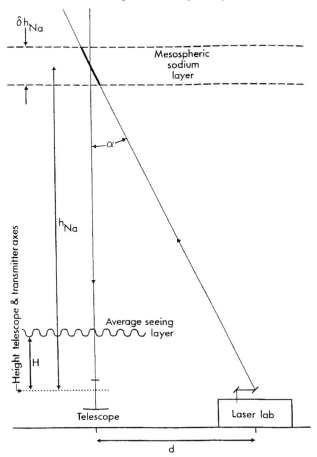

FIGURE 4. Sketch of the telescope - laser transmitter - laser guide star geometry. The plane of the figure corresponds to the plane containing these three objects. For simplicity it is shown for a zenith pointing telescope. Notations: h_{Na} = height of mesospheric sodium layer ($\approx$ 90 km); δh_{Na} = full width at half maximum (FWHM) of sodium layer ($\approx$ 11.5 km); H = average height of the seeing layer ($\approx$ 4 km); D = telescope diameter (8 -10 meter for present generation telescopes); d = distance of the telescope to the laser transmitter; α = pointing difference between telescope and laser transmitter

Three complications arise from the limited distance of the laser guide star from the telescope rather than the effectively infinite distance to the stars: (i) the laser guide star only senses the atmospheric wavefront distortions between it and the telescope and not those of the atmosphere beyond. This is only a problem for Rayleigh laser guide stars and not for high altitude sodium laser guide stars; (ii) the laser guide star focusses at a different position than the real star. Since the distance of the laser guide star to the telescope is well known, the resulting defocus signal of the wavefront sensor can be corrected for; and (iii), the rays from the laser guide star traverse a different airpath from that of the rays coming from the astronomical object. For a sodium laser guide star located on the telescope axis at a distance from the telescope of h_{Na}, the ray coming from it differs in distance from a ray coming from an on-axis star by $(h/h_{Na})r$, for a distance r from the center of the telescope pupil. For h_{Na} = 90 km, h = 9 km (the upper troposphere), and r equal to the telescope radius R of say 4 meters, this distance equals 40 cm which is larger or comparable to the Fried parameter r_0. This effect is referred to as Focus

Anisoplanatism. To remove it the use of multiple laser guide stars at slightly different sky locations has been suggested. Three to four laser guide stars may suffice. By appropriate combination of the resulting wavefront signals (also referred to as "stitching" of the wavefronts) the wavefront distortions for a star can be derived.

Tilt Problem and its Removal

The upgoing laser beam is affected by atmospheric seeing causing both some blurring and motion of the laser spot on the scattering layer. When observing the laser spot this blurring and motion will be added to the same effects resulting in the air path in the return beam. The blurring/motion effects introduced by the upgoing beam are identical for all Hartmann-Shack sub-apertures and therefore do not affect the differential wavefront tilt measurements except for a decrease in sensitivity due to the blurring. The seeing motion introduced in the upgoing beam will, however, cause a common motion signal in the wavefront sensor which causes it to be different from that of the motion of the star image caused by atmospheric seeing. In the special case where the telescope itself is used as the laser transmitter the two laser guide star motions will actually compensate each other causing the LGS to be stationary. This difference between the motion/tilt signal in the wavefront sensor is referred to as the "tilt problem". The wavefront tilt must therefore be determined from a stellar signal itself. The chances of finding a bright enough star within θ_o is however small. The area in the sky for which the image motion is small compared to the width of the diffraction limited point-spread- function obtained with the laser guide star is $\theta_m \approx 2(D/r_o)^{1/6}\theta_o$ rad in radius for $\approx 25\%$ Strehl Ratio decrease. For an 8 meter telescope and for the conditions listed in Table 4 this implies $\theta_m = 7.5$ arcsec for the V band. Within this area the probability of finding the $m_V = 21.5$ star needed for wavefront tilt sensing is 50% for the average sky (only $\approx 5\ \%$ at the galactic poles). Rigaut & Gendron therefore propose the use of dual adaptive optics to overcome the tilt determination problem. In this, LGS aided adaptive optics is used both on the astronomical object of interest as well as on the star to be used for tilt determination up to θ_m away. The latter being now diffraction limited allows the tilt determination to be done well enough to allow compensation. Their detailed analysis also shows that this results in better than 50 % sky coverage for an 8 meter telescope at visible wavelengths for very good seeing ($r_o = 20$ cm).

Perspective Elongation Effect and its Removal

When the laser transmitter is displaced from the telescopes as shown in Figure 4 (also referred to as the "bi-static" configuration as compared to the "mono-static" configuration where the telescope and the laser transmitter share the same aperture) the LGS will appear to be elongated. The amount of elongation depends linearly on the distance d of the transmitter-receiver with d = 10 meter resulting in an elongation of ≈ 3 arcsec. This elongation results in a decrease of sensitivity for wavefront sensing even if care is taken to keep the elongation less than the diameter of the isoplanatic patch ($2\theta_o$). This can be avoided, except for a small effect at the edge of large 8 - 10 meter diameter astronomical mirrors, by using a mono-static configuration. Beckers (1993) proposes removing this perspective elongation through the use of pulsed lasers.

Incorporation of LGSs into Large Telescopes and Observatories

A number of papers has been published describing the incorporation of laser guide stars in astronomical telescopes and evaluating the resulting expected performance. Most of these papers go into considerable technical detail and I refer to them for those desiring a more detailed description of the methodology using LGS-aided adaptive optics for individual large astronomical telescopes.

Because of the limited number of good astronomical sites available, and because of the expenses of developing and operating observatories, astronomical telescopes tend to occur

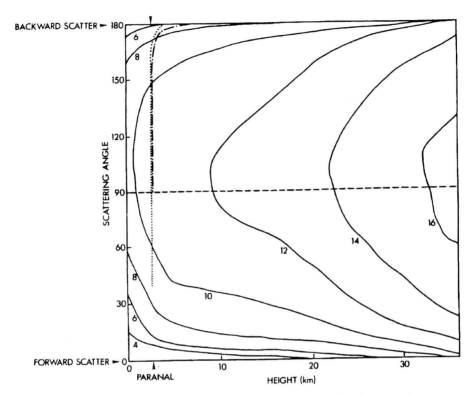

FIGURE 5. The diagram shows the magnitude difference $\Delta m_V(h, \varphi)$ (per $\square$") between the laser guide star and the laser beam scattered light as a function of height h in the atmosphere and scattering angle φ between the laser and telescope viewing direction. Dashed curve for laser pointing at zenith, dotted curve for laser pointed at 60° zenith distance. Telescope is located 100 meters from laser transmitter, pointed at the laser beam. Scattering from dust particles is not included.

in clusters. Therefore, when incorporating LGSs in telescopes one has to consider not only the effects on the performance of the telescope itself but also on all the telescopes on the observatory site. The average power and the peak power of the laser transmitter will tend to be rather high especially when laser guide stars for visible light astronomy will be used. Mie scattering on dust and aerosols and Rayleigh scattering on molecules will cause therefore an increase in sky background when looking in a direction encompassing the laser beam. The amount of average power scattered light is of concern for long duration (> laser pulse rate) exposures, the larger amount of peak power scattered light will affect high speed photometry. Figure 5 shows the difference $\Delta m_V(h, \varphi)$ in sky brightness between the laser star and the scattered laser radiation to be expected from aerosol and Rayleigh scattering as a function of height h in the atmosphere (h = 0 at sea level) and of the angle φ between the outgoing laser beam direction and the telescope viewing direction. It assumes a $1\square$" (= square arcsec) size laser guide star, a laser beam diameter of 50 cm and laser power low enough to avoid significant stimulated emission (peak power $\ll$ 5 kW). Figure 5 refers to a small telescope subaperture (< 50 cm) or to naked eye observations. For larger apertures the sky brightness background will be diluted and becomes uneven in the focal plane, its behavior depending on the telescope/laser transmitter/viewing directions configuration.

Since for visual light adaptive optics one envisages a $m_V \approx 7$ for the LGS, the sky background due to scattered laser light may amount to one $m_V = 14$ to 18 V star /□". This has to be compared with a typical sky background on moonless nights of one $m_V = 21.5$ star/□" in all radiations or of one $m_V = 24.5$ star/□" in the bright $\lambda 5577$ atmospheric OI emission line. Laser background suppression of a factor of up to $\approx 10^4$ will therefore be necessary. In telescopes which incorporate the laser guide star mono-statically the scattered light issues are much worse. It is proposed to control them with fast shutters which only allow star light in the astronomical instrument at the times when the pulsed laser is in the off-stage. Additional scattered light reduction is possible by the use of notch filters which block the laser radiation from entering the instrument. Both techniques result in light efficiency loss of the telescope, a loss which will be more than offset by its diffraction limited performance. Similar techniques can be used for the other telescopes on the site provided all lasers are pulsed simultaneously if the shutter technique is used. Notch filters are, however, likely to be sufficient to eliminate the laser scatter problems.

3.7.2. *Increasing the Isoplanatic Patch Size by Multi-Conjugate Adaptive Optics*

There have been a number of proposals to increase the size of the isoplanatic patch through the use of a number of adaptive mirrors placed at the conjugates of different layers of the atmosphere. It requires the determination of the full 3-D structure of the atmospheric wavefront distortion. This can in principle be done by means of "atmospheric tomography", using projections of the wavefront distortions at ground level and arrays of laser guide stars, or, in case of extended objects with surface structure like the sun, by wavefront sensing on different parts of the object. Since much of the seeing occurs near the earth surface and at the tropopause, a two component adaptive optics system with adaptive mirrors located at the conjugates of these layers would already help a great deal. Even a single component (conventional) adaptive optics system with the adaptive mirror located at the conjugate of the average seeing layer rather than at an image of the pupil (as is usual) would significantly increase the size of the isoplanatic patch. An additional benefit of a multi-conjugate adaptive optics system will be the reduction of the amplitude variations (scintillation).

3.7.3. *The Integration of Adaptive Optics into the Telescope*

At this moment, with adaptive optics systems still being experimental, LGSs and adaptive mirrors and their associated optics are generally combined in a package placed like an instrument at one of the focus stations of a telescope. The amount of optics is often quite staggering in order to make the package fit the generally quite limited space and in order to reimage the pupil and image respectively on the tip-tilt mirror and/or the adaptive mirror and on the detector. As a result light is lost, emissivity and scattered light is increased, and the expensive adaptive optics system is available at only one location. In the future as adaptive optics becomes more routine one might expect therefore a better integration in the telescope. It has been suggested to make the secondary mirror of the telescope adaptive. It would feed all foci except the prime focus. By making it ellipsoidal in a Gregorian telescope it could be located at the conjugate of an atmospheric seeing layer. Efforts to construct adaptive secondary mirrors are underway at a number of places. The wavefront sensor would still have to be located at the focus being used, but it and the beamsplitter feeding it would be the only components located there.

3.8. *Effect of Using Adaptive Optics on Astronomical Instrumentation*

Instrumentation on telescopes which are made diffraction limited by the use of adaptive optics looks very different from instrumentation for seeing limited telescopes. Let's examine separately spectrographs and imagers:

3.8.1. *Spectrographs*

Lets take as the slitwidth d_{slit} twice the FWHM of the star image, or $2\lambda F/r_0$ and $2\lambda F/D$ for the seeing and diffraction limited telescope respectively, where F is the telescope effective focal length and D is the telescope diameter. For a Littrow configuration the spectral resolution $R = 2\tan\beta/d\beta$, where β is the grating angle and $d\beta$ equals the angular slitwidth as seen from the grating. By straightforward analysis one obtains:

$$R_{seeing} = \tan\beta(D_{grating}/D)(r_o/\lambda) \tag{3.19}$$

and

$$R_{diffraction} = \tan\beta D_{grating}/\lambda \tag{3.20}$$

where $D_{grating}$ is the grating size.

For the Keck telescope (D = 1000 cm), $\tan\beta = 2$, $D_{grating} = 30$ cm, 1 arcsec seeing and $\lambda = 1\mu$m one obtains $R_{seeing} = 12000$ and $R_{diffraction} = 600000$, so that adaptive optics results in a greatly enhanced spectral resolution.

Equation 3.19 shows the well known problem of obtaining high spectral resolution for large telescopes. To maintain high spectral resolution one is forced to increase the grating diameter when going to large telescopes. Because of the problems in ruling very large gratings, grating mosaics have been used to do so. But also that technology has its limits. In addition, the larger the grating, the larger the spectrograph tends to be. When using adaptive optics the image size decreases at the same rate as the telescope diameter increases, hence removing the dependance of the spectral resolution on the telescope diameter. Spectral resolutions of 2×10^5 can be obtained with gratings only 10 cm in size and spectrographs of dimensions of approximately one meter in dimension.

3.8.2. *Imagers*

The higher angular resolution dictates smaller pixel sizes when expressed in arcsec. Expressed in geometrical dimensions the diffraction limited image FWHM equals λ F/D (*i.e.* λ times the telescope f-ratio). Since one needs at least two pixels to sample the image properly, one need an f-ratio for the imaging camera of at least 2 x pixel size divided by the wavelength, implying a camera slower than or equal to f/30 for 15 μm pixels and $\lambda = 1\mu$m. This is substantially slower than cameras used for seeing limited imaging.

3.9. *Coronagraphy in Solar- and Nighttime Astronomy*

There is an increasing interest to combine the diffraction limited imaging capabilities of astronomical telescopes with "coronagraphy", especially in connection with the detection of close companions (planets, brown dwarfs etc.) and faint envelopes of stars. With diffraction limited star images it is possible to observe much closer to the star than before. In addition the accompanying object, if within the isoplanatic patch, becomes sharper and gains in contrast with respect to the far wings of the stellar point-spread-function. For the most sensitive observations it is important to keep these wings as faint as possible, that's where coronagraphy comes in. Stellar coronagraphy promises therefore to become an important application of adaptive optics.

The term coronagraphy originates in solar astronomy. To detect the faint outer layers of the sun (the corona, one millionth of the brightness of the solar disk) Bernard Lyot in the 1930's invented the coronagraph, a device in which the instrumental "scattered" light was kept to an absolute minimum. It includes many ingenious concepts to do this. Its main elements consists of: (*i*) a single optical element (the telescope's *"primary"*, at that time a singlet lens), of the highest quality as far as scattered light goes, which forms the solar image, (*ii*) an *occulting disk* in that image which blocks the bright radiation from the solar disk, and (*iii*) a circular aperture placed in the image of the entrance pupil formed by an optic following the occulting disk which blocks most of the light diffracted on the edge of the circular primary. The convention is to refer to this aperture as the *"Lyot stop"*. The largest of such Lyot coronagraphs are the 52 cm aperture Russian coronagraphs and the 40 cm coronagraph at the National Solar Observatory at Sacramento Peak.

For nighttime application the term *coronagraphy* or *coronagraphic cameras* tends to refer to systems which use only part of the Lyot concept, the occulting disk and the Lyot stop. It aims at reducing the scattered light introduced in the camera itself and to reduce the amount of light diffracted from the edges of the primary and secondary mirrors as well as of the structure that supports the secondary mirror (the spiders). However, normally no attention is given to reducing the light scattering from the primary and secondary mirrors themselves. Nonetheless, it is just that light which will ultimately dominate the system's performance, especially for large telescopes. The amount of light scattered resulting from the telescope optics increases proportionally with the telescope's collecting area (D^2) whereas the diffracted light increases linearly with the aperture's diameter (D). The relative contribution of the former increases therefore proportionally with the telescope diameter. For the modest aperture solar coronagraphs the contribution of both is comparable, for large aperture stellar telescopes the former will therefore dominate if no special precautions are taken. Efforts are now underway to develop *true stellar coronagraphs* in which the telescope scattering is reduced to a minimum. They include: (*i*) reducing the imaging optics in front of the occulting disk to an absolute minimum (the primary mirror only), (*ii*) removing the secondary optics (e.g., a Gregorian secondary) and its support from the entering light beam to the telescope by making the primary mirror an off- axis paraboloid, (*iii*) using the very high quality figuring and polishing techniques developed for X-ray telescopes for optical telescopes, and (*iv*) incorporating very high quality dust control in the telescope in order to keep dust from being deposited on the optics. Ideally the primary mirror should be made adaptive in such a coronagraph. Present technology does not permit that yet, so that it will be necessary to control scattering on the surface of the (adaptive) secondary mirror as well as possible too.

4. Interferometric Imaging In Astronomy

4.1. *Principles of Interferometric Imaging*

The principle of interferometric imaging goes back to Thomas Young's experiment in 1801 in which he demonstrated experimentally the wave nature of light by generating an interference pattern using the so-called two pinhole experiment. Figure 6 shows the two aperture experiment schematically in the form which is of interest in the present context.

The configuration shown in Figure 6 is actually the one suggested by Fizeau in 1868, and later used by Stephan, for measuring stellar diameters. Its use is therefore commonly

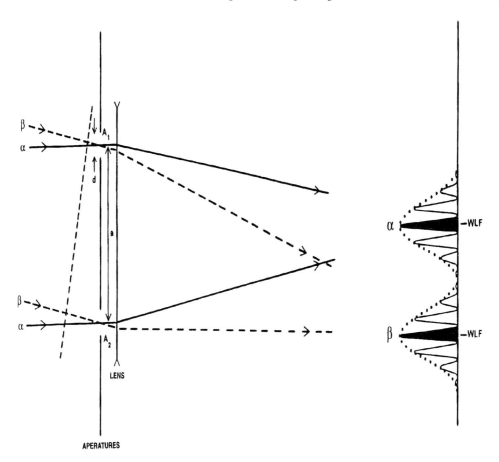

FIGURE 6. Young's fringes for a double star (α, β) with separation θ using two apertures with diameter d and separation a. The double star is imaged by a lens/mirror L. The intensity distribution in the focal plane is a combination of the Airy profile resulting from the individual apertures (width λ/d) and the interference pattern between the apertures (fringe separation λ/a). The black fringes correspond to the white-light fringes (WLF).

referred to as a *Fizeau Interferometer*. In it the fringe separation λ/a in the image plane corresponds to the actual fringe separation on the sky in radians. Also, since the external path difference in front of the apertures and internal path differences in the imaging cancel, the white-light fringe for each double star component coincides with the center of the two images. This behavior, where the white-light fringe coincides with the image of the object, is referred to in the literature as a *Wide Field Interferometer*.

In the configuration shown in Figure 7 the apparent distance a of the two apertures has been reduced optically in order to demonstrate two effects. First, because of the decreased separation $\tilde{a}$ the fringe separation has increased to $\lambda/\tilde{a}$. It does not correspond anymore to the actual fringe separation on the sky (λ/a). This increase in fringe width has been referred to as the *fluffing-out* of the fringes. Alternatively by making $\tilde{a} > a$ the fringes contract. Second, the wide field behavior of the interferometer has been lost. The white-light fringe only is centered on the image when it is located at the center of the field-of-view. Michelson and Pease used an interferometric device similar to the one shown in Fig. 7 in front of the 100 inch telescope at Mt Wilson in order to measure stellar

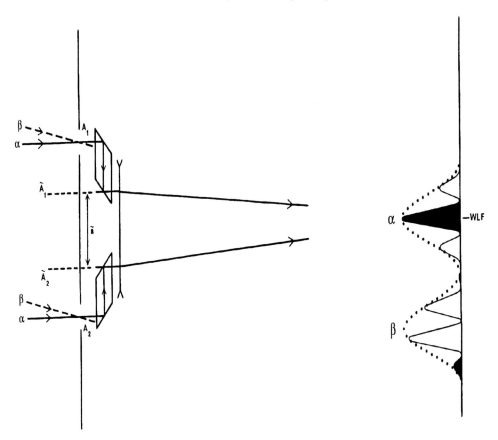

FIGURE 7. As in Figure 6 except that the two apertures have been moved closer together by means of two optical rhombs. It causes the separation a of the two apertures (A_1, A_2) to decrease to the apparent separation ã of the virtual apertures $(\tilde{A}_{-}1, \tilde{A}_{-}2)$ causing the fringes to *fluff-out* to a separation $\lambda/\tilde{a}$. Although the white-light fringe is still centered on the central α star image, it is now displaced from the β double star component.

diameters. This configuration is therefore often referred to as *Michelson Interferometry*. In it the wide field properties of the Fizeau Interferometer are lost.

    Figures 6 and 7 refer to apertures placed in front of a single telescope. The resulting behavior applies, however, also to arrays of separate telescopes in which the light of the different telescopes is combined interferometrically. It is the use of such arrays to which interferometric imaging in optical and radio astronomy normally refers. Figure 8 gives a sketch of an optical interferometer with telescopes on separate mounts. Most arrays are of such *non-monolithic* type; some like the Multiple Mirror Telescope have their telescopes on a single mount and are referred to as *monolithic interferometers*. The interferometer baseline is L, Lcosζ as seen from the stellar direction. The external path difference between the two telescopes (Lsinζ) is compensated by an optical delay line and by the location of the beam combination optics, thus forming a white-light fringe on an on-axis star.

    Table 5 list a number of optical interferometers which are currently in operation or which are being planned. Let me elaborate on a number of aspects of these interferometers.

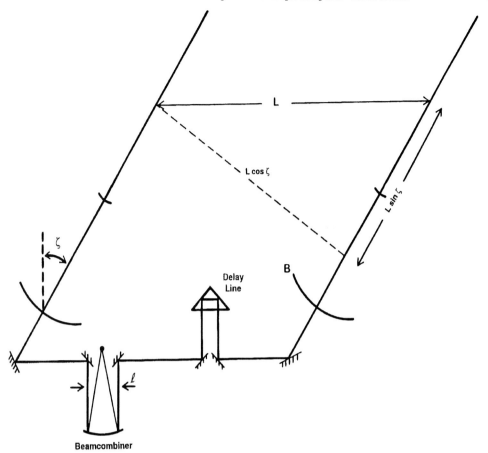

FIGURE 8. Schematic of *Non-Monolithic* Optical Interferometer consisting of two telescopes on individual mounts.

4.1.1. *Astrometric Interferometers*

Astrometric interferometers place high requirements on the precise measurement of the baseline of the interferometer. The measurement of the fringe displacement (using the variable delay line) combined with this baseline then results in very precise measurements of star distances.

4.1.2. *Imaging Interferometers*

Many two element non-astrometric interferometers measure fringe contrast as a function of telescope baseline and wavelength. From that they derive either stellar diameters or binary star separations. True imaging interferometers have at least three elements and measure both fringe contrast and the so-called closure-phases from which it is possible to reconstruct images of the astronomical object using temporal and spatial image synthesis techniques already developed for radio interferometers.

4.1.3. *Wide Field-of-View Interferometers*

In most interferometers the distance ã between the telescope pupils (Fig. 7) is made equal to zero, resulting therefore in a very small field-of-view. The field-of-view is in fact almost equal to zero, corresponding to the size of the Airy disk of the individual

Name	Location	N× D (cm)	L (m)	Comments
I2T	CERGA/F	2 x 26	12	
Mark III	Mt. Wilson/USA	2 x 5	31.5	astrometric
GI2T	CERGA/F	2 x 150	65	
SOIRDETE	CERGA/F	2 x 100	15	
ISI	Mt Wilson/USA	2 x 165	32	heterodyne
MMT	Mt. Hopkins	6 x 180	5	imaging, monolithic
COAST	Cambridge/UK	4 x 40	30	imaging
SUSI	Culgoora/AUS	2 x 14	640	
IOTA	Mt. Hopkins/USA	3 x 45	38	imaging
NPOI-I	Flagstaff/USA	4 x 12.5	38	astrometric
NPOI-II	Flagstaff/USA	6 x 12.5	437	imaging
ASEPS-0	Mt. Palomar/USA	2 x 40	100	astrometric
CHARA	*Mt. Wilson/USA*	*7 x 100*	*354*	*imaging*
VLTI	*Co. Paranal/Chile*	*4 x 800*	*130*	*imaging, wide field-of-view*
		N x 180	*200*	
KIA	*Mauna Kea/USA*	*2 x 1000*	*85*	*imaging*
		6 x 150	*165*	
LBT	*Mt. Graham/USA*	*2×820*	*14*	*imaging, monolithic*
Magellan	*Las Campanas/Chile*	*2 x 650*		

TABLE 5. Summary of Existing and *Planned (italics)* Optical Astronomical Interferometers. N = number of Elements, D = telescope diameters, L = maximum baseline.

aperture only. In monolithic interferometers like the MMT and the LBT it is quite easy to mimic the wide-field-of view interferometer case of Fig. 6 by proper combination of the beams. It requires that the interferometer's exit pupil (= the pupil configuration as seen from the combined image plane) equals its entrance pupil (= the pupil configuration as seen from the direction of the astronomical object) at all points of the field-of-view. This "Pupil In = Pupil Out" condition has to hold in detail in the sense that both the dimensions, separations, rotation, and handedness of the pupils have to be preserved. For non-monolithic interferometers it is much harder, but not impossible to achieve that same condition. The design of the VLT Interferometer in fact does that. It will be described in more detail in Section 4.5.

4.1.4. *Nulling Interferometers*

A "Nulling Interferometer" is an $\tilde{a} = 0$ configuration in which great care is taken to achieve fully destructive interference in the combined on-axis image. It implies a half wave phase shift between the (two) beams being combined and two wavefronts which interfere fully destructively over the entire pupil. It poses high demands on the adaptive optics systems (unless very small telescopes are being used). Nulling interferometers have been proposed for the detection of extra-solar planetary systems in which the light of the bright central star is suppressed interferometrically.

4.1.5. *Heterodyne Interferometers*

Radio interferometers use heterodyne techniques in which the received signal is mixed with the signal of a single local oscillator at the individual telescopes and in which the intermediate frequencies are then mixed and interfered electronically. Since the radio frequencies are relatively small to begin with, the limited bandwidth of the electronics in handling the intermediate frequencies results in rather large bandwidths in the radio signal detection. Heterodyne techniques using a 10.6 μm laser as local oscillator are being used at the ISI interferometer(see e.g., Tab. 5 The high frequency optical signal results, however, in very small bandwidths and therefore in relatively small sensitivity.

4.1.6. *Adaptive Optics and Optical Interferometers*

Many of the optical interferometers listed in Tab. 5 use small apertures, less than or comparable to r_o in diameter. In those interferometers the wavefront incident on each telescope is reasonably flat and tip-tilt control in the beam combining optics suffices to achieve good interference over the entire pupil. For larger telescopes the wavefront distortions across the apertures destroy the interference. Large telescope interferometers therefore require adaptive optics to achieve maximum sensitivity. However, even without adaptive optics, with the interference occurring in the image plane, the individual interfering speckles in the image show interference fringes of high contrast, differing in fringe position in each speckle.

4.1.7. *Atmospheric Dispersion*

In astronomical telescopes the term "atmospheric dispersion" refers normally to the displacement of the star images in the vertical direction varying for different wavelengths caused by the refraction of light in the earth atmosphere. In interferometry that is referred to as "lateral atmospheric dispersion" to contrast it with "longitudinal atmospheric dispersion." Longitudinal dispersion refers to the refractive path length differences in air for different wavelength resulting from the dispersion of the air refractive index. For an interferometer fully located in air the difference in pathlength in the air equals $L\sin\zeta$ and the longitudinal atmospheric dispersion equals hence $L\sin\zeta\{1 - n(\lambda)\}/\lambda$ waves. If broadband sensitivity is wanted, as is almost always the case, this dispersion needs to be compensated by other refractive means, generally using a combination of different glasses.

4.2. *Differences Between Interferometric Imaging in Optical and Radio Astronomy*

Although radio astronomy as a discipline is much younger than optical astronomy, interferometric imaging has blossomed in radio astronomy for years whereas the astronomical results from optical astronomy are still very limited. Because of the maturity of interferometric imaging in radio astronomy, much of optical imaging will benefit from the techniques and methodology developed there. Both because for that reason and in order to increase the understanding of those familiar with the radio astronomy interferometry it is of interest to compare both techniques. This is done in Table 6.

4.2.1. *The Closure Phase Concept*

One important concept derived from radio astronomy interferometry is the concept of closure phase. From the interference fringes both the fringe amplitude and fringe phase is determined. The amplitudes, after proper calibration, result in the contrast of the image structure at different spatial frequencies, the phase in their location on the sky. If the atmosphere, telescope, delay line, etc. introduce an unknown phase error φ_i in the electromagnetic wave coming from interferometer element i, the observed phase ϕ_{ij}^* of the interference fringe between interferometer elements i and j equals:

$$\phi_{ij}^* = \phi_{ij} + (\varphi_i - \varphi_j) \qquad (4.21)$$

where ϕ_{ij} is the true phase. In a three telescope interferometer containing three baselines, and hence three spatial frequencies in the astronomical image, the sum of the observed phases equals

Item	Radio	Optical	Consequence for Optical Interferometry
Wavelengths	0.5 to 30 cm	0.5 to 30μm	• tolerances are 10 000 times harder • baseline can be 10 000 times shorter
Frequencies	1 to 60 GHz	10 to 600 THz	• use of heterodyn techniques is very limited (Δ f/f < 10^{-4}) • direct mixing of light needed
Seeing (r_0)	$\gg$ telescope	$\leq$ telescope	• often works in multi-speckle mode • adaptive optics needed
Interference	electrically	direct	• needs optical delay lines • needs beam splitters, resulting in light loss
Number of Telescopes	any	limited (≤ 6)	• SNR losses due to beamsplitting offset gain from added baselines
Detectors	"single feed" detector	panoramic	• wide FOV detection attractive
Noise Source		vis.: photon IR: thermal	• SNR behavior very different
Imaging Algorithms	similarity for both disciplines, but also differences due to SNR and speckle presence		• can copy some of the radio methods • different algorithms often needed as well

TABLE 6. Comparison of Radio and Optical Interferometric Techniques

$$\phi^*_{12}+\phi^*_{23}+\phi^*_{31} = \phi_{12}+(\varphi_1-\varphi_2)+\phi_{23}+(\varphi_2-\varphi_3)+\phi_{31}+(\varphi_3-\varphi_1) = \phi_{12}+\phi_{23}+\phi_{31} \quad (4.22)$$

so that the unknown atmospheric/instrumental phase disturbances disappear. The observed closure phases ($\phi^*_{12} + \phi^*_{23} + \phi^*_{31}$) therefore relate the phase of the third spatial frequency to the phase of two other spatial frequencies. Provided that the phases at two spatial frequencies are known the true fringe phase at a third can therefore be determined. In generating the image these first two spatial frequencies are generally taken to be very small (short baselines). Their phases are then arbitrarily assumed to be zero, implying an assumed position of the unresolved object on the sky. By phase "triangulation" the phases at other, higher, spatial frequencies are then determined.

In addition to closure phase techniques, radio astronomy uses amplitude closure techniques as well. It allows the calibration of the gain of the different antennas. In optical interferometry that has not (yet) been done. There are probably better techniques available for gain calibration at optical wavelengths.

4.3. *Limiting Sensitivities in Interferometric Imaging*

Optical interferometric imaging is in its infancy, and experimental limits to its sensitivity have therefore not been established. Estimates of limiting sensitivities are therefore based entirely on theoretical considerations. Many factors are involved in making such estimates, and, in answering the question "What is the limiting sensitivity?", many qualifiers have to be given before the question can be answered. In this section I will describe some of these factors.

4.3.1. *Instrumental Causes for Fringe Contrast Decrease*

The basic observable in interferometry are interference fringes. It is important that their contrasts be maintained as high as possible by minimizing the effects of the many factors which contribute to their deterioration. These include the following:

Fringe Position Variations in the Exposure

These can be caused by a number of different causes, including: (i) **Vibrations** of the elements in the interferometers due to a variety of causes (e.g., microseismic activities, wind forces, man-made disturbances). Tolerances on pathlength changes due to vibrations are extremely tight ($\approx \lambda/40$ RMS) but refer to relatively short time scales set by the time scale of the temporal phase variations caused by atmospheric seeing (wavelength dependant; ≈ 50 msec at 2.2 μm), (ii) **Photon noise** when using **fringe tracking** techniques on nearby bright reference objects, (iii) **Phase variations across the pupil** resulting from imperfect optics, limited adaptive optics correction of spatial phase variations in the atmospheric wavefront, and (iv) **Phase variations within the spectral band used**. These become especially severe when the pathlengths are not closely equalized (loss of white light fringe) and when the longitudinal atmospheric dispersion compensation is inadequate.

Differential Tracking Errors

Resulting e.g. from noise in the autoguiding systems and from differential seeing motions with a nearby star used for autoguiding. It is important that the Airy disks of the telescopes coincide to a fraction of their diameter.

Polarization Effects

Differential polarization effects between the elements of an interferometer can completely destruct the interference fringes. For the light to fully interfere in the beamcombiner it is essential that either the incident polarization characteristics are preserved or that they are modified in the same way. The so-called (4 x 4) Müller Matrix which describes how the 4 Stokes parameters at the beamcombiner relate to the incident Stokes parameters has therefore to be the same for all interferometer elements. This implies careful design which takes into account : (i) identical (linear) retardation effects resulting e.g. from non-normal reflections, (ii) identical partial polarization effects, (iii) identical (rotational) retardation effects related to e.g. the rotation of the directions during the optical transfer of the beams.

Unequal Beam Intensities

Even a transmission difference of as much as 10% causes the contrast of the interference fringes to decrease only by 0.13% so that the tolerance on beam transmission equality is quite relaxed.

Detector Modulation Transfer Function

When observing closely spaced interference fringes the resolution of the panoramic detector can be a factor in reducing fringe contrast. For limited resolution detectors the fringes are therefore often "fluffed-out" until in the extreme the pupils fully overlap, and only a single detector pixel suffices.

Pupil Transfer Geometry

When a field-of-view larger than the Airy Disk is wanted the "Pupil In = Pupil-Out" geometry has to be maintained (see Section 4.1) if high fringe contrast is to be maintained for all image points in the field-of-view.

Unequal Telescope Apertures

Most interferometers use equal apertures for all their telescopes. Some interferometers in their planning phases (the VLTI and KIA) will use telescopes of unequal apertures. In that case interferometry becomes quite complex. When doing interferometry in the image plane (as in Figure 6) unequal image scales for all telescopes imply that the Airy disks of the telescopes do not correspond to each other and that the larger telescopes put much larger energy in a much smaller image area as compared to the smaller telescopes thus destroying most of the interference. Instead one should match the size of the Airy disks, at the cost, of course, of unequal image scales and a very small field-of-view.

If the interference occurs in the pupil plane it is similarly beneficial to match the two pupil sizes. In either case the intensities are sufficiently different to cause a significant fringe contrast decrease. It has been shown that in such case the effective interferometer sensitivity for two telescopes with unequal apertures A1 and A2 equals that of an interferometer with two equal sized apertures A where A is the geometrical average of A1 and A2 or $A = \sqrt{(A1 A2)}$.

Calibration

Because the sources of fringe contrast decrease are many, often seeing dependent and hence difficult to estimate, it is important to calibrate interferometers on bright unresolved stars. Even then, variable conditions will limit the precision of such calibrations.

4.3.2. *Assumptions Used in Calculating Limiting Sensitivities*

In calculating the sensitivity limits of optical interferometers, as those shown in Figure 9, a number of assumptions were made, some of which are listed on the figure (8 meter aperture telescopes like in the VLTI; 0.75 arcsec seeing, typical for Cerro Paranal; a SNR for fringe detection of 10; an interferometer emissivity of 50 % and a detector read-out noise of 100e). In addition the individual exposure times are set at r_o/V_{wind} with $V_{wind} = 10$ m/sec, the spectral bandwidth is taken as the standard bandwidth for the different spectral classes, the object is assumed to be totally unresolved (incident fringe contrast = 100%); the instrumental MTF in fringe contrast is taken as 70%, and the total exposure time as 10 minutes. It is also assumed that the telescopes are equipped with adaptive optics, although where indicated below, this may mean "partial adaptive optics" only, i.e., the adaptive optics system is designed for the K-Band but is still partially operating at shorter wavelengths. In the real astronomical sources of interest the objects are of course resolved and the fringe contrast is therefore substantially less than 100%, the amount depending on fringe frequency/interferometer baseline. These sensitivities are therefore upper limits. At these flux/magnitude limits it will just be possible to tell whether an object is resolved or not.

4.3.3. *Sensitivity for Self Referenced Wavefront Sensing and Fringe Tracking*

In case only the astronomical object is available to sense the wavefront for the adaptive optics system and in case that can only be done in the wavelength band being observed, the curve labelled I in Figure 9 gives the interferometer sensitivity. It is effectively the limit set by the adaptive optics system (Section 3.4). This sensitivity limit is extremely pessimistic since wavefront sensing can in general be done otherwise, either by using polychromatic wavefront sensing in other spectral bands (Section 3.3) or by using laser guide stars (Section 3.7). At the short wavelengths the dotted lines refer to full adaptive optics systems, the full-drawn lines to partial adaptive optics system as described above. The latter is of course more sensitive, however, performs worse.

4.3.4. *Sensitivity for Other Referenced Wavefront Sensing and Self Referenced Fringe Tracking*

In general polychromatic wavefront sensing in other spectral bands or wavefront sensing using laser guide stars will probably be available by the time large imaging interferometers like the KIA and the VLTI are placed in operation. In that case the sensitivity limits will sharply increase. Assuming that the active (white light) phasing of the interferometer still is necessary on the object being observed, the fringe sensing on this object sets the limiting magnitude. It is shown in the curve labelled II in Figure 9. The partial adaptive optics curve is, of course, less sensitive than the full adaptive optics system.

These sensitivities are of interest primarily for galactic astronomy applications, with

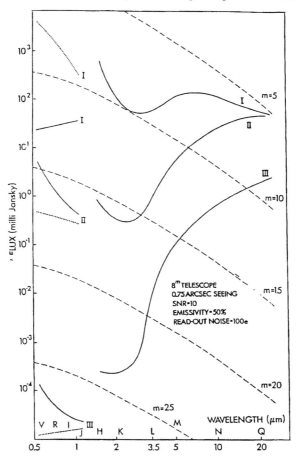

FIGURE 9. Sensitivity Estimate for the 8 meter VLTI Array. The Dashed Lines are the Apparent Magnitudes at the Wavelengths Indicated. For a Detailed Explanation See the Text.

only a few extra-galactic objects being bright enough to be observable. For most extra-galactic observations one of the two following techniques has to be used to get down to the magnitude range of interest.

4.3.5. *Sensitivity for Other Referenced Wavefront Sensing and Fringe Tracking*

Sometimes it is possible to actively phase the interferometer on a nearby object (within the isoplanatic patch for co-phasing) or even, perhaps, on the object being observed but at a different wavelength. In that case long exposures of the interference pattern can reach the very faint magnitudes shown in the curve labelled III in Figure 9. As is the case for the sky coverage for adaptive optics systems using natural guide stars, the sky coverage for interferometric imaging using this technique is very limited. Nonetheless because of their large abundance, observation of many high red-shift galaxies and a number of QSO's become possible. The VLTI has in its design included an 8 arcsec diameter interferometric field-of-view both to make this mode possible but also to co-phase the VLTI for a number of galactic objects of great interest like the supergiants αSCO and αHER as well as the galactic center.

4.3.6. *Sensitivity for Other Referenced Wavefront Sensing and "Coherencing"*

To obtain full sky coverage the so-called "blind" operation of the interferometer be-comes necessary. This can, for example, be done by co-phasing of the interferometer on a nearby "bright" unresolved object, correcting it for the differential path difference be-tween it and the object of interest and then tracking the phasing "open loop" correcting for position and systematic environmental changes. Exposure times have to be small (r_o/V_{wind}) since the atmosphere will move the fringes around. Sensitivity requires there-fore the statistical analysis of the photon-noise limited fringe observations using many observations covering a long time span. Correlation and triple correlation techniques, or their Fourier transform equivalent, are commonly used to accomplish this.

Another, perhaps better, way of blind operation consists in tracking the phasing of the interferometer on an unresolved object which lies well outside the isoplanatic patch for co-phasing. Doing so will of course still move the fringes around in the interferogram of the object, thus requiring a sequence of many short exposures. But at least, provided that the reference object is not too far away, one can be assured that there are fringes in the object interferogram. In contrast to the term "co- phasing" used above, I have referred to this mode as a "coherenced interferometer." How far away can a reference for coherencing be, depends on the spectral bandwidth used for the interferometry. The maximum bandwidth that can be used depends on the amount of atmospheric phase variations, which depends in turn on the interferometer baseline and on the upper scale of turbulence. To achieve full sky coverage one does not statistically need to go further than 0.25 degrees for finding a coherencing object. The design of the Nasmyth opto-mechanics of the VLT telescopes allows for the optical transfer of any object within such a distance to the 8 arcsec interferometric field-of-view. Calculations show that coherencing of the VLTI 8 meter telescopes will result in limiting magnitudes close to m = 20 in the V- through K-Bands.

4.4. *Results of Optical Interferometry*

So far there are no results yet from optical imaging interferometers of the non-monolithic kind. The only imaging produced so far used arrays of aperture masks in front of a single aperture telescope. With it the techniques of optical interferometric image reconstruction has been demonstrated, including the use of closure phase techniques. The current deployment of the COAST, SUSI, IOTA and NPOI-II imaging interferometers should result in interferometric imaging with arrays of telescopes in the not too distant future.

4.5. *ESO's Very Large Telescope Interferometer (VLTI)*

Because of my familiarity with the VLT Interferometer and because it probably has as much optical sophistication designed into it (it is the only wide field-of-view interferom-eter), I want to describe it in some detail. Figure 10 shows the layout of the VLTI in the 2650 meter high mountain Cerro Paranal in Northern Chile.

4.5.1. *Layout of VLTI Main Array*

The four 8 meter telescopes are arranged in a fully non-redundant configuration in order to maximize their (u,v) plane coverage. The four large telescopes in earth rotation synthesis at elevations above $30°$ result in six tracks in the (u,v) plane, each track with a considerable width because of the relatively large apertures of the individual telescopes. Figure 11 (left) shows this (u,v) plane coverage and the resulting point-spread-functions for different declinations on the sky.

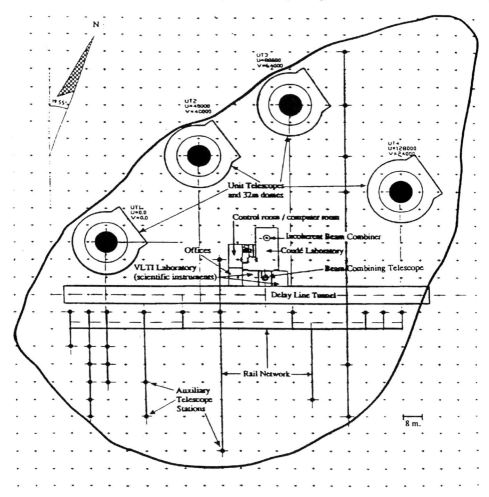

FIGURE 10. Configuration of the Very Large Telescope Interferometer on Cerro Paranal. The stationary 8 meter telescopes are shown as the large black circles, their diameter corresponding to the telescope diameter (8 meter). The smaller black circles located on the rail tracks show the stations for up to 8 mobile 1.8 meter auxiliary telescopes.

4.5.2. *Layout of VLTI Sub-Array*

In addition to the stationary 8 meter telescopes, the VLTI foresees the addition of a number of (up to 8) smaller 180 cm diameter auxiliary telescopes which can be moved around on tracks between a number of auxiliary telescope stations. In Figure 10 these stations are indicated by small black circles. The purpose of these telescopes is primarily to provide better (u,v) coverage, filling in the holes left between the 8 meter array tracks. Figure 11 (right) shows the resulting improvements for a selected set of 3 auxiliary telescopes placed on 5 selected telescope stations. The auxiliary telescopes serve other important purposes as well like : (i) they allow larger baselines (200 meters vs. 130 meters for the 8 meter telescope array), (ii) they allow for the use of the VLTI complex at any time, even when the large telescopes are being used for other purposes, (iii) they provide for a quality interferometric array when the highest sensitivity provided by the large telescopes is not needed. The VLTI array without its 8 meter telescopes will in fact still be the most sensitive interferometric array being build or planned (see Table 5).

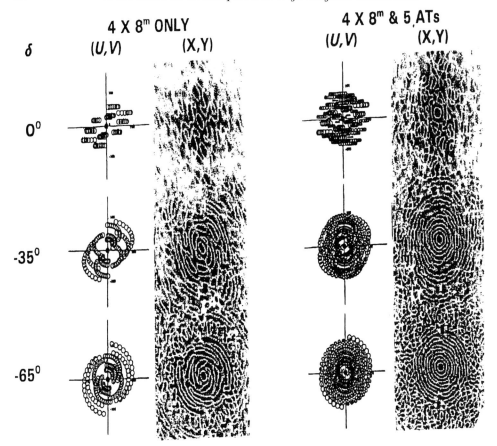

FIGURE 11. (u,v) Plane Coverage by VLTI Using the Four Large Telescopes Only (left) and Using the Large Telescopes Combined with Three or More Auxiliary Telescopes Placed on Five Auxiliary Telescope stations.

4.5.3. *VLTI Optical Configuration*

In the design of the optical paths of all telescopes, both large and auxiliary telescopes, care has been taken to arrange the optics in such a way that the image and pupil rotation and scaling and the polarization behaviors are identical. The light of all telescopes is relayed from the Nasmyth focus to a stationary coudé focus by means of 5 reflections one of which is on an adaptive mirror. Together with the three telescope mirrors feeding the Nasmyth focus, this results in 8 reflections before the coudé focus. Figure 12 shows what happens next following the ninth mirror (M9) near the coudé focus which folds the coudé beam horizontally towards the interferometer proper.

In the coudé relay optics, module M10 is located at or near the folded coudé image. The mirror is curved, causing an image of the telescope pupil to be formed near the central interferometry area (near M15). M10 has a tip-tilt mechanism to control the location of the pupil image. Since it is located at the image plane, it does not affect the location of the astronomical image. That is controlled by the tip-tilt control of the collimating mirror M11 which directs the beam to a folding flat M12 and from there to the movable delay line which consists of two mirrors (three reflections) M13/M14. M13, like M10, is located at a sky image. It is an active mirror in the sense that

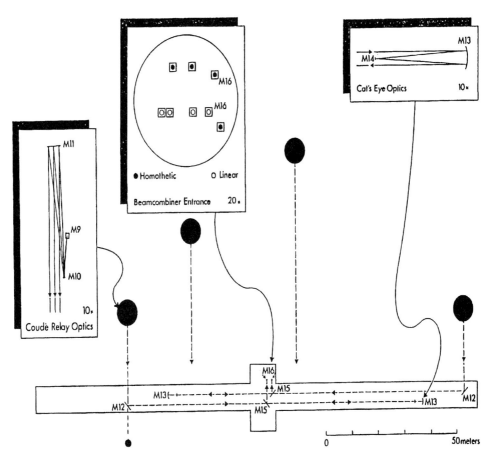

FIGURE 12. Optical Configuration of VLTI. The Insets Show More Details of the Optical Component Modules.

its curvature is variable so that the pupil image near M15 is re-imaged at that same location for the varying positions of the delay line. At the same time M14 is equipped with piston motion to allow for very small pathlength adjustments, the large adjustments being made by the delay line carriage itself. The collimated star light exiting from the delay line is then relayed to the beamcombining area via the flat mirrors M15. In Figure 12 this beamcombining is done by means of a vertically looking "beamcombining telescope" which is fed by the flat mirrors M16. Assuming two more reflections in this beamcombining telescope, one arrives at the combined focus after 19 reflections (counting M13 twice)! Large numbers of reflections are typical for optical interferometers. To maintain a reasonable high throughput and low emissivity it is therefore essential to make all coatings as reflective as possible. This is impossible at short wavelengths which is one reason why interferometers don't extend into the violet spectral regions. In the optical arrangement shown in Figure 12 the polarization properties and image/pupil attitudes are maintained throughout until the light reaches the beamcombining laboratory.

4.5.4. *Beam combination*

There are a number of different ways in which the beams can be combined in an interferometer. Figure 12 shows one option in which the images are interfered, resulting in interference fringes. The pupil images formed by M10/M14 are formed on the mirrors M16. By adjusting the positions of the mirrors M15 and M16 it is possible to arrange for any pupil configuration.

The configuration labelled "Linear" places the pupils on a line with non-redundant spacings so that the 6 fringe patterns in the combined image plane are parallel and separable in their spatial frequencies. It is the preferred configuration when doing spectroscopy of the interference pattern. It converts the six spatial frequencies with different directions on the sky into a fringe pattern which is ordered in an optimum way for spectroscopy. It has of course one major disadvantage: the "Pupil-in = Pupil-out" condition is violated, so that it has a very small interferometric field-of-view (effectively limited to the Airy disk).

The configuration labelled "Homothetic" preserves the "Pupil-in = Pupil-out" condition and is therefore the wide field-of-view configuration. The pupil image configuration equals that of the telescopes as seen from the pointing direction of the VLT. Since this configuration changes continuously with the pointing direction variation, the pupil image configuration has to change concomitantly.

4.5.5. *Instrumentation*

The instrumentation for detecting the interference in the VLTI can take many shapes depending on the science objectives. They may include: (i) spectrographs to be used, e.g., with the linear configuration, (ii) imagers using broad or narrow band filters, including, e.g., scanning Fabry- Perot interferometers to build-up a 3-D data cube (x,y, λ), (iii) fast detectors capable of very short exposures in case co-phasing of the array on a nearby reference object is not possible, and (iv) detectors and instruments covering the wide wavelength range provided by the VLTI.

5. CONCLUSION

Optical astronomy is undergoing an exciting renaissance in instrumental capabilities and in the discoveries it provides. Large ground-based telescopes result in unparalled sensitivities for our observations, but even more, they give angular resolutions which could not be achieved with smaller telescopes. The rapid developments in adaptive optics enables the best use of this high angular resolution. The next frontier will be the imaging with the super resolutions provided by interferometric arrays. These very sophisticated techniques have become possible because of advances in opto-mechanics and control techniques. Astronomy has many fascinating new discoveries to look forward to!

REFERENCES

Armstrong, J.T., Hutter, D.J., Johnston, K.J. and Mozurkewich, D., 1995, *Physics Today* **48(5)**, 42.

Babcock, H.W., 1953, *Publ. Astr. Soc. Pacific* **65**, 229.

Beckers, J.M., 1993, *Ann. Rev. Astron. & Astrophys.* **31**, 13.

Benedict, R., Breckenridge, J.B., Fried, D.L., 1994, *J. Op. Soc. Am* **11A**, 257 - 451; 783 - 945.

Cargese School on *Adaptive Optics for Astronomy*, eds D.M. Alloin & J.M. Mariotti, 1994

Christiansen, W.N. and Högbom, J.A., 1985, "Radio Telescopes", *Cambridge Un. Press*

ESO Conference & Workshop Proceedings, 1988 **29**.

ESO Conference & Workshop Proceedings, 1991, **39**.

ESO Conference & Workshop Proceedings 1993, **48**.

Fizeau, A.-H.-L., 1868, *Comptes Rendues Acad. Sci.* **66**, 932.

Hecquet and Coupinot, *J. Optics*, 1985, **16**, 21.)

Proceedings IAU Symposium 158 on "High Angular Resolution Imaging", eds. J.G. Robertson & W.J. Tango, 1994

Linnik, V.P., 1957, *Opt. & Spektrosk* (USSR) **3**, 401.

Michelson, A.A., 1891, *Nature* **45**, 160.

Noll, R.J., 1976, Zernike Polynomials and Atmospheric Turbulence. *J. Opt. Soc. America*, **66**: 207 - 11.

Roddier, F., 1981, The Effects of Atmospheric Turbulence in Optical Astronomy. *Progress in Optics* **19**: 281 - 377.

Roddier, F., 1987, Seeing and Atmospheric Turbulence : Parameters Relevant to Adaptive Optics. *Proceedings of LEST Workshop on Adaptive Optics in Solar Observations*, Merkle, F., Engvold, O. and Falomo, R., eds.,(LEST Technical Report No. 28), 7 - 15.

Roddier F., 1988, *Physics Reports* **170**, 2.

Roddier, F., 1989, Optical Propagation and Image Formation through the Turbulent Atmosphere. *Proceedings of NATO Workshop on "Diffraction Limited Imaging with Very Large Telescopes* (eds. J.-M. Mariotti and D. Alloin), pp. 33 - 52, Kluwer Academic Publishers.

Roddier, F., et.al., 1996, *in preparation*, Cambridge University Press.

Shao M. and Colavita, M.M., 1992, *Ann. Rev. Astron. & Astrophys.* **30**, 457.

SPIE Proceedings, 1990, **1237**.

SPIE Proceedings, 1994, **2200**.

SPIE Proceedings, **365**, 1983

SPIE Proceedings, **1114**, 1989

SPIE Proceedings, **2201**, 1994

Thompson, A.R., Moran, J.M. and Swenson, G.W., 1986, "Interferometry and Synthesis in Radio Astronomy", *Wiley.*

Thompson, L., 1995, *Physics Today* **47(12)**, 24.

Wohlleben, R., Mattes, H. and Krichbaum, Th., 1991, "Interferometry in Radioastronomy and Radar Techniques", *Kluwer.*

Woolf, N.J., 1982, High Resolution Imaging from the Ground. *Annual Rev. Astron. Astrophys.*, **20** , 367 - 98.

Detectors and Data Analysis Techniques for Wide Field Optical Imaging

By M. J. IRWIN

Royal Greenwich Observatory, Madingley Road, Cambridge, CB3 0EZ, UK

This contribution reviews the current status of optical wide field survey astronomy and the basic techniques that have been developed to capitalize on the large volumes of data generated by modern optical survey instruments. Topics covered include: telescope design constraints on wide field imaging; the properties of CCD detectors and wide field CCD mosaic cameras; preprocessing CCD data and combining independent digitized frames; optimal detection of images and digital image centering and photometry methods. Although the emphasis is geared toward optical imaging problems, most of the techniques reviewed are applicable to any large format two-dimensional astronomical image data.

1. Wide Field Survey Astronomy

1.1. Background

Astronomy is basically an observational science, rather than an experimental one, and the development and advancement of the subject has relied heavily on surveys of the sky at optical wavelengths to expand our knowledge of the observable Universe. Surveys form a basic foundation of observational astronomy, and provide three generic types of information:

(*a*) quantitative statistical information on the distribution of objects in our own galaxy and the Universe

(*b*) the ability to discover radically new types of object

(*c*) the means of selecting representative samples of certain types of (rare) objects, particularly the brightest examples, for further study with large telescopes.

Statistical surveys are beginning to rely ever more heavily on the wide field multi-object fibre spectroscopy capabilities of large telescopes, described elsewhere in this volume. However, surveying for rare objects, or even discovering new categories of objects, is more problematic, since often their surface densities on the sky are sufficiently low to render fibre spectroscopy approaches grossly inefficient.

We can investigate the problems in surveying for rare types of object and estimate the chances of discovering new classes of object by borrowing a statistical analogy from biological systems (Thejll & Shipman, 1988). In studies of the diversity of a population of butterflies, Fisher, Corbett & Williams (1943) proposed a simple relationship linking the sample size, N and the number of species present, S, such that

$$S = A \times ln(1 + N/A) \tag{1.1}$$

where A the diversity index, parameterizes the tendency of the population to produce separate sub-species. Although, the formula is necessarily a simplification of a complex classification process, Fisher *et al.* argued that for any procedural sub-classification scheme based on combining distinguishing factors multiplicatively, a logarithmic series provides a good practical approximation. In particular, if a similar survey has already been done, the above formula enables the diversity parameter A to be derived and subsequently used in predicting how many further observations are likely to be needed in order to, say, find one new species.

As an example, Thejll & Shipman (1988) applied this method to the Palomar–Green UVX survey covering some 10,000 deg^2 of Northern sky to B= 16 and find

$S = 11; N = 1550; \Rightarrow A = 1.6$ – stellar and

$S = 6; N = 165; \Rightarrow A = 1.2$ – non–stellar

where the categories included the first gravitational lens, ultrahot stars and new types of variables. Demers *et al.* (1986) are undertaking a similar UVX survey in the Southern sky and figure 1 illustrates how application of the Fisher *et al.* formula leads to a prediction of how much detailed follow up of candidates is required to find one or more new species. Selecting candidates in a UVX survey is relatively straightforward and cheap in terms of telescope time. On the other hand spectroscopic follow up of individual objects is a relatively costly exercise and the Fisher *et al.* formula enables researchers to assess how much telescope time is likely to be needed to reach some defined goal.

Naturally, simply looking at more objects is not the only way to find new species or find more examples of rare objects as the following points emphasize:

• Increasing N exponentially usually $\Rightarrow$ more sky coverage; this can be a problem at bright magnitudes.

• Alternatives are to survey in new wavebands, particularly outside the wavelength range normally sampled.

• Survey at different epochs, ie. time series sampling, which can give orthogonal information to conventional spectroscopy or broad-band wavelength coverage.

• Survey to fainter levels and potentially sample a different spatial volume, eg. larger look-back time for a cosmological sample.

• Higher accuracy measurements enable 'normal' object locii to be better defined, hence the parameter phase space volume for unusual objects is increased and they are easier to delineate.

1.2. *General Survey Constraints*

It is well known that reliable quantitative surveys form a vital constituent in the future success of any large telescope programme, just as the surveys based on photographic plates taken with the Palomar Schmidt and UK Schmidt have been vital to the success of the present generation of large (ie. 4m class) telescopes. Survey instruments and their associated follow up facilities must however be well matched. As an example, consider the current surveys on 1.2m Schmidt telescopes which reach limiting magnitudes equivalent to $V = 21 - 22$ in 1 hour exposures. These are well matched to follow up spectroscopy on 4m-class telescopes, where ~ 5Å resolution spectra at $V = 21 - 22$ take of order 1 hour exposure time. The "natural" survey time on these Schmidt telescopes is approximately 1 hour, since that is when most exposures are limited by sky noise rather than emulsion/fog noise. To go 1 magnitude fainter than a sky-limited plate would require an impractical $\times$ 6.25 increase in exposure. Contrast this with the equivalent numbers for a modern CCD on a 2–3m class telescope, where the limiting magnitudes would be $V = 24 - 25$ in a 10–15 minute "natural" survey time exposure. Clearly, to do spectroscopic follow up of objects detected near the limit of these types of CCD surveys will require at least 8m-class telescopes. Yet without suitably deep survey material, where are the input targets for 8m-class telescopes going to come from ?

Although the importance of next generation survey work is self-evident, it is only relatively recently that practical work has been initiated to develop survey instruments beyond the Palomar and UK Schmidt photographic surveys. The competitiveness of

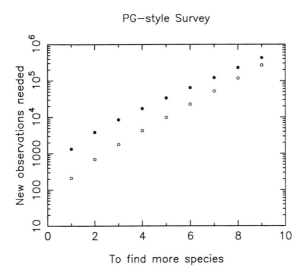

FIGURE 1. The Fisher *et al.* (1943) formula applied to the Palomar–Green survey; open symbols galaxies, filled symbols stars. Note that to significantly increase the number of new types of object found requires an exponential increase in the number of observations.

the very large telescope programmes in fundamental areas of research such as large scale structure and the formation and evolution of galaxies and AGN, critically depends on the quality of the survey work. We cannot efficiently run 8-16m telescopes in the next decade if we do not have at the same time proper survey facilities with which to select the best objects to follow-up. Fortunately, CCD mosaic cameras are now becoming available and are becoming competitive with traditional photographic survey methods. This provides a good excuse to examine the current wide field survey capabilities in optical astronomy and to assess the requirements for future optical survey material.

1.3. *Current Situation of Optical Sky Surveys*

Despite the considerable advances of detector technology within astronomy, very little improvement has been made in surveys beyond those available in the 1950's when the Palomar Sky Survey was carried out. A photographic plate taken on a 1.2m Schmidt is sky limited in about 1 hour, covers an area, A, on sky of some 40 square degrees, but is only $\approx 2\%$ efficient, η. We can characterize the overall effectiveness of the system by the product of efficiency and area covered. For photographic plates, $\eta \times A \approx 0.5$. The current best sky surveys therefore amount to no more that a 120 second glance of the Universe and all the recent dramatic improvement in exploiting the existing survey material is due entirely to the advent of the fast automatic plate measuring machines (for a review see Lasker, 1995a). By routinely processing the $\sim$ 4Gbytes of information on a Schmidt plate in a matter of a few hours, these machines have opened up new areas of quantitative survey astronomy. However, the planned and current photographic surveys (UKST second epoch R and new Palomar Sky Survey eg. Morgan, 1995) will not provide a significant improvement in this situation since essentially they will be extending what already exists.

Because of the low efficiency and relatively high grain noise of photographic plates, sky-limited exposures produce typical limiting magnitudes of 21 in the red and 22–23 in the blue. Intrinsic photometric errors are in the range $0.05 - 0.^m1$ for objects a magnitude or more above the plate limit, with the equivalent astrometric errors around 1μ, or ~0.1 arcsec. Systematic photometric and astrometric field errors are also present. Photometric field errors up to $0.^m25$ are not uncommon, though with careful processing the systematic photometric errors can generally be reduced down to the level of the random errors. Astrometrically, global distortions (some magnitude-dependent and some not *e.g.* Irwin, 1994) of up to 1 arcsec also affect plates. Again with careful processing and effective use of astrometric standards, most of these systematic errors can be reduced to the level of better than 0.25 arcsec over 2 degree scale sizes, the requirement for the 2DF facility on the AAT, and to less than the individual object random errors over the field sizes for spectroscopy on normal telescopes, such as in blind offsetting targets onto slits to 0.1 arcsec accuracy.

Impressive though these figures are for all sky surveys, they are unlikely to improve dramatically due to the relatively poor dynamic range of photographic detectors (Irwin 1992). Even the recent innovative use of 4415 emulsion by the UKST (Russell *et al.* 1992; Parker *et al.* , 1994) will not significantly change this situation. The current generation of measuring machines, including those nearing completion, fully exploit the present and planned photographic sky surveys. In order to make a significant advance (say a factor of 10) over current material any new survey must be based on detectors that are essentially photon noise limited, such as CCDs.

One area where Schmidt survey work leads the way for future CCD surveys lies in the already developed ability to handle large volumes of data. For example, the high latitude sky in the Northern hemisphere covers some 10,000 deg². If sampled at 0.4 arcsec per pixel (as proposed in the SLOAN Digital Sky Survey) this would $\Rightarrow$ 160 Mbytes / deg² per passband, or many Tera-bytes of data for a "sky" survey. Digitized photographic sky surveys have been successfully processing equivalent volumes of data for many years making use of a concept summarized in the "Kibblewhite Diagram", shown in figure 2.

The basic concepts here are:

(*a*) most of the pixels (~90%) in a survey frame belong to the smoothly varying sky background and hence can be accurately represented using **much** lower resolution sampling;

(*b*) images can be defined to be those regions of contiguous pixels lying above some user-specified threshold relative to local sky;

(*c*) most of these images are faint with little information content other than that reducible to simple descriptors such as position, intensity and shape parameters;

(*d*) all stellar images can in principle be parameterized with no loss of information using just position and intensity (although in practice the shape descriptors are necessary to judge which images are stellar);

(*e*) this leaves only a handful of bright galaxies where more detailed pixel mapping may be necessary to fully extract the information required.

This straightforward concept leads to an enormous data compression ($\times100$) with little or no loss of pertinent information. Furthermore by collecting only relevant detected image parameters, subsequent steps in the analysis become much easier.

Optical object catalogues covering large areas of sky have now been produced by several different groups (see Lasker, 1995b and references therein) and are generally available on-line at various INTERNET sites. Below is a brief summary of the main object catalogues now available and under construction.

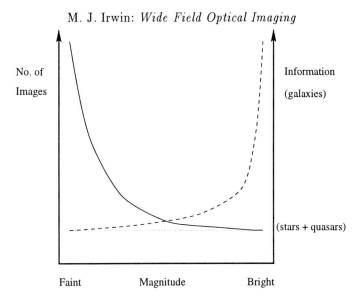

KIBBLEWHITE DIAGRAM

FIGURE 2. A Schmidt Sky Survey plate contains 4 Gbytes of data. Detecting and parameterizing objects results in a large data compression $\approx \times 100$ with only a small loss of information.

Southern Hemisphere; UKST $B_J = 22.5$ and $R = 21.0$

- B_J digitized at 2 arcsec resolution by COSMOS – catalogue
- B_J and R will be digitised by SUPERCOSMOS – pixels + catalogue
- B_J digitised at 3 arcsec resolution by STSCI – pixels + GSS
- R being digitised at 2 arcsec resolution by STSCI – pixels + new GSS
- B_J and R to $|b| > 20$ deg at 1 arcsec resolution by APM – catalogue

Northern Hemisphere; POSS1 $O = 22.0$ and $E = 20.5$; Quick $V = 19$

- O and E digitised at 2 arcsec resolution by APS – catalogue
- V digitized at 3 arcsec resolution by STSCI – pixels + GSS
- O and E digitised to $|b| > 20$ deg at 1 arcsec resolution by APM – catalogue

Northern Hemisphere; POSSII $B_J = 22.5$, $R = 21.0$, $I = 19.0$

- being digitised by STSCI, APM, SUPERCOSMOS, APS ....

There are typically 1/4–1/2 million objects detected on each photographic survey plate and between 1500 and 2000 plates required to cover the whole sky in just one passband. These object catalogues contain magnitude, colour and shape information on around 1 billion objects and in conjunction with the compressed pixel maps compiled by STSCI have become the fundamental database for most of wide field survey astronomy.

1.4. *Next Generation Optical Surveys*

Until recently CCD arrays have been physically small and the pixel sizes not well matched to the plate scales available on potential wide field survey telescopes. In addition there has been the problem of what to do with the vast amounts of data that a CCD sky survey would produce. These drawbacks now no longer limit the potential for CCD-based sky surveys - large format 2048 × 2048 and 2048 × 4096 CCDs are now available and still larger mosaics ranging from 4096 × 4096 to 8192 × 8192 pixels are being fabricated, *e.g.* Luppino *et al.* (1994), Boronson *et al.* (1994); whilst computing resources and storage media have advanced sufficiently to enable both full archiving and "real-time" processing of a CCD-based survey to take place. Thinned CCD mosaic cameras will have efficiencies $\eta \sim 70\%$ and on a medium size telescope (2-3m) will typically cover a field of area $A \sim 0.25 \text{deg}^2$, leading to an effectiveness, $\eta \times A \sim 0.2$, approaching that of photographic surveys.

Although the existing photographic surveys cover the whole sky there is no requirement that a fainter CCD-based survey should do the same. Significant scientific progress will be made by: exploiting the large gains in efficiency, particularly in the U and I-bands; by going 10 times fainter and exploiting the factor of 5 or so higher accuracy; and from the better image quality – seeing ~1 arcsec rather than the 2–3 arcsec of current survey material. In a recent review, Kron (1995) summarized the current and near–future CCD surveys that are being undertaken. Borrowing from this review and updating it as necessary, the following list indicates the current and near–future status of CCD mosaic cameras capable of doing survey work. The list is undoubtedly incomplete but hopefully provides an indication of the current state-of-the-art in this field. For more details on many of these systems consult Kron (1995) and the references therein.

• LACCD – Tyson and collaborators, Tyson *et al.* (1993); 2 × 2 array of thick Tek CCDs 2048 × 2048 with 24μ pixels; for use on Steward 2.3m, CTIO 4m, NOAO 4m; $\equiv 0.5 \times 0.5 \text{ deg}^2$ on the 4m telescopes

• Kiso Mosaic Camera – Sekiguchi *et al.* (1992); 8×5 array thick TI CCDs 1000×1018 of 12μ pixels; used on Kiso Schmidt, Swope 1m Las Campanas $\equiv 0.7 \times 0.5 \text{ deg}^2$, 4.2m WHT (March 96); $\equiv 0.6 \times 0.4 \text{ deg}^2$

• MACHO project - Microlensing toward Galactic Bulge and LMC; Stubbs *et al.* (1994); **two** 2 × 2 arrays of thick Ford/Loral 2048 × 2048 of 15μ pixels – simultaneous B and R filters; dedicated telescope, Mt. Stromlo 1.3m; $\equiv 0.7 \times 0.7 \text{ deg}^2$ twice !

• MOCAM - CFHT (1995); 2 × 2 array thick Loral 2048 × 2048 of 15μ; CFHT 3.6m common user instrument; $\equiv 0.3 \times 0.3 \text{ deg}^2$

• NOAO - Boronson *et al.* (1994); 2 × 2 array thick Loral 2048 × 2048 of 15μ pixels; next generation 8192 × 8192 array from 2k × 4k Loral CCDs; used on KPNO 0.9m, 4m; $\equiv 0.5 \times 0.5 \text{ deg}^2$

• Univ. of Hawaii - Luppino *et al.* (1994); 4 × 2 array thick Loral 2048 × 4096 of 15μ pixels; used on UH 2.2m; $\equiv 0.25 \times 0.25 \text{ deg}^2$

• SDSS - Sloan Digital Sky Survey; York *et al.* (1993); 6×5 array thin Tek 2048×2048 CCDs with 24μ pixels – drift scan simultaneous U,G,R,I,Z filters; to be based at Apache Point with dedicated 2.5m telescope; $\equiv 0.2 \times 1.4 \text{ deg}^2$ times five passbands

• INT WFC - common user instrument, see Irwin (1992); 2 × 2 array of thinned Loral 2048 × 2048 CCDs with 15μ pixels; to be used on 2.5m INT; $\equiv 0.4 \times 0.4 \text{ deg}^2$

Although at present there are no actively operating thinned CCD mosaic cameras, many of the systems in the preceding list are planned to be upgraded to thinned systems as and when such devices become readily available.

1.5. *Surveys in other Wavebands*

The last decade has seen a renaissance of wide field surveys in other wavebands. Coupled with the optical catalogues these surveys provide one of the main sources of objects for both deep single object slit spectroscopy on 4+m class telescopes and for the wider area multi-object spectrographs such as WYFFOS, 2DF, HYDRA ... Indeed many objects currently being studied in great detail on the Keck telescope were originally discovered using the wide field surveys resource.

The following list gives a necessarily brief summary of some of the major non–optical wide area surveys already completed or on the near horizon.

• X-ray surveys – all sky catalogues include Uhuru (1970-73); HEAO-1 (1977-79); Ariel V (1974-1980); Einstein/EMSS (1978-81); ROSAT WFC with $\approx$50,000 sources; in the pointed mode the ROSAT HRC has been used to survey $>$1000 deg^2 of sky; current and future missions include ASTRO-D, AXAF and XMM; see Bradt *et al.* (1992) for a comprehensive review of X-ray missions.

• Infra-red surveys – notable here are the IRAS all sky survey at 12, 25, 60, 100μ to flux limits of 0.2–1Jy, with $\approx$250,000 sources catalogued; and the recently launched ISO mission which will survey a limited region of sky but with better resolution and to much fainter flux limits than possible with IRAS.

• Near IR surveys – two major surveys are currently in progress; DENIS plans to map the southern sky in I, J, K with 1–3 arcsec sampling to a limit of $m_k \sim$14–15 using a dedicated 1m telescope; while 2MASS is an all sky survey at 2.5 arcsec sampling in J, H, K using a 1.3m telescope; both of these surveys will produce catalogues of several million sources with coordinates accurate to $\sim$1 arcsec; the review by Price (1988) covers IRAS and earlier Near IR surveys.

• Radio surveys – there are many large sky coverage radio surveys already completed or nearing completion, with flux limits ranging from $\approx$ 0.1 – 1Jy, including: Cambridge 2C – 8C and variants within; Parkes–MIT–NRAO; Greenbank87; MG and so on... Major new surveys are the NVSS 20cm survey to a flux limit of 2.5mJy over 30,000 deg^2 of sky; and FIRST, a 10,000 deg^2 survey to a flux limit of 0.7mJy; combined, these surveys will produce catalogues of several million sources with coordinates accurate to $\sim$1 arcsec (eg. Becker, White & Helfand, 1995).

1.6. *The Likelihood of an Identification*

A major requirement for all these non–optical surveys is the ability to rapidly access and identify the counterpart objects from optical sky catalogues, with minimal contamination from spurious mis-matches. In particular, when searching for intrinsically rare objects the mis-matching rate often becomes the limiting factor in the success or otherwise of the scheme. An example of this would be in looking for, say, IR detected objects that have no apparent optical counterpart to some optical flux limit. If the combined (optical + IR) positional error box is small the probability of getting an accidental match is correspondingly low. We can quantify this problem by considering the likelihood of identification.

Following Prestage & Peacock (1983) and assuming radially symmetric Gaussian errors for simplicity (generalization is relatively straightforward); the probability of detecting an object at a distance r from the expected position is given by the Rayleigh distribution

$$P(r \to \delta r | id) = \frac{r}{\sigma^2} exp(-\frac{r^2}{2\sigma^2}).\delta r \qquad (1.2)$$

where σ^2 is the combined variance in position expected. The probability of finding a confusing source at this distance is simply

$$P(r \to \delta r | c) = 2\pi r \rho.\delta r \qquad (1.3)$$

where ρ is the average cumulative surface density of sources up to some chosen limiting magnitude. This surface density could be further subdivided on say morphology (eg. stellar, non-stellar), colour and so on. Now from Bayes' theorem we know that

$$P(id|r) = \frac{P(id).P(r|id)}{P(id).P(r|id) + P(c).P(r|c)} \qquad (1.4)$$

$$P(c|r) = \frac{P(c).P(r|c)}{P(id).P(r|id) + P(c).P(r|c)} \qquad (1.5)$$

where $P(id|r)$ is the probability that the object at r is the correct identification; $P(c|r)$ is the probability of it being a confusing object; and $P(id)$ and $P(c)$ are respectively the prior probabilities that the identification is visible at the chosen limiting magnitude and that confusing objects are visible. Clearly, in general optical identification work, $P(c) = 1$ since confusing objects are both visible and greatly outnumber correct identifications. In the case of relatively bright identifications $P(id) = 1$ since it is the probability that the proposed source is reliable. For faint identifications $P(id) < 1$, since it is the combined probability of the proposed counterpart being reliable, coupled with the probability of the source being optically visible to this magnitude limit.

Therefore by rearranging the preceding equations we have that

$$P(id|r) = \frac{P(id).L(r)}{P(id).L(r) + 1} \qquad (1.6)$$

and that

$$P(c|r) = \frac{1}{P(id).L(r) + 1} \qquad (1.7)$$

where the likelihood ratio $L(r)$ is defined by

$$L(r) = \frac{P(r|id)}{P(r|c)} = \frac{exp(-r^2/2\sigma^2)}{\sigma^2.2\pi\rho} \qquad (1.8)$$

It is straightforward to see that the smaller the combined error search radius σ the easier it becomes to deal with confusing sources. In particular as non-optical surveys reach ever fainter flux limits the confusion problem becomes significantly worse, without a concomitant improvement in positional errors; since $P(id)$, the probability of the optical identification being visible, will also decrease.

2. Optical Telescope Design Constraints on Wide Field Imaging

Although almost all telescope design is now done using ray-tracing packages there are several aspects of optical telescope design amenable to simple analytic formulae. In quite general terms this approach can be used to highlight the problems of using large telescopes for wide field survey work and in particular provides a useful approximate starting point for evaluating proposed wide field imaging designs.

2.1. *Aperture, Scale and Depth of Focus*

Definitions: Effective focal length (m) - F; aperture/diameter (m) - D; focal ratio - $f = F/D$; semi-aperture angle - u'. Units as in () unless otherwise stated.

In most optical work the diameter of the entrance pupil, D, of the telescope is sufficiently large that the angular size of a diffraction limited image, ξ, of a point source given by

$$\xi = 1.2\frac{\lambda}{D} = 0.14 \times (\frac{\lambda}{5500\text{Å}}) \times (\frac{1\text{m}}{D}) \quad arcsec \qquad (2.9)$$

is significantly smaller than imperfections caused by the telescope optics or the point spread function (psf) due to atmospheric or dome seeing. The latter two effects are usually the dominant contributors to the psf.

First-order approximation (Gaussian theory) to the optical performance of a telescope assumes that apertures and field size are small compared to the focal length. With this approximation, θ, the inclination of any ray to the optical axis satisfies $sin(\theta) \approx \theta$ and $cos(\theta) \approx 1$, and we can see immediately from figure 3 that the angular scale in arcsec/mm, is given by

$$scale = \frac{d\theta}{dx} \simeq \frac{206}{Df} \qquad (2.10)$$

Therefore, two objects separated by an angle θ on the sky will appear a distance $= \theta/scale$ apart at the focal plane of the telescope. For an 8m telescope with an f/16 secondary, such as Gemini, this implies a focal scale of 1.6 arcsec/mm. Even on an 8m telescope designed for prime focus imaging, such as the f/2 prime on SUBARU, the image scale is a daunting 12.9 arcsec/mm – the proposed 30 arcmin field of view will require a detector of order 140mm × 140mm in size.

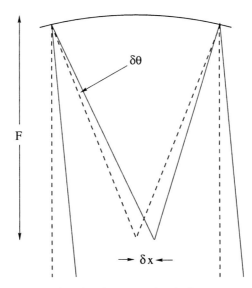

FIGURE 3. Equivalent system for deriving image scale. A change in entrance angle $\delta\theta$ gives an image shift $\delta x = F \times \delta\theta$.

Using a similar construction, illustrated in figure 4, we can derive the depth-of-focus,

δz, over the focal plane,

$$Depth - of - focus = \delta z \approx \frac{\text{FWHM}}{\text{Scale}} \times f \qquad (2.11)$$

where FWHM is the Full Width at Half Maximum of the seeing profile.

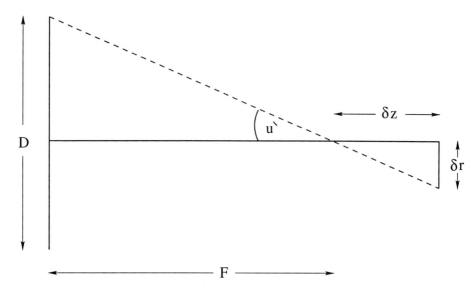

FIGURE 4. Depth of focus δz to give image spread δr. For good imagery $sin\ u' = 1/2f$ and $\delta r = \delta z \times tan\ u' \simeq \delta z/2f$.

For a telescope giving good imagery the sine condition is satisfied so that the Numerical Aperture $sin\ u' = 1/2f$. If in such a system we chose to match the pixel size, p (microns) to the expected best seeing FWHM (arcsec), to efficiently sample the sky, then

$$F \approx \frac{2p}{\text{FWHM}} \times 0.206 \qquad (2.12)$$

and therefore

$$sin\ u' \approx \frac{D \times \text{FWHM}}{4p} \times 4.85 \qquad (2.13)$$

As an example consider a pixel size, $p = 15\mu$; and a telescope situated on a good site where the average best seeing had a FWHM $= 0.5$ arcsec;

- 4-m telescope $sin\ u' \approx 0.16(f/3.1)$ – several 4m-class systems exist

- 8-m telescope $sin\ u' \approx 0.32(f/1.5)$ – corrected field difficult

- 16-m telescope $sin\ u' \approx 0.64(f/0.8)$ – impossible?

The field of view depends on the off-axis vignetting, including the effects of sky baffles, and the optical aberrations inherent in the particular telescope design. The latter limitation can usually only be adequately investigated using detailed ray tracing. However, with the current limited size of CCD detector arrays ($\approx$120mm $\times$ 120mm for a state-of-the-art 8192 $\times$ 8192 mosaic), it is often the detector that limits the useful field of view.

2.2. *Cassegrain Telescopes*

Seidel's (1856) 3rd order theory predicts the amounts of: spherical aberration, coma, astigmatism, field curvature and distortion. For example, at prime, a single paraboloidal mirror with no correcting elements will provide seeing-limited images on axis. Any other form of mirror, spherical, ellipsoidal or hyperboloidal gives fuzzy images, affected by spherical aberration. To give good image quality over a wide field of view requires the use of a prime focus corrector – generally a refractive multi-element system of low optical power. The diameter of the correcting elements is limited by glass manufacturing technology to around a maximum of 1 meter. In turn this is usually the limiting factor in the maximum possible physical size of the prime focus field, which because of the fast focal ratio ($\sim$2) and vignetting limitations, leads to a typical maximum field of order 1/2 meter diameter.

An alternative approach is to use a two mirror system and place the instrumentation at the secondary focus. In a classical Cassegrain system the primary mirror is a concave paraboloid and the secondary a convex hyperboloid. Typical secondary Cassegrain focal ratios are f/6 to f/20 and the field of view is generally modest ($\sim$10 – 20 arcmins). The field of view is limited by several factors including: aberrations, the hole in the primary, sky baffles, size of correctors at prime and so on. A schematic drawing of a classical Cassegrain telescope showing the different focii is shown in figure 5. The secondary image is usually formed a moderate distance behind the primary mirror, which needs to be perforated. The relationships between the focal lengths, radii of curvature of the mirrors, their separation and placement with the respect to the primary are all determined by simple geometrical optics.

The focal length of the secondary, F_2, is given by

$$\frac{1}{F_2} = \frac{1}{V} - \frac{1}{U} \qquad (2.14)$$

where U is the distance of the secondary inside the primary focus (see figure 5), forming a secondary image at a distance V. Usually

$$U + V = F_1 + (0.5 \sim 1.0)D_1 \qquad (2.15)$$

in order to place the secondary focus at a convenient location, and we also have from direct trigonometric construction

$$U = F1 \times \frac{D_2}{D_1} = f_1 \times D_2 \qquad (2.16)$$

where D_1 is the diameter of the primary, D_2 the effective diameter of the secondary; and

$$V = f_2 \times D_2 \Rightarrow f_2 = \frac{V}{U} \times f_1 \qquad (2.17)$$

(The real diameter of the secondary would be some $\sim$10% larger than the effective diameter to ensure no vignetting of a finite field.) One final link in the chain is to note that in the absence of astigmatism the curvature of the Petzval surface coincides with the curvature of the final focal surface, see for example Willstrop (1985). The curvature of the Petzval surface of a system of mirrors, R_p, is related to the the radii of curvature of the mirrors, R_i, by

$$\frac{1}{R_p} = \sum_{i=1}^{n} \frac{2}{R_i} = \sum_{i=1}^{n} \frac{1}{F_i} \qquad (2.18)$$

where F_i is the focal length of mirror i and the convention is to take the + sign for convex surfaces and – sign for concave surfaces. The linear magnification, m, produced by the

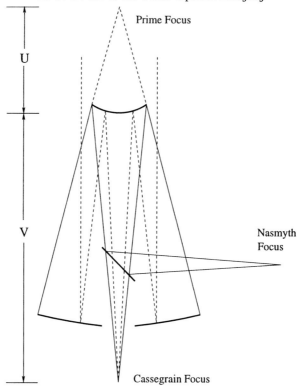

FIGURE 5. Schematic of a Cassegrain Telescope showing the different focii. The main text indicates how the primary mirror diameter, D_1, and f/ratio, f_1, and secondary mirror effective diameter, D_2, and f/ratio, f_2, are interlinked with the placement of the secondary, U, and the position of the final Cassegrain focus, V.

secondary is V/U and if the secondary image is to be free of spherical aberration then its eccentricity, $e_2 = (m+1)/(m-1)$. Images at the secondary focus are still affected by coma and astigmatism and of course field curvature. The angular image spread due to coma, $\delta\theta$, at radius θ is given by

$$\frac{\Delta\theta}{\theta} = \frac{3}{16 \times f^2} \qquad (2.19)$$

which also holds for uncorrected prime focus images. Due to the typically long Cassegrain focal lengths, f/10 – f/20, the comatic blur only reaches 1 arcsec at around 1/2 degree radius, which is normally considerably larger than the field of view used.

As an example consider the WHT: it has a primary $f_1/2.5$, $D_1 = 4.2$m, $\Rightarrow F_1 = 10.5$m; and a secondary $f_2/11$. Equations 2.15 – 2.17 imply that

$$U + V = D_2 \times (f_1 + f_2) \Rightarrow D_2 \approx 1 \qquad (2.20)$$

and therefore in conjunction with equation 2.14 this tells us that

$$D_2 = 1 \Rightarrow F_2 = -3.1; U = 2.5; V = 11.0 \qquad (2.21)$$

and the radius of curvature of the Cassegrain focal surface is 4.4m. It is straightforward to estimate if this radius of curvature is a problem by considering the change of planar depth, δz, of the surface as a function of angular distance, θ, from the optical axis

$$\delta z \simeq R_p \times \frac{\theta^2}{2} = \frac{r^2}{2R_p} \tag{2.22}$$

where r is the radial distance from the optical axis. If $R_p = 4.4$m, at a radial distance of $r = 5$cm, $\delta z = 280\mu$; cf. to the depth-of-focus $\approx 1200\mu$, ie. no problem.

2.3. Ritchey–Chrétien Telescopes

Ritchey–Chrétien telescopes are two-mirror Cassegrain-like telescopes with a relatively large secondary field of view. In this case both the primary and the secondary mirrors are hyperboloids and the secondary focus is completely corrected for spherical aberration and coma. Prime focus imaging without correcting elements is ruled out, however, it turns out that it is easier to design a wide field corrector for a hyperboloidal mirror than for a paraboloidal Cassegrain system. In practice, Ritchey–Chrétien telescopes offer some of the biggest fields available on large telescopes, both at prime and at Cassegrain. For example, the 3.9m AAT has a 40 arcmin field at Cassegrain and a new 4-element prime focus corrector gives a 2 degree field at prime.

The Gemini 8m telescopes will follow a Ritchey-Chrétien design with an $f_1/1.6$ primary. An $f_2/6$ wide field secondary option was originally considered, but note that equations 2.14–2.17 $\Rightarrow D_2 \approx 2.5$m, a huge secondary mirror; whereas the adopted $f_2/16$ secondary, which is all that is currently funded, $\Rightarrow D_2 \approx 1$m. In the latter case, equation 2.18 implies that the radius of curvature of the secondary focus will be around 1.8m without further correcting optics.

Another interesting design is provided by the 2.5m telescope under construction for the SLOAN Digital Sky Survey. This uses a secondary mirror of equal (and opposite sign) radius of curvature to the f/2.2 primary to give a flat focal plane secondary f/5 field. The relatively fast f/5 secondary is needed to provide matched sampling at the image detector plane; whilst the flat focal plane greatly simplifies both direct imaging instrumentation and the multi-fibre spectrograph design. With F_2 fixed and equal to $-F_1$, U and V are completely determined from the preceding equations, which in turn fixes the diameter of the secondary mirror to be ~ 1.3m and leads to a field of view of order 30 arcmin. This telescope will be used for imaging with CCDs in the drift scan mode and therefore has two aspheric plate refractive correcting elements, situated near the focus, to eliminate field distortion.

Although the SLOAN solution to the wide field imaging problem works well for telescopes of relatively modest aperture, the flat field constraint coupled with the relatively fast secondary focus, necessary to give a good image scale, leads to large diameter secondary mirrors, and would be impractical for 8m-class telescopes.

2.4. Field Distortion

Uncorrected field distortion at either primary or secondary focii in Cassegrain-type systems takes the form of

$$\theta_{true} = \theta \times (1 + k1 \times \theta^2 + k2 \times \theta^4 + ...) \tag{2.23}$$

As an example, at WHT prime:

$$k1 \times \theta^2 = -0.000027 \times (\frac{\theta}{arcmin})^2 \tag{2.24}$$

whilst at WHT Cassegrain the field distortion is negligible.

Field distortion can also be important in the so-called point and stare mode of operation. It is now common practice to stack sequences of offset CCD frames to reach deeper limiting magnitudes. In shifting the frames to coalign them before coaddition, it is necessary at the same time to allow for the effects of differential field distortion across the frame. For WHT prime, at a radius of 10 arcmin, the differential field distortion relative to the optical axis is over 1.5 arcsec.

2.5. *Alternative Wide Field Telescopes*

The commonest form of wide field survey telescope is the Schmidt camera. Schmidt telescopes have a spherical primary with a radius of curvature equal to the length of the telescope tube. An aspheric corrector plate placed at the entrance aperture corrects for spherical aberration and also defines the limiting diameter of the telescope. The primary advantage of a Schmidt telescope is the enormous field of view. For example the UK Schmidt telescope (UKST) has a 6 degree field which it obtains with a 1.2m aperture and a 1.8m mirror. Disadvantages are that the focal plane is curved and located internal to the telescope tube; the tube length is large for a given diameter and practical f/ratio; but the major disadvantage is that simply scaling up the design to significantly larger apertures is impractical because of the problems of both making and supporting large refractive elements.

There have been many alternative wide field designs proposed, of which one of the most promising for large aperture work is based on a three mirror system. Willstrop (1985) describes a flat-field Mersenne-Schmidt telescope which covers a field 3 degrees in diameter with images smaller than 0.5 arcsec in extreme spread; over the central 2 degrees the images are smaller than 0.1 arcsec. For more details consult the article by Willstrop (1985). More recent work by Willstrop (private communication) has pushed this design further and lead to a practical design for a 5m diameter telescope having a 5 degree field of view with images better than 0.5 arcsec over the entire field.

3. Wide Field Detectors

The traditional detector for wide field imaging over the last century has been the photographic plate. Modern IIIaJ, IIIaF, IVN (plates) and 4415 emulsions (film) provide very large area detectors, $\approx$40cm $\times$ 40cm in size, that are cheap, astrometrically stable to the level of microns and relatively simple to use. The active element, silver halide grains, are a few microns in size, and give a detective quantum efficiency between 1–5%. Although this figure is relatively poor by modern standards, as outlined in section 1., it is the product of this efficiency and the area covered, that really matters.

Historically, the earliest successful and most widely used large area electronic detectors were the image intensifier systems, of which the IPCS (Bokensberg, 1972) is probably the best known. IPCS systems were the instrument of choice for faint spectroscopic work throughout most of the last two decades. However, gradually over the last twenty or so years, Charge Coupled Devices (CCDs) have become the preferred detector for most applications. Since there are many excellent review articles on CCD development (eg. Janesick & Elliot, 1992) there is no point in repeating the whole story here. Over the last decade the active area of CCDs has grown dramatically from the typical 22μ pixels of 576×384 CCDs available in the late 1980's, through the current generation of 24μ pixel 2048×2048 thinned Tektronix CCD workhorse devices, and on to the new generation of thinned CCD mosaics of typically 15μ pixels and total array sizes of 4096×4096, or even, 8192×8192 pixels.

As we are interested in potential wide field detectors, for both imaging and spectro-

TABLE 1. Wide Field Detectors

Detector	Advantages	Disadvantages
Photography Plate IIIaJ,IIIaF & IVN Films 4415[1]	Wide area astrometry Very large area ($\approx$40cm $\times$ 40cm) Inexpensive Simple to implement Eyeball inspection	Low efficiency Grain noise high Non-linear Small dynamic range Not reproducible Needs digitizing
Image Photon Counting Systems – IPCS MIC, BIGMIC[2]	Good blue QE $\sim$25% Up to 70mm diameter Monitor in real time Zero readout noise Selectable pixel size No cooling required	Poor red QE <5% Fragile – overexposure Only low photon rates Field distortion
CCDs	Excellent QE blue – red (QE $\sim$70%) Temporal stability Large dynamic range Linear response Low readout noise Insensitive to overexposure No photon rate constraints	Still limited area No real time image Inflexible pixel sizes Cooling required

[1] See articles by Hartley (1994) and Parker *et al.*, 1994 for details of photographic emulsions.
[2] See article by Fordham *et al.* (1994) for more details of modern IPCS systems.

scopic applications, I have summarized the relative merits and disadvantages of the most commonly used wide field detectors in table 1 and indicated a few references for further reading.

High resolution spectroscopy, and usually only in the range 3200Å– 4000Å, is the only remaining area where an IPCS-type detector offers potential advantages over a thinned blue sensitive CCD. We can see why this might be the case by considering the signal:to:noise produced by the two types of detector. In a general sense, the observed counts per detector pixel, x_i, for an exposure of length t, are determined by three factors

$$x_i = s_i \times t + b_i \times t + r \qquad (3.25)$$

s_i – the signal photon rate in pixel i; b_i – the background photon rate; r – the readout noise per pixel. Assuming unit gain, ie. 1 e$^-$/γ per ADU, and ignoring spurious noise sources for simplicity, the potential signal:to:noise, $S : N_i$, for a Poisson process for pixel

i is given by

$$S : N_i = \frac{s_i \times t}{\sqrt{(s_i + b_i)t + r^2}} \qquad (3.26)$$

which translates to two limiting cases: detector noise limited where

$$S : N_i \simeq \frac{s_i \times t}{r}; \quad r^2 > (s_i + b_i)t \qquad (3.27)$$

and photon noise limited where

$$S : N_i \simeq \frac{s_i}{\sqrt{s_i + b_i}} \times \sqrt{t}; \quad r^2 < (s_i + b_i)t \qquad (3.28)$$

Since an IPCS based system produces zero readout noise it always operates in the second, photon noise limited regime. Gilmore & Walton (1993) illustrated the advantages of the IPSC-2 photon counting system over a thinned Tek CCD using the Utrecht Echelle Spectrograph on the WHT. With the maximum practical CCD exposures limited to around 1 hour (because of cosmic rays) the total signal per pixel was <10 counts and the CCD readout noise (5 e$^-$), rather than the photon noise dominates. However, the advantage is finely balanced and if the same experiment were carried out on a 10m telescope, the better blue efficiency of the best thinned CCDs, coupled with the higher photon rate at the detector, would switch the advantage back to the CCD. Figure 6 shows an example of the spectral response curves for the CCDs currently available at the Isaac Newton Group of telescopes on La Palma. With suitable treatment, eg. thinning and coating, it is feasible to have CCD QEs >50% over most of the optical passbands.

3.1. A Basic Description of CCDs

A CCD (Charge Coupled Device) is a multi-layered sandwich of various silicon- based wafers ($\sim \mu$'s deep) supported on a relatively thick ($\sim 100\mu$) silicon substrate. The basic structure is illustrated in figure 7. The conducting electrodes on the surface are insulated from the p-type silicon substrate by a thin layer of SiO$_2$. These electrodes are made from "light–transparent" polysilicon (polycrystalline silicon). Column isolators are used to define the CCD columns as buried channels, while the polysilicon electrodes define the rows. Each pixel is formed by 3 adjacent surface electrodes plus the buried channel column isolators. The central electrodes of each trio are held at a positive voltage, the outer two at earth potential, and form a potential well near the front surface of the silicon substrate. This expels the majority carriers (holes) and produces a depletion layer. Incoming photons pass through the front layer of polysilicon electrodes and are absorbed by the silicon to produce hole-electron pairs in numbers proportional to the number of incident photons. These photo-generated electrons migrate to the depletion layer formed by the potential well and collect there as single charge packets, while their counterpart holes diffuse deeper into the underlying substrate. The array of "horizontal" electrodes (rows) above "vertical" charge-transfer channels (columns) defined by the column isolators, is known as buried-channel architecture. This design permits rapid and much more efficient charge transfer (CTE ~ 0.99999) compared with surface channel CCDs (see Janesick & Elliot, 1992 for more details).

The CCD can be thought of as a photosensitive array which generates electrons that are subsequently captured and stored as small buckets of charge. At readout time the pixels form a bucket brigade; each row of charge is passed down the columns and horizontally along the final row to be measured in turn and recorded digitally. This is accomplished by parallel sequencing the triads of electrodes with clocking waveforms of the type shown in figure 8. As each row reaches the bottom of the array it enters the horizontal register where a similar triplet of electrodes and clocking waveforms moves the packets across

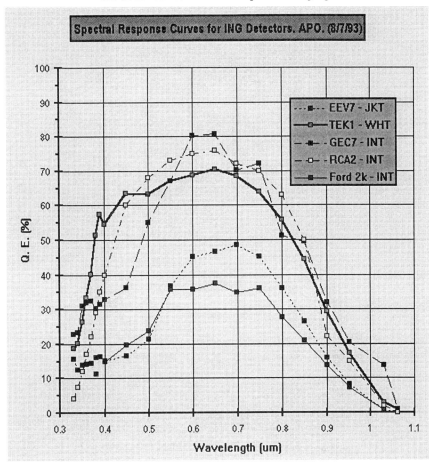

FIGURE 6. Spectral response curves for ING detectors. EEV7 is a thick blue–coated CCD; TEK1 is a standard thinned 1024 × 1024 device; GEC7 is a thinned equivalent of EEV7; RCA2 is an "ancient" thinned CCD; Ford 2k is a 2048 × 2048 Loral thick blue coated CCD. These devices are representative of the range of CCDs available on most telescopes. (Figure courtesy of Paddy Oates, Royal Greenwich Observatory)

the register. On reaching the end-of-the-line, each charge packet is detected as a voltage across a capacitance, amplified by an on-chip amplifier, and digitised for subsequent output and storage. The next row is driven down and the sequence repeated until the CCD is cleared of charge. Typical "slow" readout times for a science grade CCD are of order 10–50μs per pixel, or ≈ 40– 200s for a 2048×2048 device.

The silicon band-gap between the valence and conduction bands is 1.14eV, which corresponds to a wavelength of 10800Å. Between 1.14eV and about 5eV, each incoming photon will be potentially converted to $1e^-$ with a probability depending on the absorption depth, which varies from 0.2μ at 2500Å to 500μ at 10000Å. At still higher energies, CCDs are good X-ray detectors, the quantum yield is $1\ e^-\ /\ 3.65$ eV. The exceptional efficiency at optical wavelengths of typical CCDs is illustrated in figure 6. There are several factors that contribute to limiting the CCD efficiency. In the case of a front-illuminated (thick) CCD (figure 7), the loss of efficiency in the blue is due to absorption of blue photons by the polysilicon electrodes; whilst at near IR wavelengths ($\sim$10000Å) the capture

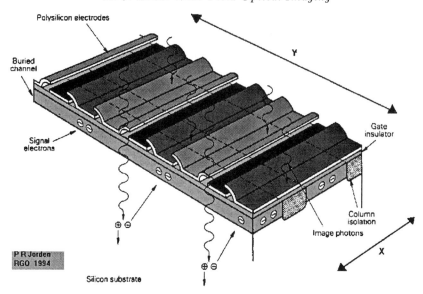

FIGURE 7. A schematic section of a CCD image sensor. (Diagram courtesy of Paul Jorden, Royal Greenwich Observatory)

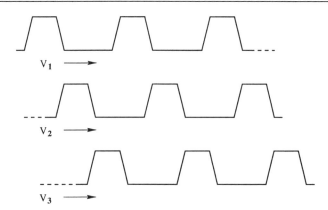

FIGURE 8. The Clocking waveform used to readout a CCD.

cross-section is the limiting factor. There is also a general loss of photons due to surface reflection losses. At longer wavelengths the penetration depth can be so large that losses due to recombination of hole- electron pairs becomes significant ($\tau_{recomb} \sim 100\mu s$). To improve the blue response of thick CCDs a fluorescent coating can be applied to the front surface. This works by absorbing blue photons incident on the surface and re-radiating them at redder wavelengths. The redder photons pass more easily through the polysilicon electrodes. However, the blue response shortward of 4500Å is only improved from typically ~5% to ~15% by this process.

Thinned CCDs, formed by either acid etching or chemical/mechanical thinning of the backside silicon, employ silicon chips that are only $10-30\mu$ thick and can therefore be illuminated from the backside. This is still an art form as can be seen from the reviews given by Lesser (1989) and Lesser (1994), and the scarcity of large format thinned devices in operation. The thinning introduces several other problems such as: oxidation of the back surface which thereby acquires a net positive charge drawing photo-generated e^-'s away from the depletion region – the effect is reduced by passivating the back surface or by UV flooding in O_2; the back surface is an excellent mirror and needs wavelength-tuned anti-reflection coating; thinned devices need some sort of mechanical support such as bonding to the front surface to a chip carrier; the thinned devices suffer from increasing fringing in the red due to internal reflection from front and back surfaces – not a problem in the blue due to the short optical path. The end result is that thinning dramatically improves the blue response, QE's of $\gtrsim50\%$ shortward of 4500Å are possible, by removing the obstacle of the electrode structure (figure 6). At the same time it might be thought that the long wavelength response, say toward 10000Å– near the silicon band-gap, should suffer because of the much reduced optical absorption depth. However, the problem/effect of multiple internal reflection of long wavelengths, that leads to fringing, also helps to compensate for the loss of silicon and in practice little degradation of red QE occurs.

A surprising feature of CCDs, given that the active depletion depth of $5–10\mu$ is a significant fraction of the pixel size, is the preservation of resolution with wavelength. There is some slight degradation of resolution due to spreading of photoelectrons before collection and some resolution loss due to imperfect charge transfer efficiency (CTE). The former effect is very difficult to detect in normal astronomical applications; while the latter manifests itself as low level smearing downstream from bright pixels. However, the internal structure of the CCD, particularly the electrode arrays and buried channels, do give rise to periodic intra-pixel variations in sensitivity (Jorden, Deltorn & Oates, 1994). These modulations are wavelength dependent and can cause problems in some situations if the data is spatially undersampled. The spatial modulation in total recorded signal is worse for front-illuminated thick devices and amounts to some $\pm10\%$ across columns, approximately independent of wavelength. As expected across rows (due to the electrode structure) the situation is worse with modulations of $\pm25\%$ at 5000Å dropping to $\pm10\%$ at 8000Å. For backside-illuminated thinned devices the spatial modulation at 5000Å is of order $\pm1\%$ in both directions increasing to of order $\pm5\%$ at 8000Å.

3.2. *Detector Requirements, Problems and Features*

Although this section is geared toward CCDs as the currently preferred detector of choice for wide field optical work, many of the requirements listed below are necessary for any potential wide field detector. For a detailed discussion about the problems caused by CCD "noise" sources and procedures to recognize and deal with them see the article by Gilliland (1992).

• Format – 2048 × 2048 pixels minimum; pixel size 8μ to 30μ; the maximum physical size of an individual CCD is limited by the size and purity of currently available silicon wafers to an area of $\approx$ 20cm^2. The yield of larger science-grade devices is very poor. Larger areas are covered by using 3-edge buttable mosaic arrays with building blocks 2048 × 4096 CCDs, eg. Luppino *et al.* (1994).

• Readout noise – 4 to 6 e^- routinely; best <2 to 3 e^- *rms*; readout noises of 1 e^- have been achieved but not routinely (Janesick & Elliot, 1992). In sky-limited imaging exposures the field photon noise per pixel $>>5$ e^-; high resolution spectroscopy is the science driver for low readout noise.

• Quantum efficiency – with anti-reflection coatings, thinning, UV flooding, etc., $\approx70\%$ QE is achievable over a large spectral range, 3000Å– 8500Å, Lesser (1994), though the best recipe remains an art form.

• Dark current – <10 e$^-$ per hour → 1 e$^-$ per hour; modern CCDs operated at ≈-110°C have dark currents around 1 e$^-$ per hour, though massive overexposure (daylight) and power cycling effects can increase this by orders of magnitude over periods up to 24 hours later (Tulloch 1995); temperature control also stabilizes the temperature -dependent pixel sensitivities. Lower operating temperatures are limited by e$^-$ mobility leading to CTE problems.

• Cosmic ray sensitivity – currently ≈2 events/cm^2/minute is the norm. This is usually the limiting factor in the length of individual exposures. Cosmic ray hits can affect many adjacent pixels making effective cleaning from the frame a non-trivial task. Best dealt with by stacking multiple exposures of the field, with the caveat that readout noise is not a problem.

• Dynamic range & linearity – >100,000 e$^-$ full well and linear over this range, ie. a true 16-bit device. Full well depths up to 500,000 e $^-$ are possible, which coupled with a selectable gain (e$^-$ / ADU) enables viable dynamic ranges (peak signal:*rms* noise) between 10^4 and 10^5:1 to be regularly obtained.

• Charge transfer efficiency – cumulative CTE's ≳0.99999 are now common but the effects of CTE are still readily noticeable downstream of saturated regions where the ambient light level is low. After all, 2048 charge transfers are required to readout an average pixel for a 2048 × 2048 device, resulting in $1 - 0.99999^{2048} = 2\%$ of the total charge left behind as low level charge smears.

• Uniformity – ≲5% large scale gradient and ≲1% pixel-to-pixel variations, make subsequent flat-fielding operations much more straightforward. A useful rule-of-thumb is that it is generally feasible to take out 90% of an effect, such as flat-fielding variations, with relatively little effort. To reduce the remainder is often impossible and **always** takes 10 × the previous effort. Problems occur with thinned CCDs due to the next effect.

• Fringing – target <10% in the I-band; thinned CCDs are prone to fringing particularly toward the red end of the optical spectrum; this is an additive effect, from the viewpoint of analyzing images superimposed on the sky, with both spatial (surface properties of the CCD) and temporal components (changing levels of night sky emission lines) and is difficult to reduce to below 1%. Its presence complicates the red-band flat-fielding procedure.

• Defect density – science grade devices should have ≲1 bad column; ≲20 bad charge traps. Low light level deferred charge (otherwise known as fat zero, charge skimming, charge depletion and low light level non-linearity) traps were more of a problem in the past but still need checking for in modern devices – see for example Gilliland (1992).

• Miscellaneous – no image retention, cross-talk between read amplifiers and/or different CCDs in mosaic should be minimal, spectral response and gain and bias should be stable over ≈week timescales. These latter requirements greatly simplify practical operation of CCDs.

3.3. *Availability of CCDs*

Most commercially available CCDs, including those manufactured by the silicon foundries are made from 4 inch silicon wafers. Some manufacturers have 5 or 6 inch wafer capability but these are at present the exception rather than the rule. Regardless of wafer size, the maximum physical size of an individual science-grade CCD is limited by the size and purity of currently available silicon wafers to an area of ≈ 10–25cm^2. The yield

TABLE 2. A Selection of State-of-the-art Large Format CCDs

CCD	Pixel Size μ	Format (pixels)	Status
DALSA	12.0	5120 × 5120	Thick
			not science grade
EEV	13.5	2048 × 4096	Thick 3 edge-buttable
			Thinned available soon
LORAL	15.0	2048 × 2048	Wafer runs in foundry
		2048 × 4096	Thick 3 edge-buttable
			Thinning by contract with
			Lesser – Steward Obs.
MIT-Lincoln Labs	?	1960 × 2560	Government funded contract
			with a consortium of
			observatories
ORBIT	15.0	2048 × 4096	Wafer runs in foundry
			Thick 3 edge buttable
			Thinning ?
PHILIPPS	12.0	7000 × 9000	Thick
			not science grade
RETICON	13.5	2048 × 2048	Thick buttable ?
			Thinned ?
TEK → SITE	24.0	2048 × 2048	Thinned packaged
			Benchmark non-buttable
	15.0	2048 × 4096	Thick 3 edge-buttable
			Thinned available soon
THOMSON	15.0	2048 × 2048	Thick
			Thinned ?

of larger science-grade devices is very poor, although non-science grade CCDs covering areas up to 100cm^2 have been made. Table 2 lists a sample of the current state-of-the-art CCD suppliers. It is instructive to compare the detective area of these devices with the "industry standard" thinned TEK 2048 × 2048 devices currently in use at many observatories. Although the active area has not improved, two important changes have occurred: first devices with significantly smaller pixels (with twice as many pixels) are now available; second these new devices can be butted on 3 sides with gaps of less than 1mm between active regions. Because of these changes the next generation of larger area CCD detectors will almost certainly be made from mosaics of individual large format CCDs. This gets around the problems of: low yield; readout limitations; and the physical limitations of the silicon wafers. The papers by Luppino *et al.* (1994) and Boronson *et al.* (1994) describe how mosaic cameras with active arrays of 8192 × 8192 pixels are being, or have been, constructed using the new "industry standard" 3 edge-buttable 2048 × 4096 CCD building block. Some of the groups using/constructing mosaic cameras were listed in section 1.4.

3.4. *Drift Scanning –v– Point-and-Stare*

Two major methods have been developed for using CCDs in wide field astronomy: drift scanning, eg. Gunn *et al.* (1987), and point-and-stare.

Drift scanning consists of carefully aligning the CCD so that the motion of the sky is exactly along the columns of the CCD and then clocking out the CCD at the same rate/pixel that the sky moves along the columns. Early versions of this technique used a stationary telescope and let the sky drift past at the sidereal rate, ie. $15cos\delta$ arcsec/s, or parked the telescope in declination (easy if equatorially mounted) and tracked at a non-sidereal rate. To minimize the distortion caused by the apparent movement of objects across the CCD, the drift tracks should ideally be along Great Circles in the sky. To see this, consider scanning along the celestial equator as an obvious example of a Great Circle route. Next, as a counter-example, consider declination tracks closer to the celestial poles, where drifting along lines of constant δ will clearly cause curved tracks across the CCD (imagine star trails near the poles or, if you prefer, consider the locii of lines of constant α or δ on a tangent plane projection eg. Smart 1956). Any tangent plane projection with a local coordinate track defined by the Great Circle passing through its centre, will provide a minimal distortion route.

The main advantages of drift scanning are:

- large areas of sky can be covered on one frame;
- the continuous readout (after ramping up and down at either end) ensures low overheads;
- the CCD systematic errors (flat-fielding) are reduced to one-dimensional problems, since each pixel in a column contributes equally to the image.

Disadvantages include:

- the problem of aligning and clocking out the CCD to minimize distortion in the output image;
- if the telescope suffers from field distortion (equation 2.23) this will also lead to distorted output images – the SLOAN survey telescope has specially designed corrector plates to remove the field distortion;
- changing observing conditions make it difficult to do precise photometry;
- it is very difficult, if not impossible, to autoguide for large area drift scanning, making long exposures too reliant on good telescope tracking. Most drift scanning produces relatively shallow surveys and as such can be well matched to wide field work.

By far the majority of imaging (and spectroscopy) work uses the point-and-stare method with the telescope tracking on the object of interest and an autoguider providing ~0.1 arcsec position lock. The advantages of this method are:

- simplicity – no complex mechanical platforms are required; optical distortions are readily dealt with;
- all the images are recorded in identical conditions enabling ~0.1% differential photometry to be routinely achieved;
- autoguiding can be used to give long integration times (up to 1 hour – limited by cosmic ray buildup);
- construction of sky-superflats enables very deep imaging to be obtained by stacking images.

The obvious disadvantages of this method are:

- a much smaller area of sky per exposure is achieved;

- readout overheads can be a significant fraction of short exposure times;
- the full two-dimensional systematics of the CCD have to be dealt with – this is not so much of a problem with modern CCDs where we expect the systematics to start at the 1% level and to be readily reduced to the 0.1% level.

4. Preprocessing CCD Images

"The only uniform CCD is a dead one." – Craig Mackay

Since there are many excellent reviews of CCD data processing (see for example, the reviews by Gullixson (1992) and Gilliland (1992); or the NOAO IRAF photometry cookbook), in this section we will concentrate solely on the steps involved in CCD data pre-processing and highlight some of the problems that may be encountered.

The first stage of any reduction of CCD images involves removing the signature of the CCD from the data. This will involve some or all of the following stages:

- Bias frames – provide both an estimate of the artificial zero-point of the digitised images (to ensure non-negative counts) and an estimate of any fixed pattern noise over the frame. The zero-point is usually monitored throughout the observing session by using the overscan region on each frame to keep track of any variation in bias level, and should generally vary by less than 1%. The "trailing" overscan region is generated by continuing to clock out the CCD after all the charge has been readout and as such solely measures the intrinsic bias level and *rms* noise of the electronics. Note that there is usually an equivalent "leading" underscan region, whose main purpose is to "ramp" up the electronics before real data is read out. Consequently, the overscan region should provide a more reliable estimate of the bias level in the frame. For a modern CCD the bias frame should be a constant, flat level, ie. characterizable by a single number. For an example of unmistakable pattern noise, see figure 9 of Hanisch (1992). If a two-dimensional bias correction is necessary, multiple bias frames need to be taken and combined to remove cosmic ray hits. All other frames are then bias-corrected and overscan regions, underscan regions and any other unwanted strips trimmed off.

- Dark current – the dark current on modern CCDs is usually $\sim$1 e^- per hour and can be ignored. Otherwise, long exposure dark frames are needed to characterize the dark current. As noted earlier, the dark current is both strongly temperature sensitive and power on/off sensitive. The dark frames need to be stacked to remove cosmic-rays and then subtracted off all bias-corrected data frames. The amount to be subtracted is proportional to the relative exposure times. A further advantage of taking a series of dark frames is that the cosmic ray sensitivity can be characterized – useful for optimizing cosmic-ray removal routines.

- Flat fields – measure the two-dimensional pixel sensitivity map, including large scale gradients and vignetting caused by dust particles on the filters or CCD window etc.. There are three basic approaches: take dome flats by attempting to uniformly illuminate either the interior of the dome or a white screen using a tungsten light source – hard to get uniform illumination but good for pixel-to-pixel sensitivity measurements; use the twilight sky – problems are short window of opportunity and sky illumination gradients can cause problems for wide field CCDs; make a "superflat" from the astronomical targets using either a shift and add technique, or a series of deep exposures on different regions – not always feasible. Whichever method is used the frames have to be first combined, to remove either cosmic rays or stars/galaxies, before being used to normalize out the sensitivity variations.

• Fringing – an awkward problem with both wavelength and spatial dependence. Only affects thinned CCDs toward red limit of optical spectrum (see previous section). Sky fringing has the added difficulty of being time-varying too, since the relative intensities of the narrow night sky emission lines (ie. those that cause the sky fringing) varies with atmospheric conditions. Continuum sources (eg. most astronomical objects and tungsten dome lamps) are not seriously affected. Note particularly that twilight sky flats will show fringing in the R, I and Z passbands and as such are not suitable for flatfielding the CCD. If fringing is a problem and the dome flats suffer from non-uniform illumination then use the dome flats to remove only the pixel-to-pixel sensitivity variations and use the sky flats to remove the large scale gradients (although even these may be affected by fringing). From the point-of-view of the astronomical objects, the sky fringing is additive, and hence is best estimated and removed by forming a "supersky" by stacking offset target exposures taken under similar atmospheric conditions.

• Cosmic rays – in target frames the effect can be reduced in three generic ways: the optimum method is to modify subsequent analysis programs to ignore them; if this is impractical cosmic-ray removers based on single frames work by variants on the theme of looking for isolated pixels with intensities/gradients "well above" the local background/gradient – helped by knowledge of the probability distribution function (PDF) of cosmic ray hits (from dark frames) and a good noise model for the detector (readout noise and photon noise); better is to use multiple frames and perform a pixel-by-pixel comparison using the noise model as previously and some form of k-sigma clipping to reject cosmic rays – see next section.

• Other problems – assorted and include hot spots, traps, dead pixels/columns, saturation, video pattern noise, CTE and smearing, deferred charge etc.. Refer to Gullixson (1992) and Gilliland (1992) and references therein for more details on recognizing and dealing with them.

To be a little more precise, we can summarize the various stages of CCD data pre-processing in a compact manner borrowing heavily from the formalism of Gullixson (1992). The raw CCD frame is related to the previous list of corrections by,

$$raw = (obj + sky \times (1 + fringe)) \times qe + dark + flash - skim + bias \qquad (4.29)$$

where most of the terms are as above, apart from *flash*, which denotes any pre-flash (low level illumination of CCD using on-board LEDs) that may be applied prior to exposing the CCD in order to reduce the non-linearities caused by deferred charge transfer/skimming, *skim*.

We first correct for the Zero Exposure Additive Spatial Systematics by forming average frames viz;

$$< dark >= dark + flash - skim + bias$$
$$< skim >= flash - skim + bias$$
$$< bias >= bias$$

and use the overscan region to monitor and remove variations in bias DC level so that,

$$raw- < .... > -(ovscn- < ovscn >) = (obj + sky \times (1 + fringe)) \times qe \qquad (4.30)$$

where $< .... >$ denotes whichever of the 3 corrections or combinations thereof is appropriate or necessary.

To correct for Multiplicative Spatial Systematics, ie. flat fielding, form

$$< flat >= qe + dark + flash - skim + bias \tag{4.31}$$

$$const \times \frac{raw - < .... >}{< flat > - < .... >} = obj + sky \times (1 + fringe) \tag{4.32}$$

where *const* is a normalizing constant chosen to make the average multiplicative *qe* correction unity.

If fringing is present, or if for some reason the twilight sky flat does not result in "flat" frames, correct for Residual Spatial Systematics – shift and stare and stack, to form

$$< sky >= sky \times (1 + fringe); \qquad < sky >= sky \tag{4.33}$$

and either additively apply, if fringing is the cause, or multiplicatively apply if problem was non-uniform flats.

Finally, although this may be necessary in the preceding steps as well, register the individual frames, including geometric distortion corrections as necessary, using variants of the general 6-plate constant linear transformation,

$$x' = a.x + b.y + c; \qquad y' = d.x + e.y + f \tag{4.34}$$

and combine. It is simplest to compute the linear transforms by matching the coordinates of pairs of images from each frame. Note that if only shifts are required, $a = 1$; $b = 0$; $d = 0$; $e = 1$; if only a small angle of rotation and shift are required, $a = 1$; $b = -d$; $e = 1$.

4.1. *Combining Images*

A common theme in the previous section for both removing the effects of unwanted sources (stars, cosmic-rays etc..) and to improve the signal-to-noise of images, involves combining image frames. Is there an optimum way to do this ? If the frames are aligned astrometrically and scaled photometrically, such that pixels on different frames can be directly combined then the problem is well defined. Otherwise, suitable frame-to-frame transformations as in equation 4.34 need to be applied and possibly some sort of intensity mapping of the form,

$$new = raw \times scale + shift \tag{4.35}$$

to correct for varying sky levels or atmospheric transmission.

Let us now consider a series of matched and scaled values of a particular pixel, with intensities in each of m frames, x_i, for $i = 1, m$. For any underlying PDF of the intensities, $P(x)$, which may include contributions from photon noise, readout noise, cosmic rays, bad pixels etc.., various estimators of central location, ie. the "average" value, can be defined. For example:

- the MEAN minimizes

$$< (x_i - \hat{x})^2 >_i = \int (x - \hat{x})^2 P(x).dx$$

- the MEDIAN minimizes

$$< |x_i - \hat{x}| >_i = \int |x - \hat{x}| P(x).dx$$

- the MODE, or a local fit to the MODE, can be used to estimate the position of the maximum of $P(x)$ but requires an impractically large number of frames to be available.

- If, and only if, $P(x)$ is exactly Gaussian (or Poisson), say with variance σ^2, is the MEAN the optimum estimator. In which case the error in the "average" values, $\hat{x}$

would be $\sigma/\sqrt{m-1}$. For such idealized cases, the MEDIAN would have error $\sqrt{\pi/2} \times \sigma/\sqrt{m-1}$, but for practical cases would have the added advantage of not being as susceptible to outlying points.

In practice, perfect Gaussian/Poisson distributions are rarely encountered so how do we generalize estimators of central location to cope with "real noise" ?

Tukey and others (eg. Hoaglin, Mosteller & Tukey 1983) have pioneered the development of so-called robust statistics, or m-estimators, to cope with exactly this problem. An m-estimator of central location can be thought of as a way of creating more robust Maximum Likelihood Estimators (MLEs) applicable to "real" noise distributions. Two of the best estimators appear to be Tukey's biweight function and Andrew's wave function. To illustrate the concept, the m-estimators are defined to be the solution of

$$\sum_i \psi(u_i) = \sum_i w(u_i).u_i = 0 \tag{4.36}$$

where $w(u_i)$ can be thought of as a weighting function; u_i is a measure of the relative deviation of the data from the solution

$$u_i = \frac{x_i - T}{c.S} \tag{4.37}$$

with T the best estimate, S the scatter (ie. a measure of the *rms* spread) and c a tuning constant. In practice equation 4.36 is solved iteratively using

$$\hat{T} = \frac{\sum_i w(u_i).x_i}{\sum_i w(u_i)} \tag{4.38}$$

which at convergence is equivalent to solving equation 4.36 directly. Tukey's biweight function is defined by

$$\psi(u) = u.(1 - u^2)^2, \quad |u| \leq 1 \tag{4.39}$$
$$= 0, \quad |u| > 1$$

and Andrew's wave function by

$$\psi(u) = \frac{1}{\pi} sin(\pi.u), \quad |u| \leq 1 \tag{4.40}$$
$$= 0, \quad |u| > 1$$

with $S = MAD$ (ie. the Median of the Absolute Deviation from the median, a popular robust estimator of sigma, equivalent to $\sigma_{rms}/0.6745$ for a Gaussian distribution) and c set to between ≈ 6 to 12. Note that both are equivalent to clipping outlying members (cf. k-sigma clipping) from contributing and both perform better than the straight median estimator, whilst remaining robust to aberrant points. For comparison, a k-sigma clipped mean estimator would be specified by $\psi(u) = u, \; |u| \leq 1; \quad = 0, \; |u| > 1$.

The condition satisfied by equation 4.36 arises naturally from minimizing a function of the form

$$\sum_i \rho(u_i); \quad \frac{d\rho(u)}{du} = \psi(u) \tag{4.41}$$

and suggests a simple equivalence with generalized MLEs. For example: consider a process whereby $x_i = \bar{x} + \epsilon_i$ and the PDF of ϵ_i is given by

$$P(\epsilon_i) = \alpha.N(0, \sigma_i) + \beta.U(-a, a) \tag{4.42}$$

where $N(0, \sigma_i)$ represents the Gaussian core of the noise distribution, and the uniform distribution, $U(-a, a)$, represents the non-Gaussian extended tail. If the range $[-a, a]$ is large compared to the core size and the fraction of outlying points is low, $\beta << 1$, then to a good approximation this distribution is equivalent to

$$P(\epsilon_i) = \frac{\alpha}{\sqrt{2\pi\sigma_i^2}}.exp(\frac{-\epsilon_i^2}{2\sigma_i^2}), \quad |\epsilon_i| \le k\sigma_i \tag{4.43}$$
$$= \beta, \quad k\sigma_i < |\epsilon_i| \le a$$
$$= 0, \quad |\epsilon_i| > a$$

or, we could equally well have postulated this form for the PDF at the beginning. Assuming independent measurements, x_i, minimizing the log-likelihood function then leads to the following estimator for $\hat{x}$,

$$\hat{x} = \frac{\sum_{i=1}^{m'} x_i/\sigma_i^2}{\sum_{i=1}^{m'} 1/\sigma_i^2} \tag{4.44}$$

where m' denotes the observations within the k-sigma clipped range. Equation 4.44 has to be solved iteratively since the clipping boundary is a function of the current estimates of $\hat{x}$ (and $\hat{\sigma}^2$). The well known k-sigma clipping is therefore equivalent to a MLE with a flat extended error tail and illustrates the general applicability of the MLE approach. Note further that equation 4.44 without the k-sigma clipping restriction, is simply the conventional inverse variance weighting scheme, usually derived directly from the requirement that the combined image signal:to:noise ratio should be a maximum. If a more realistic PDF for the non-Gaussian/ Poisson errors is known, including what fraction of pixels are likely to be affected (eg. from the PDF of cosmic rays hits on dark frames), then it is worthwhile considering a numerical solution to the MLE problem that makes direct use of this information.

5. Detecting and Parameterizing Objects

The previous section dealt with some of the rationale behind removing the detector-dependent signatures from two-dimensional CCD images. We now turn to the problems encountered in detecting and parameterizing any objects that may be present in the two-dimensional data. Since, the signature of the detector defines what device was used to create the image, once that signature has been removed, subsequent analysis techniques should be (more-or-less) device independent, ie. applicable to CCD data, digitised photographic data, infra-red array detectors, and so on.

In order to break down the overall analysis strategy into manageable portions the following main tasks can be identified (see Irwin (1985) and references therein for further discussion on these and related points):

(*a*) Estimate the local sky background over the field and track any variations at adequate resolution to eventually remove them.

(*b*) Detect objects/blends of objects and keep a list of pixels belonging to each blend for further analysis.

(*c*) Parameterize the detected objects, ie. perform astrometry, photometry and some sort of shape analysis.

In deciding on a rationale for these three stages, natural questions to consider are: is it feasible to do the whole analysis automatically – after all wide field optical astronomy has the potential to generate Gbytes of data per night; how does a human observer interactively interpret two-dimensional maps and what can we learn from this; how accurate are any derived image parameters likely to be; and closely related to this last question, which methods of analysis will give the best results at reasonable computational cost ?

5.1. *Sky Background Analysis*

To track any potential background variation, the whole field can be partitioned up into an array of suitably-sized "background regions". (Here, suitable means, a compromise between small enough size to follow real variations and large enough size to reduce subsequent error in the estimate.) For each region compute the pixel intensity histogram and derive various statistics from it, as illustrated in figure 9. A straight mean is clearly hopeless, since even in regions free of faint objects, the odd pixel defect can be enough to seriously bias it. The median is a significant improvement, but owing to the innate skewness of the histogram will always tend to overestimate the local sky level. A popular method, and one which I recommend, is to fit a Gaussian to the core (ie. within say ± 1 sigma) of the histogram, as a robust estimator of the mode, an estimate of sigma is a useful byproduct. This works well in relatively uncrowded regions but again systematically overestimates the sky level as the image number density goes up (see Irwin (1985) for examples). Finally, modeling the skewness of the histogram explicitly (Bijaoui, 1980) can in principle give the most unbiased answer, although unfortunately the strong correlation between the asymmetry in the data (usually a free-parameter in the model) and the sky location can lead to instability problems.

A problem with histogram methods, particularly with full 16-bit data, is selecting a sensible bin size – too fine a sampling and the mode is ill-defined; too coarse and resolution is lost due to quantization. A useful rule of thumb is to select a bin size such that the noise level at sky, $\sigma_{sky} \approx 10$ units. Gaussian fitting to the modal core region can readily recover information at a 1/10th bin size level. This is usually much smaller than either the systematic errors caused by contaminating objects, or the random error in the sky estimate $\approx N_{pixels}^{-1/2} \times \sigma_{sky}$; where N_{pixels} is the number of pixels in the histogram within $\pm 1\sigma_{sky}$ of the sky level.

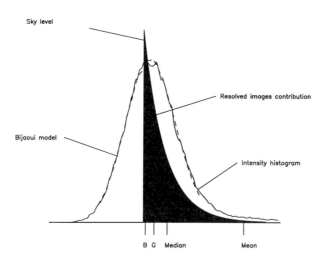

FIGURE 9. Sky background estimation from assorted statistics of the pixel intensity histogram. The histogram of the shaded region is convolved with the PDF of the noise to give the observed intensity distribution. Note in particular how the most widely used estimators of sky level are ordered: Mean > Median > mode (from Gaussian fit - G) > Bijaoui model fit (B), due to the skewness caused by resolved images. A popular variant, the k-sigma clipped mean computed directly from the data, can be as good as the Gaussian fit to the mode.

Combining all the background region sky estimates then gives a two-dimensional array of "background" values that can be further processed to help remove regions affected by, for example, the halos of bright stars or galaxies. A local sky value at any point in the original field can then be obtained by interpolation from the background array. This is the approach adopted by the APM facility in automatically processing photographic plates (see Kibblewhite *et al.* , 1984 and Irwin & Trimble, 1984). Note that the process of estimating the background, or sky level, has been completely decoupled from estimating the parameters of an object.

The preceding approach can be thought of as a variant on the scheme of digitally filtering the full two-dimensional pixel map to remove "slowly varying" features, ie. background variations. A popular alternative based on this analogy, is to use something like a two-dimensional median filter (Tukey, 1971) followed by a linear filter (to reduce the step-like appearance median filtering leaves behind), to track the slowly varying features. These are then subtracted off to leave a sky-less map. This method itself is of course a variant of unsharp-masking (eg. Pratt, 1978).

A final alternative is to estimate the background on an image-by-image basis (eg. Newell, 1982). To do this, define an "annulus" around each image, with inner radius several psf radii away, and an outer radius as large as possible. Within the annulus compute either a k-sigma clipped mean estimate, or use the modal estimator $mode = 3 \times median - 2 \times mean$. The latter formula is a good working approximation for a generalized Gaussian distribution of the form $N(0,1) \times (1 + \alpha.(x^3 - 3x))$, but I would recommend the clipped mean because of its generality.

There are three main problems with the image-by-image method. First, setting appropriate annulii boundaries when dealing with mixtures of stars and galaxies of unknown profile (plus the problems of neighbouring images), is non-trivial. Second, although the combination of spatial clipping, inherent in the annulus approach, and intensity clipping seems like an improvement on the histogram, remember that the psf of star images is much larger than a few core radii and it is precisely these same low level, but large area, intensities near sky, that cause the trouble with either the k-sigma clipping algorithm or the histogram method. A final criticism, is that while a histogram-based method attempts to make full use of all the pixel information and the continuity of the background values, this is not true of the local image method.

5.2. *Optimal Detection of Images*

The classical approach for optimum detection of signals (images) in the presence of random noise is to apply a matched detection filter to the raw data prior to setting detection criteria (see for example Pratt, 1978). To a good approximation the random noise present on the detector comes from a Gaussian/Poisson process and is easy to deal with. However, the presence of other images in the data, particularly faint stars and background galaxies, and possible systematic artefacts due to the detector, complicates things by introducing a "systematic" noise component. Here we define any artefact/image outside of the image of interest, to be a potential systematic noise problem. A good example of this arises in the problem of detecting Low Surface Brightness Galaxies (LSBGs). Since LSBGs often have scale sizes of ≈ 1 arcmin, multitudes of other faint images will generally be superimposed and make the identification task more complex.

Woodward & Davies (1958) presented a useful insight to the problem of optimal detection of images using a straightforward application of Bayes' theorem. The argument runs as follows: consider the question of detecting the reflected echo of a radar pulse of known shape, $x_{t+\tau}$, but unknown lag τ, buried within a noisy, discretely sampled, time series, y_t, such that

$$y_t = x_{t+\tau} + \epsilon_t \tag{5.45}$$

where ϵ_t is assumed to be random Gaussian noise of variance σ_t^2 †. The aims are to locate the signal and in doing so accurately estimate the position, τ. We first construct the likelihood function of observing the sequences, x_t, y_t; $t = 1, ...N$, for a particular value of τ,

$$L(\mathbf{x}, \mathbf{y} \mid \tau) = \prod_{t=1}^{N} P(\epsilon_t \mid \tau) \tag{5.46}$$

and then use Bayes' theorem to argue that in the absence of prior information, finding the maximum of the likelihood function is equivalent to finding the most probable value of τ given $\mathbf{x}$ and $\mathbf{y}$. Now

$$P(\tau \mid \mathbf{x}, \mathbf{y}) \propto exp(-\sum_{t=1}^{N} \frac{\epsilon_t^2}{2\sigma_t^2}) \tag{5.47}$$

and maximizing the probability is equivalent to a weighted least-squares minimization of

$$\sum_t \frac{\epsilon_t^2}{\sigma_t^2} = \sum_t \frac{1}{\sigma_t^2} \cdot (y_t - x_{t+\tau})^2 \tag{5.48}$$

or in the case of $\sigma_t^2 = constant$, finding the maximum of the cross- correlation of y_t with $x_{t+\tau}$. Not only does this illustrate how a cross-correlation function is related to the solution we seek, it shows how to generalize the traditional cross-correlation approach to deal with more complex forms of noise contamination.

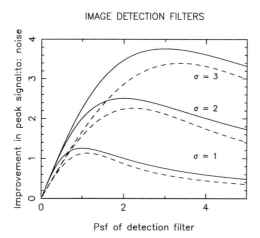

IMAGE DETECTION FILTERS

FIGURE 10. Performance of various detection filters for target Gaussian images of $\sigma = 1, 2, 3$. Solid lines are for Gaussian filters and dashed lines are for circular top-hat filters. Note: how the improvement peaks for filters of equivalent scale size to the target objects; how insensitive the improvement is to the exact size of the filter; and how insensitive the performance is to the exact shape of the filter.

---

† It is straightforward to generalize to non-independent noise using a generic multivariate Gaussian distribution, which leads to a "least-squares" estimator based on a Mahanoublis metric.

To go back to our original problem of detecting images in noise; since the only images difficult to locate are those marginally above the sky noise, a constant noise approximation is a valid assumption and certainly for a CCD, the noise in adjacent pixels is to all intents and purposes random. In other words, to optimally detect images, cross-correlate ‡ the entire frame with the appropriate filter; the psf in the case of point-like sources and possibly an exponential function of chosen scale length for larger faint galaxies.

From a practical viewpoint a matched filter maximizes the resultant peak signal:to:*rms* noise. The same criteria can indeed be used to derive it. Figure 10 illustrates this improvement for a series of Gaussian target objects using both Gaussian and top-hat detection filters of various sizes. Defining the two-dimensional target profile as $I = I_p.exp(-r^2/\sigma^2)$ and the normalized detection filter as $I = exp(-r^2/\sigma_g^2)/\pi\sigma_g^2$ changes the peak intensity, $I_p \rightarrow I_p \div (1 + \frac{\sigma_g^2}{\sigma^2})$, and changes the noise, $< \epsilon^2 >^{\frac{1}{2}} \rightarrow < \epsilon^2 >^{\frac{1}{2}} \div \sqrt{2\pi\sigma_g^2}$. A similar exercise can be carried through for normalized circular top-hat filters of radius, a, to show that the improvement becomes $\sqrt{\pi a^2} \times \sigma^2/a^2 \times exp(-a^2/\sigma^2)$. Because the performance gain in peak signal:to:*rms* noise is relatively insensitive to the exact size and shape of the filter, it is then feasible to use non-linear variants of the preceding filters to detect large LSBGs in the presence of contaminating foreground/background objects. See Irwin *et al.* (1990) for more details and examples of the galaxies detected.

5.3. *Detecting Using Pixel Connectivity*

The main purpose of the detection filter is to enhance the peak signal:to:noise to optimize the performance of object locating algorithms. Most of these algorithms are some variant on the theme of peak (eg. Herzog & Illingworth, 1977) or island searches (eg. FOCAS – Tyson & Jarvis, 1979). Both methods work by first prescribing a pixel detection threshold above background, otherwise known as a detection isophote, and then proceed to look for either isolated peaks, or islands of contiguous pixels above this threshold. Since most peak-based methods examine the intensities of neighbouring pixels too, there is in practice, little difference between the two approaches and I will stick to describing the pixel connectivity approach.

Figure 11 illustrates the main concepts. Potential images are defined as regions of simply-connected pixels above the detection isophote, relative to the local sky background. Two passes through the data are recommended. On the first pass the sky background is estimated, as outlined previously, and on the second pass the sky level (interpolated if necessary) plus the threshold, define those pixels that require further examination. Spurious images or noise are rejected using a combination of the requirements for a pixel to be above threshold and for it to be connected with enough neighbours to meet a chosen minimum size criterion. The threshold is usually set to be some fixed multiple (≈ 2) of sky noise above sky. Once a potential image has been located, the original pre-detection filter intensities of the pixels are used for subsequent analysis. This latter requirement ensures that any derived image parameters are unaffected by the blurring action of the detection filter. Of the two most commonly used isophotal connectivity paths, the four nearest neighbour route (N,S,E,W) seems to give better performance than the eight-fold neighbour approach (N,S,E,W + NW,SE,NE,SW). The latter type of connectivity tends to produce more spurious pixels in the outer parts of an image leading to somewhat noisier image parameters.

---

‡ This is the same operation as convolution for symmetric profiles – desirable since they induce no systematic shift in image position. Most detection filters are also normalized to unit area under the curve, to ensure average intensities remain the same.

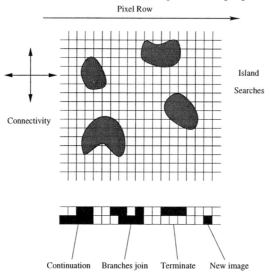

FIGURE 11. The basis of pixel connectivity algorithms

The operations required to do all this, using only one pass through the data and minimal memory requirements, ie. enough rows of pixels in memory to do the detection filtering, can be summarized as follows (see Lutz, 1979 for more details):

(*a*) Set up parent, link-list and pixel information arrays/stack. The parent list, a label for potential new images, points to the position of the 1st link in the link-list for this image; the link-list entry then points to the next pixel for this image and so on; while the pixel information arrays store the i,j position and original intensity of the pixel in the location specified within the link-list.

(*b*) Read in a new row of pixels; do filtering and save original intensities. The filtering is of necessity out of phase with the row of data currently being analyzed for objects.

(*c*) Set the background level and add on threshold and flag pixels above this level.

(*d*) Is pixel i,j neighbour to any active pixels in the current parent/link lists or not? It must be either: a continuation of an image to add to the linked-list; a joining of two branches of previously separate images (see figure 11) requiring updates of parent and link-list arrays; or a potential new image.

(*e*) If within a row no pixels join to a previously active parent image, terminate that image, process pixels to generate parameters and return free slots to stack.

5.4. *Estimating Image Parameters*

The isophotal connectivity method of the previous section only makes use of the information stored in those pixels that passed the threshold and contiguity criteria. The advantages of this approach are that it is computationally fast and straightforward to implement; and that a wide variety of useful and robust image parameters can be derived. The following list demonstrates the range of typical image parameters that can be computed from the connected pixel list in a single pass. These are the image parameters that are widely used by the measuring machine community for astrometry, photometry and image shape description (see for example MacGillivray & Stobie, 1985). It is worth stressing that a single photographic Schmidt plate, the raw material for the measuring machines, contains up to 4 Gbytes of data, recorded in a highly non-linear fashion, and that the currently available sky surveys total well over 20 Tera-bytes of data. The ability to process large volumes of data quickly is also highly relevant for processing wide field CCD data. Examples of alternative strategies, usually involving significantly more computational effort, will be given in the next section. In all the following equations, $I(x_i, y_i)$ denotes the intensity of the ith pixel found in the island search phase and the sum is over all detected connected pixels. We shall assume for simplicity that all detected images are single objects. For more details on dealing with overlapped blended images see Irwin (1985).

Isophotal Intensity – the integrated flux within the boundary defined by the threshold level; ie. 0th moment

$$I_{iso} = \sum_i I(x_i, y_i) \tag{5.49}$$

For Gaussian images, this is related to the total intensity by the factor $(1 - I_t/I_p)^{-1}$, where I_p is the peak flux and I_t the threshold level relative to sky. This implies that for images close to the detection limit, $I_p \approx I_t$, isophotal magnitudes will be around 1 magnitude fainter than total magnitudes.

Position – computed as an intensity-weighted centre of gravity; ie. 1st moment

$$x_o = \sum_i x_i . I(x_i, y_i) / \sum_i I(x_i, y_i) \tag{5.50}$$

$$y_o = \sum_i y_i . I(x_i, y_i) / \sum_i I(x_i, y_i)$$

Note that the intensity weighting and the isophotal connectivity criteria automatically reduce the sensitivity of the estimator to outlying points.

Covariance Matrix – the triad of intensity-weighted 2nd moments is used to estimate the eccentricity/ellipticity, position angle and intensity-weighted size of an image

$$\sigma_{xx} = \sum_i (x_i - x_o)^2 . I(x_i, y_i) / \sum_i I(x_i, y_i) \tag{5.51}$$

$$\sigma_{xy} = \sum_i (x_i - x_o).(y_i - y_o).I(x_i, y_i) / \sum_i I(x_i, y_i)$$

$$\sigma_{yy} = \sum_i (y_i - y_o)^2 . I(x_i, y_i) / \sum_i I(x_i, y_i)$$

The simplest way to derive the ellipse parameters from the 2nd moments is to equate them to an elliptical Gaussian function having the same 2nd moments. It is then straightforward to show (eg. Stobie, 1980) that the scale size, $\sqrt{\sigma_{rr}}$, is given by, $\sigma_{rr} = \sigma_{xx} + \sigma_{yy}$; the eccentricity, $ecc = \sqrt{(\sigma_{xx} - \sigma_{yy})^2 + 4.\sigma_{xy}^2} / \sigma_{rr}$; and the position angle, θ is defined by, $tan(2\theta) = 2.\sigma_{xy}/(\sigma_{yy} - \sigma_{xx})$. Higher order moments have been used as shape descriptors (eg. Tyson & Jarvis, 1979), but are generally much more prone to corruption by outlying noisy data pixels.

Areal Profile – an interesting variant on the radial profile, which measures the area of an image at various intensity levels. Unlike the radial profile, which needs a prior estimate of the image centre, the areal profile provides a single pass estimate of the

profile

$$Areal\ Profile \rightarrow \quad T + p_1, T + p_2, T + p_3, ...T + p_m \qquad (5.52)$$

where $p_j; j = 1, ...m$ are intensity levels relative to the threshold, T, usually spaced logarithmically to give even sampling.

The peak height, I_p, is a useful related addition to the areal profile information and is defined as

$$I_p = max[I(x_i, y_i)]_i \qquad (5.53)$$

or alternatively measured by extrapolation from the areal profile if the image is saturated (straightforward for photographic plates, but charge leakage causes problems with CCD data).

Other useful measures of image size are those defined by Kron (1980) based on moments of the radial profile.

$$r_1 = \int r.I(r)dr / \int I(r)dr; \quad r_{-2} = \int r^{-2}I(r)dr / \int I(r)dr \qquad (5.54)$$

These involve more than one pass through the data but are good shape discriminators for morphological classification. There are many other methods of parameterizing the shape information as the front end to an object classification scheme (eg. star/galaxies/noise). A recent review of the field is given by Odewahn (1995).

6. Alternatives Methods for Astrometry and Photometry

Maximum likelihood theory enables us to devise optimum methods for estimating image parameters and also specifies the minimum attainable error it is possible to achieve for images of a given profile and signal:to:noise (eg. Irwin, 1985). Isophotal methods are not generally optimal, but they are robust, require no prior knowledge of the image profile, and are simple to compute. A serious drawback in their direct use is that the derived intensities are biased, since a varying fraction of the total flux is included. Although there are assorted tricks to get around this problem for stellar images (eg. Bunclark & Irwin, 1983), recovering the lost flux for an unknown galaxy profile is more difficult. From an information-theoretic viewpoint the loss of information in neglecting the outer parts of the profile is not so serious, since most of the information regarding image location and total intensity comes from the higher signal:to:noise inner part of the profile. Consequently, bias notwithstanding, isophotal methods actually come quite close, certainly within <50%, of the minimum attainable error bound.

In principle, a stellar image is completely specified by its coordinates and intensity. In practice, variations in psf across the field and lack of prior knowledge about which images are stellar conspire to make life more complicated. The problem is much worse for galaxies, since we have no prior knowledge of their profile. These problems have lead to a whole range of methods, particularly photometric methods, to circumvent the lack of knowledge.

6.1. *Aperture Photometry*

There are two basic approaches for estimating total intensities. The first is to integrate the flux within an aperture. The second is to fit some sort of profile to the image and then renormalize the profile-based intensity to be equivalent to a total one. Apertures are either fixed in size (usually for bright stellar photometry only) or allowed to "grow" automatically out to some pre-specified maximum area.

Aperture photometry is defined by

$$I_{ap}(r) = \sum_i^N I_i - N \times sky \tag{6.55}$$

such that all pixels indexed by i are within a radius, r, of the image centre †. An isophotal estimate of the image centre, as in equation 5.50, is usually adequate. Intensity contributions from pixels crossing the boundary are usually added in on a pro-rata basis. Defining I_i to be the object flux + sky flux in pixel i, clarifies the dependence of $I_{ap}(r)$ on the accuracy of the sky estimate – assumed constant over the image. The problem with aperture photometry is the necessary tradeoff between capturing all the light, = large radius; adding in too much sky noise = $\sqrt{N} \times \sigma_{sky}$; and being mislead by systematic errors in the sky level = $N \times \Delta_{sky}$. The usual approach to minimize these problems is to "examine" the curve-of-growth, $I_{ap}(r)$ -v- r, as illustrated in figure 12. Another problem, not shown in the figure, is the presence of neighbouring images, which will also lead to an upturn in the curve. In this example, the systematic errors in the sky estimate correspond to $\pm 1\%$ errors in the absolute sky level for a typical scale length galaxy image at a similar magnitude to a 1 arcsec^2 patch of sky. Figure 2. of Irwin (1985) shows how typical estimators of sky can readily be off by 1% in even moderately crowded fields.

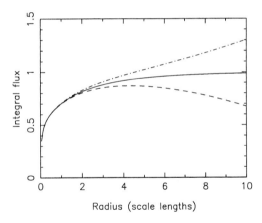

FIGURE 12. Idealized curves-of-growth for aperture integration of an exponential profile. If the sky level is exactly correct, $I_{ap}(r)$ will asymptotically stabilize to the desired level. If the sky estimate is too high the curve will turn over (dashed), too low and the curve continues rising (dot-dash). The effect of noise in the data is not shown.

For stellar images, assuming the psf does not vary rapidly around the field, it is possible to minimize the preceding problems by calibrating the curve-of-growth using isolated bright stellar images. Fainter stars can then be integrated out to a compromise radius (usually a few core radii) and then a well defined correction applied. Clearly this is not possible for galaxy images and some sort of empirical fit to the curve-of-growth (eg. 3rd order polynomial and look for zero derivative out to some maximum radius) seems to be

† Elliptical apertures, parameterized by semi-major axis radius, were used by Irwin & Hall (1982) for galaxy photometry to minimize the area of sky included.

as good a method as any of automating the decision and minimizing the effects of sky noise (see Irwin & Hall (1982) for more details).

6.2. *Profile Fitting*

Profile fitting, usually limited to stellar images, is now well served by several software packages such as: ROMAFOT – Buonanno & Iannicola (1989); DoPHOT – Mateo, Saha & Schecter (1990); and DAOPHOT – Stetson (1987). These references address the main problems and issues involved in successful profile fitting; whilst the work by King (1983), Irwin (1985) and Goad (1986) investigates the theory underpinning optimal fitting and shows how this can be used to predict the likely errors of the derived parameters. Stellar psf fitting is the only reliable technique to use in crowded regions such as those found toward the centre of globular clusters. By iterative use of difference maps (data – psf model) and by simultaneous fitting of overlapping components in a blended image, results very close to the theoretical error bound can be achieved.

Very briefly, the main decisions to be made in profile fitting are:

(*a*) is the profile, ie. psf, adequately constant across the frame or does some allowance for geometric variation need to be folded in (eg. HST data; ROSAT data) ?

(*b*) can the psf be approximated using an analytic function ? Usually some mixture of Gaussian-exponential-Lorentzian or a Moffat profile, $(1 + r^2/a^2)^{-\beta}$, are used.

(*c*) if the images are noticeably trailed is an elliptical deformation of one of the above profiles adequate ? or,

(*d*) does an empirically constructed profile need to be formed from isolated stars ? If the seeing psf is not well sampled then some sort of sub-pixel interpolation will be necessary to reconstruct the model profile at non-integral pixel locations. This is easy with analytic profiles, harder for empirical profiles.

(*e*) how large a region around each image should the psf be fitted over ? With a complex blend of several images this is not necessarily straightforward to answer. A simple solution is to only use those pixels picked out by some isophotal detection algorithm.

(*f*) should a local estimate/update for the background level, caused by unresolved images etc., be made simultaneously with the psf parameter fitting ?

(*g*) if the profile is an exact match for the psf, total intensities = profile intensities. In practice this is rarely the case and isolated bright stars are used to estimate the offset between the profile and total intensity systems.

For more details on these and other points consult the preceding references.

There is a very simple link between isophotal intensity estimation and profile fitting for isolated images, which goes some way to show why isophotal estimates are so popular. With the usual Gaussian/Poisson noise encountered in optical work, all profile fitting methods reduce to weighted least-squares methods of the form of minimizing

$$F = \sum_i \frac{(d_i - m_i)^2}{\sigma_i^2} \tag{6.56}$$

where d_i and m_i are respectively the data and model at pixel i, and σ_i^2 is the expected noise variance $< \epsilon_i^2 >_i$, such that $d_i = m_i + \epsilon_i$. † For a single profile, with assumed known position, $m_i = k.\phi_i$, where k is the unknown scaling factor and ϕ is the normalized psf (ie. $\int \int \phi(x, y) \, dx dy = 1$) shifted to the appropriate location. Solving for the intensity

---

† For those condemned to work with few photons per resolution element, as is often the case in X-ray or γ-ray astronomy, the equivalent expression is, $F = \sum_i d_i.ln(m_i) - m_i - ln(d_i!)$.

parameter then gives

$$k = \frac{\sum_i \phi_i.d_i/\sigma_i^2}{\sum_i \phi_i^2/\sigma_i^2} \qquad (6.57)$$

which is nothing more than a weighted sum over the pixel intensities belonging to the image; although subtle differences between the data–model overlap integral in the numerator, and the model–model overlap integral in the denominator, are the cause of the intensity renormalization problem mentioned previously. There are two interesting limiting cases: for bright images well above sky, Poisson noise dominates, so that $\sigma_i^2 = \phi_i$ (if the gain is unity, $\propto$ otherwise) and the profile intensity is simply $\sum_i d_i$, the isophotal intensity; for a faint image dominated by constant sky noise the profile intensity is just the profile-weighted sum of the pixel intensities. This analogy can be extended to multiple overlapping images, using more complex weighting factors.

6.3. *Astrometry – Image Centering*

In the late 1970's the most popular method for locating the centre of an image involved using marginal sums of the image intensity distribution. This reduces a two-dimensional problem to a pair of one-dimensional problems, with subsequent savings in computer time. The marginal sum method involves some variant along the lines of the following steps. First, form $\rho(x_i)$, and equivalently $\rho(y_i)$, where

$$\rho(x_i) = \sum_{j=-a}^{j=+a} I_{i,j} \qquad (6.58)$$

and then find the centre of the one-dimensional distribution(s) using either: the median, a Gaussian fit to the distribution (often including a sloping background component rather than a constant), or a variant of the centre of gravity such as

$$\bar{x} = \frac{\sum_i [\rho(x_i) - \bar{\rho}].x_i}{\sum_i [\rho(x_i) - \bar{\rho}]} \qquad (6.59)$$

using only those pixels with $\rho(x_i) \geq \bar{\rho}$. Iterate through these two steps if $\bar{x}$ is more than 1 pixel from the window centre. Although computationally fast, there are several shortcomings of the method: the window has to be centred fairly accurately to start with or the iterative solution can diverge; neighbouring images can corrupt the marginal distribution and bias the answer away from the true centre; if the window $[-a, a]$ is too large excessive noise is folded in, too small and much of the information is lost. Irwin (1985) outlined how this potential loss of information degrades the performance of the algorithm and suggested ways of minimizing the problem by using Maximum Likelihood theory to establish the optimum window to use.

In many ways, the marginal sum method was superseded by the centre-of-gravity estimates derived from pixel connectivity algorithms – ie. isophotal centre-of-gravity estimators such as equation 5.50. If we generalist such an estimator to include a weighting term, w_i, (cf. equation 6.57 for intensity estimation) along the lines of

$$\bar{x} = \frac{\sum_i w_i \times x_i \times I_i}{\sum_i w_i \times I_i} \qquad (6.60)$$

where the summation is over all isophotally connected pixels, is there an optimum weighting factor ? Put another way, what does Maximum Likelihood theory suggest ? For an image with underlying normalized profile, $\phi(x, y)$, the optimum weight is given by

$$w_i = \left[\frac{1}{x - \bar{x}} \times \frac{\delta\phi}{\delta x} \right]_i \times \frac{1}{\sigma_i^2} \qquad (6.61)$$

Clearly, this involves an iterative solution for $\bar{x}$ (and $\bar{y}$). Irwin (1985) shows how to extend this scheme to overlapping blended images.

As an example, for a radially symmetric Gaussian profile, this simplifies to $w_i \propto \phi(x_i, y_i)/\sigma_i^2$. As in the previous section, the two limiting cases are very revealing. For bright images, w_i is a constant, implying that for such cases isophotal centre-of-gravity methods are close to optimal. For faint images, $w_i \propto \phi(x_i, y_i)$, and profile weighting, with iteration, is the answer.

For more details of the derivation of optimum estimators and a full error analysis see the papers by, King (1983), Irwin (1985) and Goad (1986).

7. Acknowledgements

In preparing the section on telescope optics I have benefited from many fruitful conversations with Roderick Willstrop and Ian Parry and would like to thank them for their help. I am also indebted to Richard McMahon for letting me borrow from his notes of sky surveys in other wavebands. Finally I would like to acknowledge the financial support provided by the Instituto de Astrofísica de Canarias, and to thank them for inviting me to take part in the Winter School.

REFERENCES

Becker, R.H., White, R.L. & Helfand, D.J., 1995, *Astrophys. J.*, **450**, 559.

Bijaoui, A., 1980, *Astr. Astrophys.*, **84**, 81.

Boksenberg, A., 1972, in *Proceedings of the ESO-CERN Conference on Auxiliary Instrumentation for Large Telescopes*, eds. S. Lausen & A. Reiz, 295.

Boronson, T., Reed, R., & Wong, W.Y., 1994, in *Instrumentation in Astronomy VIII*, Proc. SPIE, **2198**, 877.

Bradt, H.V.D., Ohashi, T., & Pounds, K.A., 1992, in *Ann. Rev. Astron. Astrophys.*, **30**, 391.

Buonanno, R., & Iannicola, G., 1989, *Publ. astr. Soc. Pacific*, **101**, 294.

Bunclark, P.S., & Irwin, M.J., 1983, in *Proc. Statistical Methods in Astronomy*, ESA SP-201, 195.

CFHT MOCAM User Manual, 1995.

Demers, S., Kibblewhite, E.J., Irwin, M.J., Nithakorn, D.S., Beland, S., Fontaine, G., & Wesemael, F., 1986, *Astr. J.*, **92**, 878.

Fisher, R.A., Corbett, A.S., & Williams, C.B., 1943, Journal of Animal Ecology, **12**, 42.

Fordham, J.L.A., Bone, D.A., Michel-Murillo, R., Norton, T.J., Butler, I.G., & Airey, R.W., 1994, in *Instrumentation in Astronomy VIII*, Proc. SPIE, **2198**, 829.

Gilliland, R.L., 1992, in *Astronomical CCD Observing and Reduction Techniques*, ed. S.B.Howell, ASP Conf. Series, Vol. 23, 68.

Gilmore, G., & Walton, N., 1993, *Gemini*, **40**, 16.

Goad, L., 1986, in *Instrumentation in Astronomy*, SPIE Vol. **627**, 688.

Gullixson, C.A., 1992, in *Astronomical CCD Observing and Reduction Techniques*, ed. S.B.Howell, ASP Conf. Series, Vol. 23, 130.

Gunn, J.E., *et al.*, 1987, *Optical Engineering*, **26**, 779.

Hanisch, R.J., 1992, in *Astronomical CCD Observing and Reduction Techniques*, ed. S.B.Howell, ASP Conf. Series, Vol. 23, 285.

Hartley, M., 1994, in *Digitised Optical Sky Surveys*, eds. H.T.MacGillivray and E.B. Thomson, Kluwer Academic Publishers, Dordrecht, p.117.

Herzog, A.D., & Illingworth, G., 1977, *Astrophys. J. Suppl.*, **33**, 55.

Hoaglin, D.C., Mosteller, F., & Tukey, J.W., 1983, *Understanding Robust and Exploratory Data Analysis*, Wiley, Chp 11.

Irwin, M.J., & Hall, P., 1982, *Occ. Repts. R. Obs. Edin.*, **10**, 111.

Irwin, M.J., & Trimble, V., 1984, *Astr. J.*, **89**, 83.

Irwin, M.J., 1985, *Mon. Not. R. astr. Soc.*, **214**, 575.

Irwin, M.J., Davies, J.I., Disney, M.J., & Phillipps, S., 1990, *Mon. Not. R. astr. Soc.*, **245**, 289.

Irwin, M.J., 1992, in *Digitized Optical Sky Surveys* , eds. H.T.MacGillivray and E.B. Thomson, Kluwer Academic Publishers, Dordrecht, p.43.

Irwin, M.J., 1994, in *Working Group on Wide Field Imaging – Newsletter No. 5*, 25.

Janesick, J., & Elliot, T., 1992, in *Astronomical CCD Observing and Reduction Techniques*, ed. S.B.Howell, ASP Conf. Series, Vol. 23, 1.

Jorden, P.R., Deltorn, J., & Oates, A.P., 1994 in *Instrumentation in Astronomy VIII*, Proc. SPIE, **2198**, 836.

Kibblewhite, E.J., Bridgeland, M.T., Bunclark, P.S., & Irwin, M.J., 1984, in *Proc. Astronomical Microdensitometry Conference*, NASA-2317, 277.

King, I., 1983, *Publ. astr. Soc. Pacific*, **95**, 163.

Kron, R.G., 1980, *Astrophys. J. Suppl.*, **43**, 1.

Kron, R.G., 1995, *Publ. astr. Soc. Pacific*, **107**, 766.

Lasker, B.M., 1995a, in *Future Utilisation of Schmidt Telescopes*, IAU Symposium No. 148, eds. R.D.Cannon *et al.* , ASP, San Francisco (in press).

Lasker, B.M., 1995b, *Publ. astr. Soc. Pacific*, **107**, 762.

Lesser, M.P., 1989, in *CCDs in Astronomy*, ed. G.H. Jacoby, ASP Conf. Series, Vol. 8, 65.

Lesser, M.P., 1994, in *Instrumentation in Astronomy VIII*, Proc. SPIE, **2198**, 782.

Luppino, G.A., Bredthauer, R.A., & Geary, J.C., 1994 in *Instrumentation in Astronomy VIII*, Proc. SPIE, **2198**, 810.

Lutz, R.K., 1979, in *Image Processing in Astronomy*, eds. G. Sedmak, M. Capaccioli, R.J. Allen, Trieste, p. 218.

MacGillivray, H.T., & Stobie, R.S., 1985, in *Vistas in Astronomy*, **27**, 433.

Mateo, M., Saha, P., & Schecter, P., 1990, DoPHOT Users Manual.

Morgan, D.H., 1995, in *Future Utilisation of Schmidt Telescopes*, IAU Symposium No. 148, eds. R.D.Cannon *et al.* , ASP, San Francisco (in press).

Newell, E.B., 1982, *Occ. Repts. R. Obs. Edin.*, **10**, 111.

Odewahn, S.C., 1995, *Publ. astr. Soc. Pacific*, **107**, 770.

Parker, Q.A., Philipps, S., Morgan, D.H., Malin, D.F., Russell, K.S., Hartley, & Savage, A., 1994, in *Astronomy from Wide-Field Imaging*, IAU Symposium No. 161, eds. H.T.MacGillivray *et al.* , Kluwer Academic Publishers, Dordrecht, p.129.

Pratt, W.K., 1978, *Digital Image Processing*, Wiley.

Prestage, R., & Peacock, J., 1983, *Mon. Not. R. astr. Soc.*, **204**, 355.

Price, S., 1988, *Publ. astr. Soc. Pacific*, **100**, 171.

Russell, K.S., Malin, D.F., Savage, A., Hartley, M. & Parker, Q.A., 1992, in *Digitised Optical Sky Surveys* , eds. H.T.MacGillivray and E.B. Thomson, Kluwer Academic Publishers, Dordrecht, p.23.

Seidel, L., 1856, Astronomische Nachrichten, **43**, Nos. 1027,1028,1029.

Sekiguchi, M., Iwashita, H., Doi, M., Kashikawa, N., & Okamura, S., 1992, *Publ. astr. Soc. Pacific*, **104**, 744.

Smart, W.M., 1956, *Spherical Astronomy*, CUP, Cambridge.

Stetson, P.B., 1987, *Publ. astr. Soc. Pacific*, **99**, 191.

Stobie, R.S., 1980, *Journal British Interplanetary Society*, **33**, 323.

Stubbs, C.W., *et al.* , 1994, in *Charge Coupled Devices and Solid State Optical Sensors III*, Proc.

SPIE, 1900.

Thejll, P., & Shipman, H.L., 1988, *Publ. astr. Soc. Pacific,* , **100**, 398.

Tulloch, S., 1995, *ING La Palma Technical Note No. 99.*

Tukey, J.W., 1971, in *Exploratory Data Analysis*, Adison Wesley.

Tyson, J.A., & Jarvis, J.F., 1979, *Astrophys. J.*, **230**, L153.

Tyson, J.A., Bernstein, G.M., Blouke, M.M., & Lee, R.W., 1993, in *High-Resolution Sensors and Hybrid Systems*, Proc. SPIE, 1656, 400.

Willstrop, R.V., 1985, *Mon. Not. R. astr. Soc.*, **216**, 411.

York, D., *et al.* , 1993, A Digital Sky Survey of the Northern Galactic Cap, NSF proposal.

Modern methods of image reconstruction

By R. C. PUETTER[1]

[1]Center for Astrophysics and Space Sciences,
University of California, San Diego
9500 Gilman Drive
La Jolla, CA, 92093-0111, USA
rpuetter@ucsd.edu

Chapter 1 reviews the image restoration/reconstruction problem in its general setting. We first discuss linear methods for solving the problem of image deconvolution, i.e. the case in which the data is a convolution of a point-spread function and an underlying unblurred image. Next, non-linear methods are introduced in the context of Bayesian estimation, including Maximum-Likelihood and Maximum Entropy methods. Finally, the successes and failures of these methods are discussed along with some of the roots of these problems and the suggestion that these difficulties might be overcome by new (e.g. pixon-based) image reconstruction methods.

Chapter 2 discusses the role of language and information theory concepts for data compression and solving the inverse problem. The concept of Algorithmic Information Content (AIC) is introduced and shown to be crucial to achieving optimal data compression and optimized Bayesian priors for image reconstruction. The dependence of the AIC on the selection of language then suggests how efficient coordinate systems for the inverse problem may be selected. This motivates the selection of a multiresolution language for the reconstruction of generic images.

Chapter 3 introduces pixon-based image restoration/reconstruction methods. The relationship between image Algorithmic Information Content and the Bayesian incarnation of *Occam's Razor* are discussed as well as the relationship of multiresolution pixon languages and image fractal dimension. Also discussed is the relationship of pixons to the role played by the Heisenberg uncertainty principle in statistical physics and how pixon-based image reconstruction provides a natural extension to the Akaike information criterion for Maximum Likelihood estimation. The lecture then discusses practical implementation of pixon-based reconstruction, and how these techniques can be applied to a variety of problems, including problems outside of astronomical imaging.

Chapter 4 presents practical applications of pixon-based Bayesian estimation to the restoration of astronomical images. It discusses the effects of noise, effects of finite sampling on resolution, and special problems associated with spatially correlated noise introduced by mosaicing. Comparisons to other methods demonstrate the significant improvements afforded by pixon-based methods and illustrate the science that such performance improvements allow.

Chapter 5 explores the application of pixon-based methods to the reconstruction of complexly encoded images. The special problems sometimes associated with complexly encoded data sets, i.e. an extremely complex and convoluted solution space, are discussed, and numerical methods for dealing with these problems, e.g. simulated annealing, genetic algorithms, and evolutionary programing, are presented. Several sample reconstructions provide examples of reconstructions of complexly encoded images using both standard multi-dimensional minimization techniques and stochastic search methods.

---

1. The Image Restoration/Reconstruction Problem

1.1. *The Mathematics of the Image Reconstruction Problem*

Image reconstruction is becoming a more and more common technique in optical and infrared astronomy. Of course, the problem of reconstructing images from Fourier domain data has been a long standing problem in radio interferometry (see Thompson *et al.* 1986;

Cornwell and Perley 1991; Wohlleben *et al.* 1991). Indeed, in this discipline image reconstruction is a "bread-and-butter" application. The meaning of one's data simply cannot be interpreted from the raw data except in the very simplest of cases. In optical and infrared imaging, one distinguishes between image reconstruction and image restoration. As the names suggest, image restoration generally is a less drastic extrapolation from the data to the final image than is image reconstruction. Normally, image restoration refers to modifying data which already is in the form of an image and applying corrections to achieve an image which is superior in some way (e.g. has higher resolution or has reduced distortion). In the case of image reconstruction one is normally dealing with data that is not in the form of an image. The image is then literally reconstructed from this data. The example of radio interferometry is an excellent one here. In the case of radio interferometry the collected data is a sparsely sampled interferogram. While the data set (if complete) would fill a 2-dimensional image, the spatial relationships of features in the interferogram are very different from the spatial distributions of sources in the sky, and the image and the data are related to each other through a mathematical relationship that encodes the image in a complex manner.

While image restoration and reconstruction have many differences, both are examples of a larger class of problems called the "Inverse Problem". In the general inverse problem one seeks to evaluate, or estimate, the values of a certain set of parameters or functions from the known properties of other parameters or functions, e.g. one has the relationship:

$$\mathbf{L}\left(\{f_i\}, \{g_j\}\right) = 0 \quad , \tag{1.1}$$

where $\mathbf{L}$ is an operator (which can be quite general), the parameters (or functions), $\{f_i\}$, are sought, and the values of the parameters (or functions), $\{g_j\}$, are known.

In the context of the inverse problem, one often discusses the existence and uniqueness of the solution. In physical problems, the existence of a solution is assured, provided the problem is well posed. Furthermore, in the presence of noise, it is virtually assured that the solution is not unique. After all, any two solutions differing only by properties within the noise limits (i.e. statistically indistinguishable solutions) are equally valid. A more interesting question is what are the properties of the solution space for various physical problems. The answer to this is rather complex, and varies from problem to problem. To gain some intuition, let us ask the question first in the image restoration framework. As already mentioned, any two statistically indistinguishable solutions are equally acceptable. That means that another solution can be generated from any known solution by adding a bit of noise. More precisely, to account for noise we would write:

$$\mathbf{L}\left(\{f_i\}, \{g_j\}\right) = \mathbf{N} \quad , \tag{1.2}$$

where $\mathbf{N}$ is the noise which may be a scalar or multi-dimensional noise vector (note that this functional form may not be appropriate for all situations), then any solution, $\{f_i\}$, can generate new solutions, $\{f_i + \Delta f_i\}$, provided that

$$\mathbf{L}\left(\{f_i + \Delta f_i\}, \{g_j\}\right) \approx \mathbf{N} \quad . \tag{1.3}$$

In fact, if $\mathbf{L}$ were a linear operator, we would talk about the null space of $\mathbf{L}$ and say that any operator contained in the null space could be added to any solution to construct new solutions. However, since in general we are not dealing with linear operators, this idea is not exact. Nonetheless, for the image restoration problem (which is often linear) it is clear that any small perturbation can be added to a known solution to produce another solution, provided that the perturbation produces an effect which is small relative to the noise.

We can also carry this exercise through for the radio interferometry problem. In this

Sample Reconstructions

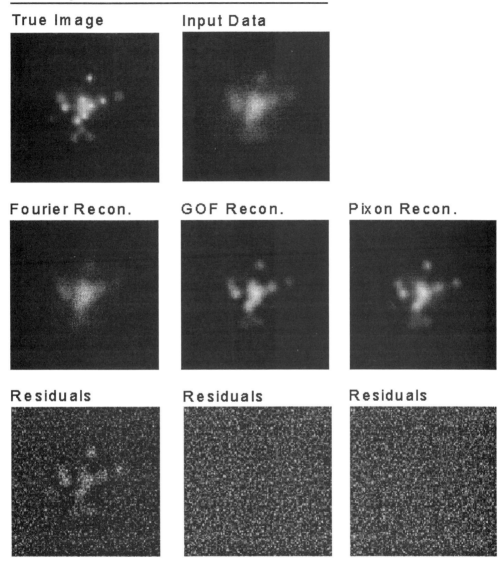

FIGURE 1. Comparison of image reconstruction techniques for a mock data set. Shown at the top is the true, underlying, high resolution, noise-free image. This has been convolved with a Gaussian PSF with FWHM=8 pixels and Gaussian noise added to make the input data image. The peak signal-to-noise ratio in the data is about 30. Shown at the bottom of the figure are reconstruction results for a direct Fourier deconvolution, a pure GOF deconvolution, and a pixon-based deconvolution. Beneath each deconvolution are the resulting residual errors, i.e. the result of subtracting the reconstruction convolved with the PSF from the input data.

case, any solution can have added to it a function whose Fourier transform is small relative to the measured noise in the interferogram. Even more sinister for this case is the fact that huge regions of the Fourier plane are unsampled. Hence one is free to add functions with arbitrarily large Fourier components in these regions without violating the collected data, simply because no information is available at these frequencies. Clearly, in this situation there is a huge solution space. What is more, this solution space contains a wealth of unphysical solutions. Indeed, the typical solution is unphysical. It is clear in this case that in order to ensure the selection of a physically realistic solution, other conditions must be imposed, and indeed, this is the case—see below.

1.2. *Linear Methods*

While most inverse problems are non-linear in nature, an important linear inverse problem commonly occurs in astronomy. This is the case of image restoration of point-spread function (PSF) blurred object. In this case we have the situation:

$$D(\vec{x}) = \int dV_{\vec{y}}\, H\left(\|\vec{x} - \vec{y}\|\right) I(\vec{y}) + N(\vec{x}) \quad , \tag{1.4}$$

where D is the data collected, H is the PSF, I is the underlying, true, unblurred image, and N is the noise. The integral in equation (1.4) is over volume in image space and accounts for the general, multi-dimensional nature of this space. Further, we have assumed that the PSF is a function of the difference in position between $\vec{x}$ and $\vec{y}$, i.e. a convolution. Since equation (1.4) is a convolution, we can use the Fourier convolution theorem to solve the problem. In the absence of noise:

$$F(D) = F(H * I) \quad , \tag{1.5}$$

$$= F(H)F(I) \quad , \tag{1.6}$$

$$I = F^{-1}\left(\frac{F(D)}{F(H)}\right) \quad , \tag{1.7}$$

where $F(f)$ is the Fourier transform of f. In the absence of noise, equation (1.7) provides an exact inversion of equation (1.4). This is a commonly used trick in many early attempts at image reconstruction. It is still often used for the reconstruction of bright objects in which the error due to noise is minimal. A good example of this is speckle image reconstruction (e.g. Ghez *et al.* 1993, Jones 1983) and Fourier transform spectroscopy (e.g. Serabyn and Weisstein 1995; Abrams *et al.* 1995; Prasad and Bernath 1994; also see the review by Ridgway and Brault 1984). Unfortunately, because of the noise contribution to $D(\vec{x})$, we cannot exactly solve for the image in this manner. Fourier methods are notorious for their poor noise propagation properties. This can be understood through simple noise propagation arguments. As can be seen from equation (1.7), the image at each point will be obtained by calculating the Fourier transform of the data, dividing by the Fourier transform of the PSF and performing an inverse Fourier transform. Each of these operations requires a large number of adds and multiplies. If the original numbers are noisy, each add and multiply increases the noise. By the time the final answer is achieved, the propagated noise is quite large, giving rise to the characteristic "ringing" and spurious sources plaguing inverse Fourier methods. If this isn't bad enough, the PSF may have natural "zeroes" in its Fourier transform. These zeroes amplify the noise in the reconstruction. Furthermore, even if the PSF has no "natural" zeros in its Fourier transform, if the PSF is experimentally measured and suffers from noise, then many artificial zeroes may arise in the PSF's Fourier transform, causing noise problems.

Figure 1 presents a comparison of the performance of direct Fourier deconvolution to two non-linear techniques, i.e. straight GOF deconvolution and pixon-based deconvolution —see Chapter 3 for a description of pixon-based methods. For the purposes of this comparison, a mock data set was constructed. The true underlying image is displayed at the top of Figure 1. This is the goal image of the reconstructions. The input data was created from the true image by convolving it with a Gaussian PSF with FWHM=8 pixels. To this image Gaussian noise was added to give a peak signal-to-noise ratio of

30. This was the input data for each of the deconvolutions. As can be seen from the reconstructions, the direct Fourier inverse is by far the poorest. There is a great deal of noise in the reconstruction and the achieved resolution is poor. In addition, the residual errors are quite large. By comparison the non-linear methods give better results with excellent residuals and considerably better resolution. Note, however, that in the straight GOF reconstruction a number of spurious sources have been created. This is indicative of overfitting of the data. By far the best reconstruction is given by the pixon-based technique. Here the residuals are excellent, the resolution is high, and there are no spurious sources.

In summary, Fourier methods provide a simple inversion technique for convolution problems. However, these techniques are only effective if the data are virtually noise free and the PSF is well known and noise free. Because of these difficulties, most modern efforts in image reconstruction methods have turned to non-linear methods. Such an approach holds significant benefit for controlling the propagation of noise in the solution.

1.3. Non-Linear Methods

Most non-linear methods can be interpreted in terms of a Bayesian estimation scheme in which the hypothesis sought is in some sense the most probable. To derive a suitable goal function with which to judge the relative merits of various hypotheses, Bayesians use conditional probabilities to factor the joint probability distribution of D, I, and M, i.e. $p(D, I, M)$, where D, I, and M are the data, unblurred image, and model respectively— we now explicitly assume that the data plays the role of $\{g_j\}$ in equation (1.3) and the image plays the role of $\{f_i\}$. The model, M, includes all aspects of the relationship between D and I such as the physics of the image encoding process, the details of the measuring instrument, e.g. pixel size, noise properties, etc., and the mathematical method of modeling the data, e.g. that equation (1.3) might be discretized, etc. As we shall see, all aspects of the model are important and can affect the quality of the solution to the inverse problem.

To derive equations useful for inverting equation (1-3), Bayesians factor $p(D, I, M)$ as:

$$p(D, I, M) = p(D|I, M)p(I, M) \quad , \tag{1.8}$$
$$= p(D|I, M)p(I|M)p(M) \quad , \tag{1.9}$$
$$= p(I|D, M)p(D, M) \quad , \tag{1.10}$$
$$= p(I|D, M)p(D|M)p(M) \quad , \tag{1.11}$$
$$= p(M|D, I)p(D, I) \quad , \tag{1.12}$$
$$= p(M|D, I)p(D|I)p(I) \quad , \tag{1.13}$$

where $p(X|Y)$ is the probability of X given that the value of Y is known. By equating various terms, these equations give rise to the formulae

$$p(I|D, M) = \frac{p(D|I, M)p(I, M)}{p(D|M)} \propto p(D|I, M)p(I|M) \quad , \tag{1.14}$$

$$p(I, M|D) = \frac{p(D|I, M)p(I|M)p(M)}{p(D)} \propto p(D|I, M)p(I|M)p(M) \quad . \tag{1.15}$$

The formula of equation (1.14) is the typical starting place for Bayesian methods. This equation essentially assumes that the appropriate model, M, is known and will not be varied during the reconstruction. This assumption is made explicit in the proportionality of equation in (1.14) where the term $p(D|M)$ is dropped (it is assumed to be constant—D and M are not varied during the reconstruction). Our preferred formulation, however, is given by the formula of equation (1.15). Here we do not assume that all of the parameters associated with the model belong to the so called "nuisance parameters", i.e. parameters which might require estimation, but which are of no interest to the scientist. In pixon-based reconstruction, however, the model parameters associated with the local smoothness scale of the image (see later chapters) are extremely important.

The significance of the terms on the far right hand side of equations (1.14) and (1.15) is readily understood. The term $p(D|I, M)$ is a goodness-of-fit (GOF) quantity, measuring the likelihood of the data given a particular image and model. The terms, $p(I|M)$ and $p(M)$, are "priors", and incorporate our prior knowledge (or expectations) about the measurement situation or the nature of suitable selections for I and M. The term "prior" is used since each of these terms make no reference to the data, D, and hence can be decided on *a priori*, i.e. before the act of making the measurement.

In GOF (or maximum likelihood) image reconstruction, $p(I|M)$ and $p(M)$ are assumed to be constant, i.e. there is no prior bias concerning the image or parts of the model that might be varied. The standard choice for $p(D|I, M)$ is to use $p(D|I, M) = (2\pi\sigma^2)^{1/2n_{pixels}} \exp\left(-\chi^2/2\right)$, i.e. the joint probability distribution for n_{pixels} independent pixels with normally distributed (Gaussian) noise. This assumes that (1) within the chosen space of image/model pairs, a statistically acceptable fit to the data is obtainable, and (2) that near this solution, image/model space uniformly fills data space so that the resulting probability distribution of data realizations is dominated by the probability density of the Gaussian distributed measurement noise, i.e.

$$p(\text{Blurred Image+Noise}|I, M) = p(\text{Blurred Image}|\text{Noise}, I, M)p(\text{Noise}|I, M) \quad (1.16)$$
$$\cong const \times p(\text{Noise}|I, M) \quad . \quad (1.17)$$

The "constant" $p(\text{Blurred Image}|\text{Noise}, I, M)$ is then generally dropped from the expression. This is generally not too bad an assumption, but it is by no means guaranteed, especially if the chosen space of image/models is selected to be restrictive in some way.

In Maximum Entropy (ME) image reconstruction, the image prior is based upon "phase space volume" or counting arguments, and the prior is expressed as $p(I|M) = \exp(\alpha S)$, where S is the entropy of the image and α is an adjustable constant which weights the relative importance of the GOF and image prior. Again, usually the model is considered to be known and held fixed, i.e. $p(M) = constant$, and the usual assumption is to choose $p(D|I, M) = (2\pi\sigma^2)^{1/2n_{pixels}} \exp\left(-\chi^2/2\right)$. While many specific formulations for α and S appear in the literature (Kikuchi and Soffer 1976; Bryan and Skilling 1980; Narayan and Nityananda 1986; Skilling 1989), all ME methods capitalize on the virtues of incorporating prior knowledge of the likelihood of the image into the image restoration algorithm.

Each of the equations (1.14)-(1.15) allows the calculation of the Maximum *A Posteriori* (M.A.P.) image (or image/model pair), i.e. the image (image/model) which maximizes $p(I|D, M)$ [or $p(I, M|D)$]:

$$I_{MAP} = \arg\{\max_I [p(I|D, M)]\} \quad , \quad (1.18)$$
$$\{I_{MAP}, M_{MAP}\} = \arg\{\max_{I,M} [p(I, M|D)]\} \quad . \quad (1.19)$$

The M.A.P. image or image/model, of course, is only one possible choice of what one might select as the "best" solution. This choice selects the arguments that maximize the posterior probability density function. Another sensible choice, however, would be to choose the mean image or image/model pair:

$$< I > = \int dM \; dD \; I \; p(I|D, M) \quad , \quad (1.20)$$

$$< I, M > = \left(\int dD \ I \ p(I, M|D), \quad \int dD \ M \ p(I, M|D) \right) \quad . \tag{1.21}$$

1.4. The Bayesian Embodiment of Occam's Razor

While most scientists have heard of Occam's Razor, and most of the rest have unknowingly embraced the concept as a working rule of thumb in their research, it is not widely understood that Occam's razor can be justified on solid probabilistic grounds. (The principle of Occam's Razor states that the best hypotheses, or explanation, of a data set is the simplest one.) One of the best discussions of the Bayesian embodiment of Occam's Razor is that presented by David MacKay (MacKay 1994). Our discussion shall parallel MacKay's discussion.

In order to judge the relative merits of two hypotheses, H_1 and H_2, for explaning the data set D, Bayesian's would form the relative probability ratio:

$$\frac{p(H_1|D)}{p(H_2|D)} = \frac{p(H_1)}{p(H_2)} \frac{p(D|H_1)}{p(D|H_2)} \quad . \tag{1.22}$$

The first ratio involves hypothesis priors, i.e. hypothesis probability densities with no functional dependence on the data. These can be used to express our prior conceptions of the relative likelihood of hypothesis H_1 relative to H_2. As pointed out by MacKay, this corresponds to the motivations behind Occam's razor. However, it is unnecessary to make any prior assumptions about the relative probabilities $p(H_1)$ and $p(H_2)$. Equation (1.22) will still give rise to an Occam's Razor like effect. This is because the simpler hypotheses, say H_1, make the most precise predictions. Complex hypotheses embrace a much larger data space and hence spread their predictive probability $p(D|H_2)$ more thinly over this larger data space. Thus, in the case in which both hypotheses H_1 and H_2 are equally compatible with the data, the simpler hypothesis H_1 will be favored by equation (1.22), without having to express any subjective dislike for one hypothesis relative to the other. Of course even if one assigns an equal prior probability to certain aspects of the hypothesis, e.g. the model in image reconstruction, one can usually make sensible arguments for appropriate priors for the other aspects of the hypothesis, e.g. the image prior. For example, in the image reconstruction case, the hypothesis for a given data set is the pair (I, M). The prior is thus $p(I, M) = p(I|M)p(M)$. So even if one expresses no prior bias for a given model, e.g. $p(M) = constant$, one might easily assign a sensible prior probability distribution for $p(I|M)$, e.g. a ME prior. Such an assignment also acts in the cause of simple models, since any sample state drawn from a simple model typically has a higher probability simply because it represents a larger fraction of all possible states. Thus at every turn probability theory favors Occam's Razor.

1.5. Successes and Failures of Classical Methods

The very fact that image restoration/reconstruction has been performed continually for many years is testimony to its basic success as a method. For PSF blurring problems, its primary advantage is its ability to provide enhanced resolution. As discussed above, for linear problems, Fourier methods are able to provide such enhancement if noise is absent or inconsequential. In the low noise case, only digitization noise (i.e., the fact that the data or equation (1.1) is discretized in some manner for computer solution) and natural "zeroes" in the Fourier transform of the PSF limits our ability to achieve infinite resolution. While Fourier techniques are widely used for high SNR data, these techniques fail badly for moderate to low SNR data.

Non-linear methods provide an improvement over linear methods in moderate to low SNR situations. Unlike Fourier methods, non-linear methods achieve significantly greater control of noise propagation. This is done by insisting (through the GOF criterion) that the modeled data (e.g. the data predicted by the inferred image and model) match the measured data within the uncertainty specified by the noise. This is quite different from Fourier methods were no such demand is made. Of course, insisting that the predicted data "fit" the measured data within the noise limits does not guarantee that the inferred image is without significantly larger noise. Indeed, as discussed above, each image reconstruction problem can have a significant "null space". [Quotes are used around "null space" since there need not be a strict null space for the data encoding problem under consideration, but large Δf_i relative to f_i may exist that still satisfy equation (1.3).]

Common problems with GOF (or Maximum Likelihood) and ME methods are over-fitting of the data, the production of signal correlated residuals, and spurious sources. Overfitting of the data occurs when the GOF criterion is driven below expected values. For example, suppose an image is acquired with a $n \times n$ pixel CCD system. Further suppose that the CCD plate scale finely samples the PSF (say, m pixels across the PSF FWHM). Normally GOF reconstruction tries to drive the value of χ^2 to zero to obtain the best fit. Indeed, this is the proper prescription for obtaining the best estimate of n^2 independent pixel values and provides a consistent and unbiased estimate of the resid-uals. Consequently this procedure would provide an excellent image estimate if a large sample of images were taken, the reconstructions performed, and the average value of the reconstructed image evaluated. However, this is usually not the situation one is faced with. Normally one wishes to estimate the best image from a single data frame. In this case, attempting to drive the value of χ^2 to zero often causes overfitting of the data. After all, the expected value of χ^2 is not zero for n^2 independent, Gaussian distributed variables, but is equal to $n^2 - 2$ (= peak, or mode, of the χ^2 distribution). Another way of stating this is that if one does a perfect job of estimating the underlying image, after convolving with the PSF and subtracting the result from the data, one expects to see a random noise frame. The expected (modal) value of χ^2 for this noise frame is not zero, but is $n^2 - 2$ with a standard deviation of $\sigma_{\chi^2} = \sqrt{2n^2}$. In this case, the likelihood function for the data (given the image and mode) is provided by the χ^2-distribution for n^2 independent variables.

As already mentioned, the production of signal correlated residuals and spurious sources are common problems with GOF and ME methods. Both problems are re-lated to using an inappropriate number of degrees-of-freedom (DOFs) to fit the data and consequently are largely overcome by pixon-based methods—see later chapters in this series. Briefly, however, signal correlated residuals are produced by a "diluted" GOF criterion. Suppose we have imaged a galaxy in a large CCD frame. Assume that the CCD has dimensions 1024×1024 pixels. Further suppose that the galaxy only fills the central 100×100 pixels. If the astronomer were to use the entire data frame in his reconstruction and chose to represent the image as a rectangular grid of numbers with the same spatial frequency as the data (as is commonly done), then the reconstruction would use roughly 10^6 DOFs. If a standard GOF criterion were used, then one would stop the iterative procedure when there were less than 10^3 residuals larger than 3σ (0.1% of the residuals are 3σ or larger in a population of Gaussian distributed variables). Since iterative procedures that minimize GOF functions normally spend most of their time fit-ting the bright sources, they will be working on adjusting the bright source levels when the stopping criterion is met, resulting in the vast majority of the large residuals lying under the bright sources.

The origin of spurious sources in standard image reconstruction techniques is even easier to understand. Using the example above, it should be clear to most readers that the 10^6 DOFs used in the reconstruction are probably far more than are needed to describe the information present in the data. In fact, the data probably is inadequate to constrain the values of this many DOFs. Hence the vast majority of these DOFs are free to produce whatever bumps and wiggles they like, with the only requirement being that after they are smoothed by the PSF [or acted on by L in equations (1.2)] they average to zero within the noise. Hence their amplitude can be very large as long as they have a spatial scale small relative to the PSF and are arranged in such a manner that hills and valleys tend to cancel each other out. Since this large number of unconstrained DOFs represents a huge phase space and the typical member has numerous bumps and wiggles, spurious sources are guaranteed.

One of the main motivations for developing Maximum Entropy methods was to help overcome the spurious source problem. After all, the ME prior acts to make unsmooth images undesirable. However, ME methods, in their zeal to flatten spurious sources, can over flatten the true sources. This affects the ability of ME methods to perform statistically unbiased photometry. In fact, while ME methods are significantly more immune to spurious sources (they more effectively reject situations with large adjacent hills and valleys than GOF methods) they still tend to systematically underestimate source brightnesses (Sibisi 1990; Cohen 1991) because of using too many DOFs in the reconstruction.

1.6. *The Promise of Pixon-Based Methods*

Most of the problems facing standard GOF and ME methods are over come by pixon-based methods. Pixon-based methods are a generalization of ME methods and perform the Bayesian estimation problem in a more natural language. The pixon language is information-based. Consequently it concisely and accurately describes the key variables in the problem and avoids the use of too many DOFs in the reconstruction. Because the number of DOFs is tuned to the number required to accurately describe the image, problems with stopping the reconstruction because a "diluted" GOF criterion is used are avoided. In addition, spurious source production are avoided since the pixon-basis automatically rejects spurious sources as being unneeded in the reconstruction. The generalization of the pixon to arbitrary data is the *informaton*, i.e. a quanta of information—see later chapters in this series, and has a very large range of application to a variety of problems. The informaton is directly related to a data set's Algorithmic Information Content—see Chapter 2—and is the appropriate coordinate system for the general inverse problem and optimal data compression. The value of using a particular language for image reconstruction will be motivated in Chapter 2 of this series, and pixon-based methods will be developed in Chapter 3.

2. Language and Information content

2.1. *Introduction*

The many problems that arise with classical image reconstruction, *e.g.* Goodness-of-Fit (GOF) methods (= Maximum-Likelihood methods), and Maximum Entropy (ME) image reconstruction, all stem from the reconstructed image being too complex. This is seen in the severe over-fitting of the data by these methods and in their production of spurious sources. Indeed, the basic improvement of ME methods over GOF methods results directly from attempting to reduce the amount of information contained in the reconstructed image—one attempts to maximize the entropy, or disorder in the image. This strongly suggests that controlling the image information content is a key issue for successful image reconstruction. That this is true should be especially apparent when the collected data have associated measurement noise. In this case the data contain much more detailed information than is relevant to the image—e.g. the precise values of the pixels in a digital image are unimportant since noise contributes to their value. Hence, it is clear that there should be some smaller, but quantifiable, amount of information in the image, and that the ability to identify this information and ignore the other details of the data would be valuable. Other clues that controlling the information content of the image is highly desirable come directly from Bayesian estimation theory. Here, the Bayesian prior always favors the simplest (and hence the *a priori* most likely) hypothesis (=image or image/model) for explaining the data. This is the Bayesian embodiment of Occam's Razor.

It is the goal of this lecture to find a suitable method for controlling the information content of the image and to use this to constrain the image reconstruction process. Looking to the information sciences, we find that there is a rich literature on the nature of information and the role of language in the specification of information. Furthermore, from the field of image processing, we find that multiresolution languages and operators provide powerful tools for describing the structure of generic images. These insights will allow us to develop concise and powerful languages to handle generic images and develop highly optimized Bayesian priors for our image reconstruction applications. The concepts crucial to these goals are described below.

2.2. *Types of information*

The concepts of information and entropy have been discussed in numerous works. While the thermodynamic concepts of entropy have a very specific meaning and apply to the state of an ensemble of particles, the concepts of information have a much broader scope. Nonetheless, the laws of statistical physics can be derived directly from information theory concepts (Jaynes 1957; Jaynes 1963; Hobson 1971), and information and entropy (or negentropy) are often talked about as the same quantities. The most common definition of information, now known as Shannon information, was introduced in 1948 (Shannon 1948; Shannon and Weaver 1949) and specifies the average information contained in a string of symbols transmitted across a communication line. In this definition, the information per symbol is given by

$$H = - \sum_{i=1}^{n} p_i \log_2 p_i \quad , \tag{2.23}$$

where n is the total number of possible symbols, and p_i is the probability of occurence of the ith symbol. There are many reasons why the Shannon information is a suitable definition for information. One of the most convincing arguments is given by the *Noiseless Source Coding Theorem* (see, for example, Rabbani and Jones 1991), which states that for an ergodic source with an alphabet of n characters, that as the size of the transmitted string approaches infinite length, it is possible to construct a unique random length binary code for the string that has an average length in bits per encoded symbol equal to, but no less than, the Shannon entropy.

While the *Noiseless Source Coding Theorem* in and of itself speaks strongly for Shannon entropy as a sensible information measure, there are still other motivations. For example it can be proven that the Shannon information is the unique and self-consistent measure which satisfies conditions which would reasonably be expected of a measure of information (Khinchin 1957; Feinstein 1958; Hobson 1971). Using Khinchin's formulation of the properties uniquely defining an information measure, it is found that only the function H satisfies the conditions that (1) H takes it's largest value when all the p_i are equal, (2) H doesn't change it's value when additional impossible events are added—i.e. events for which $p_i = 0$, and (3) the information in two not necessarily independent events is given by the information of the first event plus the expectation value of the additional information provided by the second type of event after the first event has occurred, i.e. if A and B are sets of events, then

$$H(AB) = H(A) + \sum_i p_i H(B|A_i) \quad , \tag{2.24}$$

$$H(AB) = H(B) + \sum_j p_j H(A|B_j) \quad , \tag{2.25}$$

where $H(X|Y)$ is the additional information in X given that Y has occurred.

Readers who have had classical training in physics will be interested that the formula of equation (2.23) is essentially equivalent to the physical definition of entropy, $\sigma = k \ln W$, (modulo a constant to convert logarithm types) where k is Boltzman's constant and W is the total number of available states. In fact, equation (2.23) is the Boltzman (or Gibbs) definition of entropy. Kittel (Kittel 1969—see page 118) gives a particularly fluid derivation of the Boltzman definition of entropy starting from the Boltzman factor, i.e. $p_i = Z^{-1} \exp(-E_i/kT)$. Thus Shannon information is seen to be not merely a mathematical construct of theoretical interest, but has a substantial practical "bite"— statistical physics describes the real world and works exceedingly well.

While we have now provided motivation for the merits of Shannon information, of what use is this concept for solving the inverse problem in general or image reconstruction in particular. To be sure, ME methods make use of the entropy or Shannon information definition. However, Shannon information doesn't get to the heart of the problem. This definition of information doesn't allow us to describe the *information content* of an image or data set. As has been pointed out by several authors (Wicken 1987; Chaiten 1982), Shannon information is an ensemble notion. It is a measure of ignorance concerning possible realizations of events which have a given *a priori* probability distribution, and measures the statistical rarity or "surprise value" of a particular realization (Cherry 1978). It does not deal with the information content expressed by a given realization. For this we need a different concept, i.e. that of algorithmic information content (AIC), algorithmic randomness, or algorithmic complexity (Solomonoff 1964; Kolmogorov 1965; Chaiten 1966). As commonly used in information or computer science, the AIC of a string of characters is defined to be the size of the minimum computer program required for a universal computer to produce the specified string as its output. (A computer U is universal if for any other computer M there is a prefix program μ such that the program μp makes U perform the same tasks as performed by computer M running the program p, i.e. the program μ makes computer U simulate computer M.) As suggested by the name, AIC describes the information content of the specific item under scrutiny. In fact it does this in a very practical way, i.e. algorithmically. It provides a prescription for how the data can be reproduced. It should also be clear that the AIC represents the optimal compression of the data. There is no shorter description that will allow a complete reconstruction of the original information. In fact, it is this optimal compression of the information that will allow us to optimally extract an image in the image reconstruction problem in a practical manner.

2.3. The Role of Language in AIC and Bayesian Estimation

The term "randomness" comes into the definition of AIC since random strings of characters have maximum complexity or information content. This is because if the pattern of the string is non-random, there is some way of describing the string which is typically shorter than listing the string itself. For example, in strings of 1s and 0s, a string with

Character Expressions of a Number

Character String	Language
11111010011100111011	binary
3723473	octal
102581	decimal
FA73B	hexadecimal

TABLE 1. The character representation of a given number in different languages. The simplest language, i.e. binary, with two characters in the language, requires 20 characters to express the number. The richest language, hexadecimal, with 16 characters in the language, requires only 5 characters to express the number.

10^9 copies of the number 1 would take a lot of paper to write down, but can be described with a single sentence. Similarly this same string with a 0 placed in the millionth place can also easily be described without wasting paper. However, a totally random string of 10^9 digits cannot be so easily described, and if no rule is known (and this is what we mean by random), only the actual string will impart all of the information. It is well known that quantification of the AIC is a function of the "richness" of the language used to describe the character string or data set. In the example used above, we demonstrated how the English language could be used to briefly describe large strings of digits. Other languages, e.g. the binary digits 0 and 1, decimal digits, or even hexadecimal digits, do not provide such a terse description. An example of the character strings used to express a given number in various languages is given in Table 1.

The above discussion makes it quite obvious that rich, i.e complex, languages allow a terse description of a data set—the AIC is low for rich languages. This seems like an obvious advantage. It says that the data can be compressed to a higher degree and that the information can be more precisely located and identified. Similar concepts occur in Bayesian estimation, e.g. the concept of Occam's Razor. Clearly, in the Bayesian estimation problem, it is highly beneficial to develop concise, simple hypotheses (images or image/model pairs) to explain the data. Such simple hypotheses give rise to optimized priors [$p(I|M)$ or $p(I|M)p(M)$—see Chapter 1 of this series]. Consequently, they produce optimal reconstructions and a more likely M.A.P. image or image/model. Thus it is seen that minimizing the AIC of the image/model, optimally compressing the information in the data, and performing highly optimized Bayesian image reconstruction are directly related. Hence the goal for improved image reconstruction is clear. One must improve the language in which the reconstruction is performed, or equivalently the language in which the image is described. The language must be rich so that the description can be concise, the AIC low, and the Bayesian prior optimized. The fact that the description will be concise suggests that the language for describing the image must be "natural" in some sense, and we turn now to a discussion of possible languages for image description.

2.4. Languages for Images

There are many "languages" for images in current use. The most common of these are usually not even thought of as languages since they seem so natural. For example, in electronic or digital imaging we essentially expect to see every image in the form of a rectangular pixel array. This is so natural that we don't even consider the consequences of this language for the image. However there are consequences. The pixels are almost always rectangular and so is the picture format. However this may not be optimal for image reconstruction. Certainly in the case of astronomical imaging, many objects are not square, e.g. stars. Also images of star clusters or galaxies, etc., seldom uniformly fill a rectangular picture to its corners. Hence these clearly are not an optimal format or pixel shape for astronomical situations. There are usually too many pixels in the image and many pixels are used to describe single objects such as stars.

Another common basis for images is the Fourier basis. This, of course, is quite natural in the case of radio interferometry where the collected data are actually points in the Fourier domain. This basis is often also used when image deconvolution is attempted for high signal-to-noise (SNR) data—see Chapter 1 of this series. Astronomers also often think in terms of the Fourier basis when considering the fundamental limits on spatial frequency response of a telescope, e.g. the sharp cut off in frequency space giving rise to the telescope's Airy diffraction pattern. However, there are many reasons why the Fourier basis often is inappropriate (and even misleading) for astronomical imaging. This is seen quite easily once one realizes that most objects in astronomical images are spatially localized and different objects usually bear no causal relationship with each other. By contrast, each Fourier component of an image contains information from all parts of the image. Hence every location in the image becomes intertwined in the Fourier transform. This is quite unnatural and is at the root of the difficulties associated with Fourier deconvolution in the moderate to low SNR case—noise from all over the image gets mixed together.

Thinking about images in the Fourier domain can also be misleading. A common misunderstanding, for example, is that the finite size of a telescope prevents the detection of star pairs more closely separated than the full-width-half-max (FWHM) of the diffraction pattern. This is simply not true. If the SNR of the data is high enough one can easily tell the difference between a single Airy pattern and the sum of two closely spaced Airy patterns. What is really happening is that the Airy pattern prevents information on the Fourier components with frequencies higher than a certain cut off frequency. This does not mean, however, that high spatial resolution imaging is prevented. What it means is that there is some ambiguity in the image. One can add, for example, any image that has only components with frequencies higher than the cut-off frequency and not change the collected data. Usually, however, images such as this are unphysical and can be discriminated against by other means, e.g. by probabilistic methods.

In addition to the above languages there are two more languages that should be discussed and which are in many ways more suitable that either the pixel or Fourier bases. They are the wavelet basis and the pixon basis. Both of these languages are largely motivated by the structure of picture information. Indeed, while the pixel and Fourier bases are extreme in their view point (one being very local, the other being completely global), the wavelet and pixon bases take a middle-ground approach. Before discussing these bases further, let us first ask the question: "What is the nature of picture information?" This, of course, can have many answers, and the answer changes, depending on the object at which the photographer points his camera. We are interested, here, in "generic" images. So let use first ask the general question: "What do we really expect when we take a picture?" Thinking about this in the broadest sense, all we can really say is that we expect the picture to contain some limited amount of information. We expect, in particular, that the manufacturer of the film or the designer of the camera will have made the grain or pixel size fine enough to do service to the picture, i.e. not to leave out any important detail. We expect interesting features, perhaps people, with open spaces between them. In other words, we expect that at each point in the image there is a finest spatial scale of interest and that there is no information content below this scale. Indeed, this is how photographic grain sizes are chosen and why data is not sampled finer than the Nyquist frequency when pixel elements are at a premium.

How does one capture this prior expectation in mathematic form and incorporate it into a set of image basis functions? Again, the key comes from thinking about photographic grain sizes or Nyquist sampling. We would do just as well at recording the picture information with large photographic grains in portions of the image with coarse structure. We need only have fine grains when we need to record fine spatial structure. This means that the picture information can be dealt with by using variable sized cells, with the cell sizes set so as to capture the spatial information present. This is the fundamental idea behind wavelets and multiresolution techniques in general. These approaches seek to localize the information in different parts of the image. Wavelets attempt to localize the frequency information, while multiresolution techniques seek to analyze or localize the information on different spatial scales.

Having had the foregoing discussion, we are ready to discuss in more detail both wavelets, multiresolution methods, and our current implementation of pixon-based methods. Wavelets are basically an extension of Fourier methods in which the frequency information is localized. A complete description of wavelet theory is beyond the scope of this lecture, but suffice it to say that wavelets are a linear, orthogonal set of basis functions that do a much better job (are more concise) of describing images that contain localized structure. However as we have shown for the reconstruction problem, arguments based on optimization of the Bayesian prior indicate that the languages that are the most concise will provide the best reconstruction. So we are left with the question is the wavelet basis the most concise language for describing generic images?. The Fourier basis failed because images with highly localized structure and large empty spaces have Fourier components at all frequencies and so are very verbose. Pixel bases fail because they may use many more pixels than necessary to describe a large smooth structure in the image. Wavelet bases fall somewhere in the middle. They localize the information. Yet they still describe the localized information in terms of a number of oscillating basis functions. High frequency oscillating functions are required to describe small scale structure and many components are required to "cancel-out" the resulting wavelet oscillations at the edges of fine bumps.

Like the wavelet approach, multiresolution techniques attempt to quantify the information present in an image at a variety of scales. Unlike wavelets, however, these techniques do not require the basis functions to be linear or orthogonal. They simply try to quantify the information. Our pixon-based methods currently use a multiresolution language. We would like to stress here that our concept of a pixon—see Chapter 3—is that each pixon can be identified with a unit of information (in the AIC sense) in the image, and that the collection of an image's pixons are the most concise description possible in the chosen language, i.e. the pixon basis is the AIC of the image. We have chosen to implement pixon-based methods using a multiresolution language since it seems that this language is "natural" for generic images—generic images have a variety of structure on a variety of spatial scales. It is clear from this that our pixon bases need not be either linear nor orthogonal, since having these extra constraints on the selection of the pixon basis might compromise the all important goal of the basis being concise.

While we have now made it clear that a basis that allows concise descriptions is important, there is, of course, another important property of an image basis. It must be complete (or near complete) to be useful. This property is easily seen to be an attribute of the pixel, Fourier, and wavelet bases, at least for images that are definable on the pixel grid. We must be careful, however, to select a pixon basis that is complete. Since we have used a multiresolution language, completeness can be easily insured if single (or effectively single) pixels are included in the set of basis functions. This is clearly desirable anyway since one may wish to describe structures that are this fine in spatial scale.

2.5. *Conclusions*

This lecture has described the basic information properties in an image. We have pointed out that each image has a finite information content and that the information we are interested in is the Algorithmic Information Content as defined by the computer and information sciences. We argued that if we were able to identify this information then we would be in a position to develop a scheme for its optimal extraction. We discussed the role of language in determining the AIC of a data set and the direct relationship of a concise (rich) language and small AIC values, optimized Bayesian priors, and consequently high quality solutions to the inverse problem. We then discussed appropriate languages for image description, stressing that the languages should be concise and complete. We finally argued that for generic images a multiresolution description should be superior to the direct pixel basis and the Fourier and wavelet bases. In the next lecture we shall see how to explicitly use the ideas developed in this lecture to construct practical pixon bases.

3. Pixon-Based Image Restoration/Reconstruction

3.1. *Introduction*

As discussed in previous chapters of this series, while Goodness-of-Fit (GOF), or Maximum-Likelihood, and Maximum Entropy (ME) methods have improved dramatically on the performance of linear inversion methods for image deconvolution, they still suffer from signal correlated residuals and spurious source production. One of the fundamental sources of the problems with these methods is their selection of the coordinate system in which to represent the image. This lecture will introduce pixon-based methods which use an information theoretic coordinate system which quantifies the image's Algorithmic Information Content (AIC), and uses this to optimally constrain the reconstruction process (see Piña and Puetter 1993, Puetter and Piña 1993a, Puetter and Piña 1993b, Puetter 1994, and especially the review article Puetter 1996a). We shall also see that pixons (or *informatons*–their natural extension to general data sets) are in effect quanta of information, and comprise the smallest collection of degrees-of-freedom (DOF) required to specify the data within the accuracy allowed by the noise.

3.2. *Practical Methods for Implementing Pixons*

How one might practically implement pixon-based methods is discussed in a number of previous papers. The material in this section follows closely the discussion in (Puetter 1996b).

3.2.1. *A Multiresolution Pixon Language for Image Description*

In Chapter 2 we argued that a multiresolution image description language would be suitable (i.e. concise) for describing generic images. The success of multiresolution languages is not surprising. In fact, these ideas are familiar to most scientists and are fundamental to simple, well known concepts, e.g. the use of fine grain photographic film to capture fine detail. So our multiresolution pixon language will use the idea that generic images can be concisely described by using fewer degrees-of-freedom (DOFs) per unit area in portions of the image which are smooth and a greater density of degrees-of-freedom where there is greater detail. Each of these degrees-of-freedom might then be likened to a single photographic grain or a generalized pixel. The value of this pixel represents the average brightness in a given region. This, in fact, is the origin of the name "pixon". Each pixon is a single DOF used to describe the image in a particular region. The "pix" part of the name recognizes its pixel heritage, while the "on" suffix recognizes its more fundamental nature (the pixon is fundamental to the image, not to the instrument that took the picture). However, since any scheme which controls the local density of DOFs used to describe the local image information is suitable for constraining the image reconstruction and optimizing the Bayesian prior, rather than using the image signal contained in cells with hard boundaries, we have chosen to use a local correlation scale formulation to control the DOF density. We call these "fuzzy pixons". To formalize the definition, at each point in the image, $\vec{x}$, we write the image, I, as

$$I(\vec{x}) = (K \otimes I_{pseudo})(\vec{x}) \tag{3.26}$$

$$(K \otimes I_{pseudo})(\vec{x}) = \int dV_{\vec{y}} \, K(\vec{x}, \vec{y}, \delta(\vec{x})) I_{pseudo}(\vec{y}) \tag{3.27}$$

$$K(\vec{x}, \vec{y}, \delta(\vec{x})) = K\left(\frac{\|\vec{x} - \vec{y}\|}{\delta(\vec{x})}\right) \quad ; \text{ for radially symmetric pixons} \quad , \tag{3.28}$$

where equations (3.26) through (3.28) show that the image is a local convolution of a "pseudo-image" with a blurring function with a given local scale, $\delta(\vec{x})$. Note that the local scale varies with position $\vec{x}$ in the image and the integration in equation (3.27) is carried out over volume in pseudo-image space. We have also indicated in equation (3.28) that one suitable functional form for the pixon kernel function, K, might be a radially symmetric function that depends only on the distance between the kernel center and the image position relative to the local scale. We have found this functional form to work quite well for centrally peaked kernel functions with a finite foot-print. We normally use truncated paraboloids, i.e.

$$K(\vec{x}, \vec{y}, \delta(\vec{x})) = \begin{cases} \left(1 - \frac{\|\vec{x} - \vec{y}\|^2}{\delta(\vec{x})^2}\right) / \int dV_{\vec{y}} \left(1 - \frac{\|\vec{x} - \vec{y}\|^2}{\delta(\vec{x})^2}\right) & ; \; \|\vec{x} - \vec{y}\| \le \delta(\vec{x}) \\ 0 & ; \; \|\vec{x} - \vec{y}\| > \delta(\vec{x}) \end{cases} \tag{3.29}$$

Of course AIC concepts suggest that a richer language, e.g. elliptical pixons, would yield a more concise image description, which Occam's razor then says should have a more optimized prior. In fact, we have used elliptical pixons recently to perform some image restorations—see Chapter 4. Nonetheless, for generic images it seems clear that we have reached a point of diminishing returns, and it is unlikely that pixon kernels more complicated than ellipses are warranted for generic images.

3.2.2. Solving for the M.A.P. Image/Model Pair

Now that we have described a suitable language in which to solve the problem, we shall move on to the details of finding the M.A.P. (Maximum *A Posteriori*) image/model pair, i.e. the image/model pair that maximizes $p(I, M|D)$. To maximize $p(I, M|D)$, we need to maximize the product of $p(D|I, M)$ and $p(I, M)$. The common choice for $p(D|I, M)$ in the case of pixelized data with independent Gaussian noise is $p(D|I, M) = (2\pi\sigma^2)^{1/2n} \exp(-\chi^2/2)$ where χ^2 is the standard chi-squared value, [i.e. $\chi^2 = \sum_1^n (x_i - \langle x_i \rangle)^2/\sigma^2$], σ is the standard deviation of the noise, and n is the number of independent measure parameters (here the number of pixels). As already mentioned in Chapter 1, this choice of the Goodness-of-Fit criterion assumes that (1) within the chosen space of image/model pairs, a statistically acceptable fit to the data is obtainable, and (2) that near this solution, image/model space uniformly fills data space so that the resulting probability distribution of data realizations is dominated by the probability density of the Gaussian distributed measurement noise, i.e.

$$p(\text{Blurred Image+Noise}|I, M) = p(\text{Blurred Image}|\text{Noise}, I, M)p(\text{Noise}|I, M) \quad (3.30)$$
$$\cong const \times p(\text{Noise}|I, M) \quad . \quad (3.31)$$

The "constant" $p(\text{Blurred Image}|\text{Noise}, I, M)$ is then generally dropped from the expression. This is generally not a bad an assumption, but it is by no means guaranteed, especially if the chosen space of image/models is selected to be restricted in some way.

While the functional form for the GOF criterion given above is almost always adopted, the various choices for $p(I, M)$ are far from standard. The choice of Maximum Entropy enthusiasts is something like

$$p(I|M) = \frac{N!}{n^N \prod_{i=1}^{n} N_i!} = e^S \quad (3.32)$$

$$p(M) = constant \quad , \quad (3.33)$$

where n is the number of pixels, N is the total number of counts in the image, N_i is the number of counts in pixel i, and S is the entropy. This choice is made on the basis of counting arguments. The arguments are basically sound, but operate within a fixed language, e.g. the pixel basis for the image. As discussed previously, a more appropriate choice for the image basis would be multiresolution pixons. This more critically models the data and provides a superior prior.

For our pixon prior we could use the same prior as used by ME workers with the pixons substituted for the pixels. The *a priori* probability arguments for the prior of equations (3.32)-(3.33) remain valid. It is just that we now recognize that the pixons are a more appropriate coordinate system and that we will use vastly fewer pixons (i.e. DOFs) to describe the image than there are pixels in the data. To make the formulae explicit, we define the pseudo-image on a pseudo-grid (we loosely refer to this as the "pixel grid" when talking about the pseudo-image) which is as least as fine as the data pixel grid (we use this finer "pixel" grid for the image too) and then use the following substitutions:

$$p(I|M) = \frac{N!}{n^N \prod_{\substack{i=1, \\ i \, \epsilon \, Image}}^{n_{pixons}} N_i!} \quad (3.34)$$

$$N = \sum_{\substack{i=1, \\ i \, \epsilon \, Image}}^{n_{pixons}} N_i = \sum_{\substack{j=1, \\ j \, \epsilon \, Image}}^{n_{pixels}} N_j \quad (3.35)$$

$$N_i = \int dV_{\vec{y}} \, k(\vec{x}, \vec{y}, \delta(\vec{x})) I_{pseudo}(\vec{y}) \quad (3.36)$$

$$\prod_{\substack{i=1, \\ i \, \epsilon \, Image}}^{n_{pixons}} N_i! = \prod_{\substack{j=1, \\ j \, \epsilon \, Image}}^{n_{pixels}} (N_j!)^{p_j} \quad (3.37)$$

$$p(\vec{x}) = 1 \, / \int dV_{\vec{y}} \, k(\vec{x}, \vec{y}, \delta(\vec{x})) \quad (3.38)$$

$$p(M) = constant \quad (3.39)$$

where N_i is the number of counts in pixon i, N_j is the number of counts in pixel j, p is the pixon density, i.e. the number of pixons per pseudo-pixel (in image space), and

$k(\vec{x}, \vec{y}, \delta(\vec{x}))$ is the pixon shape kernel normalized to unity at $\vec{y} = \vec{x}$. [Note that while the formula of equation (3.34) can be easily justified, it is not the only justifiable expression. Other expressions might be more appropriate in certain situations.]

To obtain the M.A.P. image/model pair, one can now proceed directly by minimizing the product of $(2\pi\sigma^2)^{1/2n_{pixels}} \exp(-\chi^2/2)$ and equation (3.34) with respect to the local scales, $\{\delta(\vec{x_j})\}$—the pixon map, and the pseudo-image values, $\{I_{pseudo,j}\}$. However, this is not what we do in practice. Instead, we divide the problem into a sequential chain of two repeated steps: (1) optimization of the pseudo-image with the local scales held fixed, (2) optimization of the local scales with the pseudo-image held fixed. The sequence is then repeated until convergence–see Figure 3. To formally carry out this procedure, in step (1) we should find the M.A.P. pseudo-image, i.e. the pseudo-image that maximizes $p(I|D, M)$—note that we are using the notation here that the local scales belong to M, while the pseudo-image values are associated with I. In step (2) we would then find the M.A.P. model, i.e. the scales that maximize $p(M|D, I)$.

While the above procedure is quite simple, we have made still further simplifications. In neither step do we evaluate the prior. We simply evaluate the GOF term $p(D|I, M)$. So in step (1) we essentially find the Maximum Likelihood pseudo-image with a given pixon map. In step (2) we must take into account some knowledge of the pixon prior, but we simply use the fact that the pixon prior of equation (3.34) or any sensible prior increases rapidly as the number of pixons are decreased. So at each pseudo-grid point, j, we attempt to increase the local scale until it is no longer possible to fit the data within the noise. In other words at each pseudo-grid point we progressively smooth the current pseudo-image until the GOF criterion is "violated". We then select the largest scale at each point which was acceptable and use these values in the next iteration of step (1).

There is one more practical matter to consider. As with any interactive method, there can be convergence problems. With the approach outlined above, we have noticed that if small scales are allowed early in the solution, then these scales become "frozen-in", even if they would have later proved inappropriately small. To solve this problem, we start out the pseudo-image calculation with initially large pixon scales. We then use our pixon-calculator [the code that performs step (2)] to determine new scales. The pixon-calculator, of course, will report that over some of the image the initial scales are fine, but over other parts of the image smaller scales are required. At this point, however, we do not allow the smallest scales requested. Instead, we let the scales get somewhat smaller over the portion of the image for which smaller scales were requested and proceed to step (1) of the next iteration. We repeat this process, letting the smallest scales allowed get smaller and smaller until the method converges. This procedure has proven to be very robust.

3.3. *The Relationship of Pixon Concepts to Other Fields*

Pixon-based concepts have a direct relationship to other fields of thought. Most of the subject matter in this section has already appeared in print (Puetter 1996a), but is repeated here for completeness.

3.3.1. *The Pixon and Fractal Dimensional Concepts*

Previous studies have introduced several different pixon schemes. The first scheme proposed (Piña and Puetter 1993) was called the Uniform Pixon Basis (UPB) since the AIC of each pixon was equal. The second scheme (Puetter and Piña 1993a) was called the Fractal Pixon Basis (FPB). In fact, the pixon approach described in section 2 is essentially that used when calculating an FPB basis. In the FPB approach the pixon kernel function is radially symmetric. Because the FPB approach analyzes the AIC of

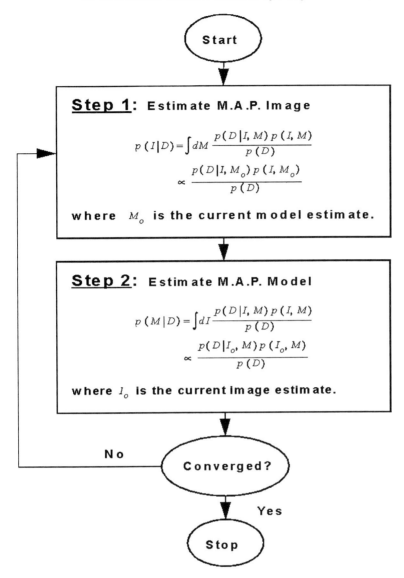

FIGURE 2. Iterative scheme for obtaining the M.A.P. image/model pair. This figure shows the pixon algorithm as currently implemented. The procedure is divided into two steps: (1) GOF optimization step in which the image is optimized while holding the pixon map (local smoothness scales) fixed, (2) Pixon optimization step in which a new pixon map is calculated while hold the current image estimate fixed. The method alternatively iterates between these two steps until convergence is obtained.

the image in terms of a single spatial scale, it is conceptually similar to the definition of fractal dimension. While there are numerous definitions for the fractal dimension of a geometric object, all definitions have one thing in common. Each calculates a quantity that changes as a scale (or measurement precision) changes. For example, the compass dimension (also referred to as the divider or ruler dimension) is defined in terms of how the measured length of the line varies as one changes the length of the ruler used to make the measurement. The commonly used box-counting dimension is defined in terms of how many cells contain pieces of a curve as the sizes of the cells are changed. As can be seen from the above discussion, these ideas are closely related to the pixon representation since in order to calculate the local pixon scale, we ask: How does $p(I, M|D)$ change as the local size of the pixons are varied? In other words, $p(I, M|D)$ acts as our measure of length and the pixon scale acts as our ruler. (This is only true for radially symmetric pixons in which the only variable quantity is scale. For elliptical pixons, for example, we would also be interested in how $p(I, M|D)$ changes as the ellipticity and major axis angle changes.) So in this sense radially symmetric pixon kernel methods share much in common with fractal concepts. They each analyze and quantify geometric structure by how things change when varying a single spatial scale.

3.3.2. *Pixons and Data Compression*

Since pixons quantify the AIC of an image (or in the general case a data set), they are related directly to data compression. In fact, once a language has been chosen, the pixon representation is the most concise description of the image. Hence pixons provide optimal data compression. Indeed, it is precisely the fact that pixons are the natural and most concise coordinate system for the image, that gives pixon-based methods their computational power to provide an optimal solution to the inverse problem. So in effect, the image reconstruction and image compression problem are intimately related. Both can be optimized by using Bayesian techniques to maximize the fidelity to the data, while simplifying the model with which to represent it.

When one discusses image (or data) compression schemes, it is also of interest to understand whether or not the process is lossless or lossy. As should be clear, any data compression scheme based on pixon bases can be as lossless as desired. To adjust the fidelity to the data one need only adjust the GOF criterion. Such an adjustment will allow a uniform degradation or increase in the information content over the entire image. Alternatively, if it was desired to preserve certain sections of the data with higher fidelity, one need only to express this fact in the GOF criterion. Hence pixon-based methods provide an optimal method for data compression. As before, however, selection of a language suitable for the compression is still a key issue. Nonetheless, pixon-based compression with fuzzy, radially symmetric (or perhaps elliptical) pixons should produce excellent results for generic images.

3.3.3. *Pixons and Wavelets*

A current popular method for performing both image compression and image reconstruction is to use wavelets (see, for example, Press *et al.* 1994). Since the time a number of years ago when wavelets were first introduced, the theory of wavelet transforms has blossomed. This is with good reason. Wavelets provide many practical advantages over straight Fourier methods. Nonetheless, from the Bayesian image reconstruction theory developed above, it should be clear that the performance of wavelet data compression and image reconstruction will be inferior to pixon-based methods. There are several ways to see this. First, since standard wavelet bases are orthogonal, the basis functions have positive and negative excursions. Thus in order to construct part of a positive definite image, a number of wavelet components are required. This violates the principle of Occams Razor. Such a representation cannot hope to be minimal. Hence its Bayesian prior will be inferior to the pixon prior and it will be less likely from a probabilistic point of view. If this is not satisfying enough, the additional degrees of freedom represented by the many wavelet components needed to specify a local bump in the image may be inadequately constrained by the data. As with the pixel basis of more standard methods, this will give rise to spurious sources and signal correlated residuals. Hence pixons in any language which has positive definite pixon shape functions will provide a more optimal (i.e. more concise) description of the image.

3.3.4. *The Pixon and the Akaike Information Criterion*

In the 1970s, Akaike (Akaike 1973) introduced a penalty function for model complexity for maximum likelihood fitting. This formulation has come to be known as the Akaike Information Criterion (see Akaike 1973; Akaike 1977; Hall 1990; Veres 1990). The Akaike Criterion (AC) takes the log-likelihood of the fit and subtracts a term proportional to the number of parameters used in the fit. In this sense, the AC acts in the same manner as the pixon prior, i.e. acts as an *Occam's Razor* term and works for the cause of simple models. One problem with the AC approach is that it is rather *ad hoc*. Each new parameter that is added to the model invokes an identical penalty independent of the innate merit of the parameter. In effect, the AC criterion uses a uniform prior for each new variable. This, however, often has serious flaws. For example, if it is known that one is fitting a data set that can be described by polynomial dependence on the variable, then introducing a new polynomial power in to the set of basis functions should be viewed with a different weight than adding an exponential function to the basis set. Furthermore, the AC method gives no suggestion as to the appropriate model variables that should be used. A pixon-based Bayesian approach does both, i.e. (1) an appropriate selection of the prior [e.g. that of equations (3.34)-(3.39)] invokes different penalties for each parameter of the model, and (2) pixon-based methods suggest exactly which DOFs are required to most succinctly model the data. In this sense, pixon-based methods are a direct generalization of the Akaike Information Criterion.

3.3.5. *Relationship to Statistical Mechanics/Heisenberg Uncertainty Principle*

The pixon has a very close connection to the concept of "coarse graining" in physics. Pixons, in fact, directly describe the "natural graininess" of the information in a data set due to the statistical uncertainties in the measurement. This also has a direct relationship to the role of the Heisenberg uncertainty principle in statistical mechanics. As is well known, the description of a system in statistical mechanics introduces a phase space to describe the state of a system. This is akin to the "language", basis, or coordinate system used to perform image reconstructions. Phase space in statistical mechanics is what is used to specify information about the system in the AIC sense, i.e. to completely specify the state of the system. Furthermore, all of the methods of statistical mechanics are in essence Bayesian estimation applied to physical systems. The partition functions used to make statistical predictions about the system are uniform priors (each volume of phase space is given equal probability), while the GOF probability distributions are normally taken to be delta functions since all of the macroscopic parameters of the system are usually given and assumed to be known exactly. This means that the state of the system is localized to a hyper-surface in phase space, e.g. that of constant energy, temperature, or particle number. Schematically, this might be written

$$p(Prop|Sys) \propto p(Sys|Prop)p(Prop) \quad , \tag{3.40}$$

where we have used the short hand notation that *Sys* stands for the state of the system and *Prop* stands for the set of system properties in which one is interested. If, then, $Prop = \{T, E, N, \ldots\}$, i.e. a set of macroscopic variables that might include temperature, T, total energy, E, total particle number, N, etc., then specification of the system temperature, T_0, would be equivalent to specifying the GOF term as

$$p(Sys|Prop) = \delta(T - T_0)p(E, N, \ldots) \quad . \tag{3.41}$$

In this language all of the familiar concepts of temperature, Boltzman factors, etc., arise from the prior through definitions of the change in the volume of phase space with respect to extensive variables (e.g. $\frac{1}{kT} = \frac{\partial}{\partial E}\ln W$ where W is the volume of phase space and E is the energy exchanged between system and reservoir). Thus we would write

$$p(Prop|Sys) \propto \delta(T - T_0)p(E, N, \ldots)\frac{1}{Z}Z(T, E, N, \ldots)\,, \tag{3.42}$$

$$\propto \delta(T - T_0)p(E, N, \ldots)\frac{\exp(-\Delta\sigma(T, E, N, \ldots))}{\sum_{States,\, i}\exp(-\Delta\sigma(T_i, E_i, N_i, \ldots))}\,, \tag{3.43}$$

$$\propto \delta(T - T_0)p(E, N, \ldots)\frac{\exp\left(-\left(\frac{\Delta E}{kT} + \ldots\right)\right)}{\sum_{States,\, i}\exp\left(-\left(\frac{\Delta E_i}{kT} + \ldots\right)\right)}\,, \tag{3.44}$$

which contains the normal expression for the partition function.

What, then, is the role of uncertainty in statistical mechanics? The best known example of this is the role of the Heisenberg uncertainty principle. This principle declares that states within a particular hyper-cube [e.g. $(\Delta p\Delta x)^{3N} \sim (h/2\pi)^{3N}$ in the case of N free particles] are indistinguishable. This puts a natural graininess (or degeneracy) on phase space and directly affects calculation of the number of available states. Being unable to distinguish between states, however, can arise from other causes. The one of interest for the image reconstruction problem is uncertainty due to noise. The source of uncertainty, however, is inconsequential to the Bayesian estimation problem. Hence it is seen that the $(\Delta p\Delta x)^{3N} \sim (h/2\pi)^{3N}$ chunks appropriate to quantum phase space are nothing more than pixons induced by the uncertainty associated with the fundamental laws of physics.

Both the Heisenberg Uncertainty Principle and the uncertainty produced by measurement error in the image reconstruction problem cause the scientist to reevaluate the appropriateness of standard coordinate systems. This uncertainty in what might be an appropriate *coordinate system* can be seen from the above discussion to be a GOF induced problem. We might, for example, consider the effect on the statistical mechanics partition function if there were uncertainty in the system temperature. The direct outward effect of this would be to replace the delta functions in equations (3.41)-(3.44) with a broader distribution. However, we would then experience the same sort of effect that we see with pixons in the image reconstruction problem. Our old *coordinate system* for describing the properties of the system, i.e. that including the system temperature, becomes less appropriate. We can no longer use T_0 directly in our calculation of $p(Sys|Prop)$ or the Boltzman factor. The temperature T_0 has become an *inappropriate variable*. This is not because there is no appropriate value of the temperature. Temperature has become inappropriate because of uncertainty in its value. Just as in our use of pixons to smooth together adjacent pixel values (since having separately adjustable values with such a dense matrix was unjustifiable), a similar thing must be done with the statistical mechanics coordinate system (e.g. phase space). Phase space must be divided up into larger chunks, and this time the size of the chunks will have something to do with the uncertainty in the temperature. The specific size of the chunks will be determined by whether or not the states within the chunks (perhaps states with different temperatures) can be distinguished from each other.

3.4. *Application of Pixon Based Methods*

Pixon-based reconstruction has now been used by the astronomical community to perform a variety of image reconstruction problems. These include HST spectroscopy (Diplas et al. 1993), coded mask X-ray satellite imaging (Metcalf et al. 1996), far-IR airborne imaging (Koresco et al. 1995), IRAS far-IR survey data (Puetter and Piña 1994,

Puetter 1996a), and mid-IR ground-based imaging (Smith et al. 1994). We shall concentrate on such applications in the next 2 chapters. Here, however, we would like to point out that pixon-based methods have application to a wide range of problems outside of astronomy. These include radar imaging, seismology, communication, medical imaging, laboratory spectroscopy, satellite earth-resource surveys, audio and video recording and play-back, and data compression, just to name a few. Most of these applications are obvious extensions of astronomical image reconstruction techniques to areas outside of astronomy, e.g. medical imaging. The data compression and communications aspects, however, are also quite obvious since pixon-based reconstruction begins by selection of a minimal representation, or coordinate system, in which to perform the reconstruction. This is effectively performing an optimal compression of the information contained in the data and so is directly applicable to image compression and compression of signals for efficient communication.

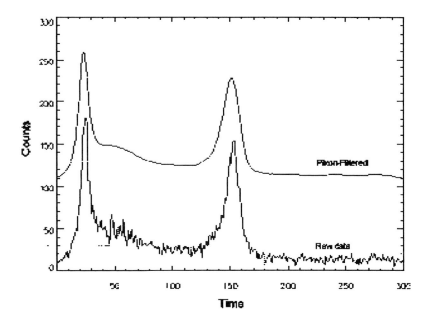

FIGURE 3. Pixon-based spatially adaptive filtering of X-ray timing data from the Vela satellites. Lower curve: Vela satellite data. Upper curve: pixon-filtered data with a constant added to provide a vertical shift. The noise in the data is given by the counting statistics in the signal. The pixon-filtered image displays only those structures calculated to be statistically significant relative to this noise.

In the present section we would like to give 2 examples of pixon-based estimation that are not directly related to astronomical image reconstruction. In fact, in both of these examples, no reconstruction is done. Pixon-based methods have been used solely to perform an optimal spatially adaptive filtering of the data. The first example is presented in Figure 3. This example is again drawn from the field of astronomy. However this time we are interested in the temporal behavior of an X-ray source. Figure 3 displays a sample of timing data from the Vela satellites, along with the pixon-based estimate of the true underlying signal. Again, as already mentioned, in this case no reconstruction has been performed, i.e. we have not attempted to remove the temporal response of the Vela satellite X-ray detector. We have simply used pixons to filter the data and report only the statistically significant structure present in the data.

The second example is presented in Figure 4. This is a mammogram of a standard medical phantom. In this case a fiber of thickness 400 microns has been placed in a piece of material with X-ray absorption properties similar to the human breast. Again, in this example pixon-based methods were used solely to filter the data. No attempt was made to sharpen the image by removing the X-ray beam spread. As can be seen from this example, the pixon filtered image provides vastly superior feature detection contrast and would allow an X-ray technician to scan mammograms much more rapidly and with greater sensitivity to finding features such as cancerous tumors. It is obvious, for example, that even though the fiber present in this image is barely at the detection threshold (as evidenced by the fact that it breaks up into a chain of features due to noise), the pixon filtered image easily detects the presence of whatever statistically significant structure is present in the image.

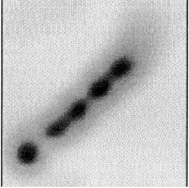

FIGURE 4. Pixon-based spatially adaptive filtering of a standard mammogram phantom. Left panel: raw phantom image. Right panel: pixon-filtered image. The phantom in this case is a 400 micron thick fiber placed in a piece of material with X-ray absorption properties similar to the human breast. The pixon-filtered image reports only those structures calculated to be statistically significant. Note that in this case the fiber is near the detection threshold–the fiber breaks up into a string of features because in some places the signal is statistically insignificant because of the large noise level. Nonetheless, the pixon-filtered image easily detects the remaining statistically significant structure.

3.5. *Conclusions*

The present lecture has shown how one might practically implement a multiresolution pixon-based image reconstruction algorithm. The relationship of pixons to other fields of study has been discussed, and it was shown that pixon-based (or pixon-like) concepts are already being used in a variety of fields. This lecture has also shown how pixon-based concepts can be used to extend the usefulness of these ideas. We have also argued that pixon-based (or more generally informaton-based) methods can be applied to a variety of fields outside of astronomical imaging. In the following chapters we shall present a number of examples of pixon-based methods applied to astronomical studies. Chapter

4 will concentrate on applications to the image restoration problem, and Chapter 5 will concentrate on reconstruction of images from complexly encoded data.

4. The restoration of Astronomical Images

4.1. *Introduction*

Image restoration is the improvement of the quality of image-like data. This includes noise filtering and resolution enhancement. Image restoration is to be distinguished from image reconstruction in which the underlying image is estimated from data which is not in image-like form, but which is complexly related to the underlying image. An example of image reconstruction is radio interferometry in which one deduces the underlying spatial structure of sources from a sparsely sampled interferogram. A common example of image restoration is Point Spread Function (PSF) deconvolution, and we shall concentrate on this application in the present lecture.

Before presenting examples of pixon-based restorations, however, we discuss several important issues which limit our ability to perform perfect PSF deconvolutions. These include the presence of random noise, the finite sampling represented by digital detector arrays, and spatial noise correlation caused by the co-addition of multiple image frames.

4.2. *The Effects of Noise on Image Restoration*

In the absence of noise, the PSF image deconvolution problem is completely (or nearly completely) invertible. Non-uniqueness of the solution only results because of pathologies in the PSF. If viewed in Fourier space, for example, one can see that there is a straightforward solution to the deconvolution problem, i.e. that given by equation (2.7) of Chapter 1. In the absence of noise, this solution is well defined and completely un-ambiguous so long as the Fourier transform of the PSF has no zeroes. In this case, each Fourier component of the underlying image can be recovered, and since the Fourier basis is complete, the solution is known exactly. If the Fourier transform of the PSF has only a few zeroes, then the problem is not too much worse. Then nearly all of the Fourier components of the underlying image are known and there is only small ambiguity in the solution. In the presence of noise, however, things change dramatically. Then even if the Fourier transform of the PSF has no zeroes, one is unable to accurately determing the Fourier components of the underlying solution over large regions of Fourier space. This occurs whenever the product of the Fourier transforms of the PSF and data is small relative to the noise.

As discussed in Chapter 1, directly using the Fourier deconvolution formula is a bad idea for performing image reconstruction. While the above considerations are valid for any deconvolution algorithm, the Fourier deconvolution method suffers greatly from noise propagation problems. Calculation of the Fourier components of the data and PSF is an inherently noisy process, and results in an estimate of the Fourier components which is much noiser than what actually generated the measured data and PSF. In addition, in order to estimate the underlying image, this noise propagation is further enhanced by taking the Fourier inverse of the noisy quotient $F(data)/F(PSF)$.

As can be seen, we can place some fundamental limits on the information content of a PSF blurred image, i.e. we know at what level we begin to lose information concerning certain Fourier components in the data. However this is not very useful for several reasons. First, knowing that certain Fourier components of the data and PSF are noisy, does not give an intuitive handle on the nature of the limitations to one's knowledge of localized features in the image, since such features are comprised of a wide range of Fourier components. Second, we have already noted that taking the Fourier transform of the data and PSF is a noisy process. Hence a good image restoration algorithm should be able to recover an image with Fourier components significantly smaller than the noise level in the computed Fourier components. This leaves us with an unsatisfactory feeling, and we are left searching for a more practical method for assessing the limitations placed by noise on a given image restoration.

A direct solution to this problem is offered by multiresolution pixon-based image reconstruction. Since we are normally interested in the properties of localized features in an image, multiresolution pixon-based methods are a natural choice. The language itself is phrased in terms of structures of various scales, and the pixon map informs us of the level of believable structure based on a statistical analysis of the noise. In more general terms, we can use Bayesian estimation to answer any question we like about the solution. For example, if we wanted to know what the relative likelihood was of the image having properties $\{x_i\}$ relative to it having properties $\{y_i\}$, we would simply calculate $p(\{x_i\}|D)/p(\{y_j\}|D)$. The properties under discussion could be the width of a given structure, its peak value, its shape, or any combination of a variety of properties.

What, then, has Bayesian estimation in general, and pixon methods in particular, taught us? Multiresolution pixon methods have taught us that at any given place in the image, the resolution limit is given roughly by the smallest statistically justifiable scale. Hence such structure must be more than a 1-σ event. A good rule of thumb is that one can see the presence of a lump above a smooth background only if it contains roughly 3-σ of signal above the background. Figure 5 illustrates the nature of the detectability of structure in the presence of noise. Displayed is a particularly noisy data set along with the underlying true structure. On the figure is a feature which is labeled as detectable structure. This feature is detectable because relative to the underlying background, the signal contained in the bump is statistically significant, i.e. greater than 3-σ. Hence if one were to use multiresolution pixons to assess the local smoothness scale at this position, one would find Bayesian evidence for structure on the size scale of this bump.

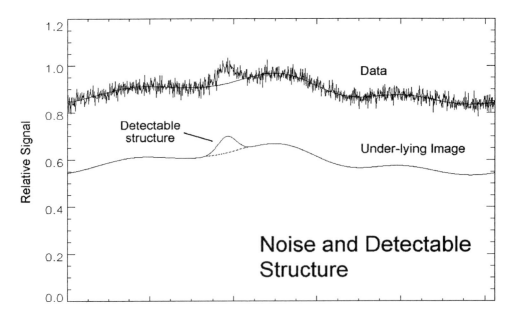

FIGURE 5. Ability to detect structure in the presence of noise. Pixon-based methods detect the structural scale present in an image by looking for Bayesian evidence of this scale.

4.3. *Finite, Discrete Sampling and Image Restoration*

A typical assumption made by many workers in the field of image reconstruction is that the finest resolution obtainable with a digital image is given by 1 pixel (some would say 2 pixels). This, however, is incorrect. There have been a number of studies that have demonstrated that digital images contain information at sub-pixel scales (see, for example, Weir and Djorgovsky 1990). The ability to estimate sub-pixel scale structure is illustrated in Figure 6. In this figure we present 3 sample situations. In the first case (illustrated by the underlying source and data set in the lefthand column), the true source uniformly fills a single pixel. When convolved with the PSF, the resulting data sampled with a resolution of 1 pixel is shown in the lower lefthand panel. If the underlying source had smaller spatial extent, e.g. filled a $1/3 \times 1/3$ pixel corner of a pixel, yet retained the same total flux, the situation would be as illustrated in the central column of Figure 6. In this case the resulting data looks significantly different, and is easily distinguishable from the uniform source case. Thus the data set produced by the $1/3 \times 1/3$ pixel source contains information about sub-pixel sized structure. In some situations, however, sources with clear sub-pixel scale structure produce data which are ambigous. Such a case is illustrated in the righthand column of Figure 6. Here the underlying source has sub-pixel sized structure, but the resulting data is identical to that produced by the uniform source.

While it appears that some sources are unable to "express" any evidence of sub-pixel scale structure, the real problem has to do with the particular placement of the sampling function. For example, for the cases presented in Figure 6, the ambiguity between the situations given in the lefthand and righthand columns can be removed by changing the sampling centers. Simply put, in each of these the source distributions behaves differently as a top-hat sampling function is dragged over them; each case has a distinguishable response function, $R_i(\vec{x}) = \int dV_{\vec{y}} H(|\vec{y} - \vec{x}|) S_i(\vec{y})$, where H is the top-hat sampling function of 1 pixel width and S_i are the different source functions. Ambiguity is only present if the values for R_i are given at only a few values of $\vec{y}$, or if the sampling function is pathological—deduction of sub-pixel scale structure is mathematically identical to the deconvolution problem which is usually invertible. This is quite valuable information. It tells us how to improve our ability to deduce sub-pixel scale structure. We can remove sub-pixel scale structure ambiguities by dithering our sampling function over the image. Indeed, if the signal-to-noise is high enough, the sampling dense enough, and the sampling function is not pathological, this approach should result in infinite resolution.

While we can now see that digitally sampled data contains sub-pixel scale information, and we have identified a manner in which to extract this information (i.e. dithering the image sampling), in practical situations the spatial resolution typically is limited to being on the order of the pixel scale or perhaps a factor of a few smaller. This is because noise counteracts our attempts to remove ambiguity. It makes each of the dithered images look similar, except in the case of the very brightest sources. Because high signal-to-noise ratios are required to resolve fine scale structure, it is usually only practical to measure a few dithered images. However, application of Bayesian methods makes the most use of this information. Bayesian methods allow us to select the most probable solution consistent with all of the information at our disposal—in this case the dithered images. So unlike direct inversions methods, which typically suffer terribly when information is incomplete, Bayesian methods work exceedingly well. Pixon-based methods, being Bayesian in their formulation, share these properties. In addition, since the structural scale is part of the estimation process in pixon-based reconstruction, this method reports

Point Spread Function

Examples of data with sub-pixel information content and ambiguity

- Each source in contained in a single pixel
- Each source has the same integrated flux

Uniform, 1 pixel source

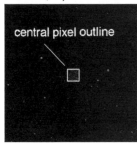

central pixel outline

1/3 pixel dia. source

4 sources, 1/3 pixel dia.

Resulting pixelated data

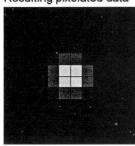

Resulting pixelated data

Resulting pixelated data

FIGURE 6. Illustration that pixelized data contains sub-pixel information. Upper left panel: PSF (Gaussian, FWHM = 2 pixels). Bottom two rows: sources and resulting data sampled at the pixel scale. Lefthand column: source uniformly fills a pixel with unit integrated flux. Center column: source fills the lower-left 1/9th of a pixel and has unit integrated flux. Right-hand column: source fills the 4, 1/9th pixel corners of a pixel and has unit integrated flux.

(through the pixon map) the resolution achieved at each point in the image. Hence one is always aware of the level of spatial structure ambiguity since this is directly estimated.

4.4. *Spatially Correlated Noise*

A final problem, often associated with digital imaging, is the introduction of spatially correlated noise by mosaicing a number of independent data frames. Of course, the best method of performing an image restoration is to use each of the data sets directly, and not to use an intermediate data product such as a mosaiced data frame. If this prescription is followed, then the noise properties of the data are simple and clear. Each data frame has independent (and hopefully random) noise. This can be dealt with easily. However, often dealing with a large number of data frames in a restoration is computationally too expensive. This is the major reason for coadding or mosaicing all of one's images into a

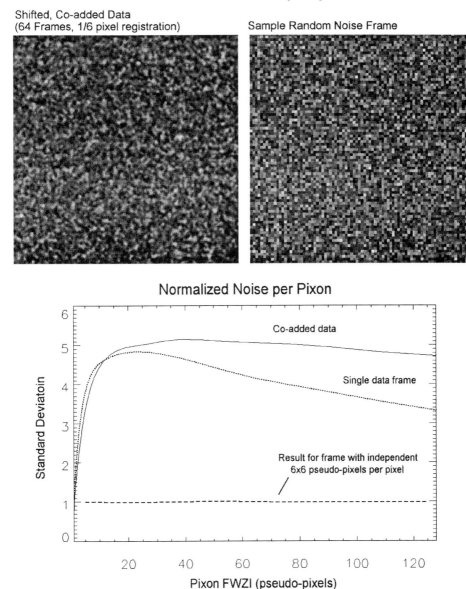

FIGURE 7. Illustration of the effects of sub-pixel registration on the spatial correlation of noise. Top left panel: Noise generated by randomly registering 64 independent noise frames to 1/6 of a pixel. Top right panel: A sample of one of the 64 noise frames used to make up the top left panel noise frame. Bottom plot: Noise per pixon as a function of pixon FWZI.

single data frame. One must clearly understand, however, that if this is done, then the noise properties of the resulting mosaic are very complex.

To illustrate the non-Gaussian nature of mosaiced or frame-coadded data, we present the results of Figure 7. In the example illustrated here, we have taken 64 independent, random Gaussian noise frames, shifted them randomly on a grid with 1/6th pixel centers (call this the pseudo-pixel grid) and coadded them. The maximum displacement of one data frame relative to the other was 108 pseudo-pixels (108 pseudo-pixels = 6 pseudo-pixels/pixel × 18 pixels), and the 64 displacements were randomly and uniformly distributed over these 108 positions in both the x and y directions. The result is presented in the upper lefthand image of Figure 7. A sample of one of the 64 Gaussian noise frames used to make up the mosaiced image is presented in the upper righthand image. The full pixel grid size is apparent in this image.

As can be seen from the mosaiced image, the noise is far from Gaussian, i.e. a random Gaussian noise field at the pseudo-pixel scale. Despite the randomness of coadding a large number of images, the noise is clearly spatially correlated. Furthermore, this spatial correlation does not go away as the number of coadded frames goes to infinity. It is a feature of coadded noise. The properties of this noise can be analyzed in a number of ways. One could, for example, look at the Fourier power spectrum of the noise. However, this is not what we are interested in here. We are interested in the effects that this noise has on our ability to detect localized structure. We are not interested in how this noise affects our ability to measure sine waves. Hence the appropriate analysis is to study the noise as a function of structural scale, i.e. a pixon-based analysis. This is what we have done in the bottom plot presented in Figure 7. In this diagram we plot the noise contained in a given radially symmetric, truncated parabolic pixon as a function of pixon full width at zero intensity (FWZI), i.e. we plot $\sigma_{N,pixon}/\sigma_{G,pixon}$ where

$$\sigma_{N,pixon} = \sqrt{\sum_i \frac{1}{N_{pseudo-pixels}} \left(\sum_j k(\vec{x}_i, \vec{y}_j, \delta(\vec{x}_i)) N(\vec{y}_j) \right)^2} \quad , \qquad (4.45)$$

$$k(\vec{x}, \vec{y}, \delta(\vec{x})) = 1 - \frac{|\vec{x} - \vec{y}|^2}{\delta(\vec{x})^2} \quad , \qquad (4.46)$$

$$\sigma_{G,pixon} = \sigma \sqrt{\sum_j k(\vec{x}_i, \vec{y}_j, \delta(\vec{x}_i))} \quad , \qquad (4.47)$$

where summations are over the pseudo-pixel grid, N is the noise frame in question, δ is the pixon FWZI, $\sigma_{G,pixon}$ is the value of $\sigma_{N,pixon}$ for the case in which the noise is a random Gaussian noise field on the pseudo-pixel grid, and σ is the standard deviation of the noise at the pixel scale.

Note that in the diagram at the bottom of Figure 7, we have presented results for both a random Gaussian field over the pseudo-pixel grid (dashed line—the value is expected to be 1 since the noise is normalized to this case), as well as for the shifted co-added data, and the single sample Gaussian noise frame at the pixel resolution. Note that both of these cases have reduced noise below the pixel scale (i.e. 6 pseudo-pixels). This property is directly apparent from visual inspection of the images. Again, this is valuable knowledge. Co-adding data affects the spatial structure spectrum of the noise. In calculating a pixon map, this noise spectrum must be acknowledged. There is less noise at small scales. Of course at sub-pixel scales there usually is also less signal—the signature of sub-pixel scale structure is relatively weak in noisy dithered, mosaiced images. Use of the appropriate noise level for each pixon scale, however, is necessary if one wishes to achieve an optimum restoration of the structure present in the data at each scale.

4.5. *Examples of Pixon-Based Reconstruction*

In this section we present 2 pixon-based image reconstructions. The first example compares pixon-based reconstructions to Maximum Entropy methods for a mock data set, i.e. a completely made up data set in which the underlying image, PSF, and noise frames are manufactured and hence known precisely. Mock data sets are useful in comparisons of image restoration/reconstruction methods since there can be no argument about the goal of the reconstruction. The true, underlying image is known perfectly *a priori*. Furthermore, the noise and all parameters of how the input data was made are completely specified. The second example presents an elliptical pixon-based reconstruction of Keck $2\mu m$ imaging data for the gravitational lens of FSC 10214+4724. This example illustrates the use of non-radially symmetry pixon bases. We shall go into some detail in the discussion of this example principally because most pixon reconstructions presented in previous work have used radially symmetric basis functions and because the science allowed by the reconstruction is quite interesting.

4.5.1. *Mock Data*

We present in this section pixon-based and MEMSYS 5 restorations of a mock-data set. For the comparisons presented here, the ME algorithms chosen are those embodied in MEMSYS 5, the most current release of the MEMSYS algorithms. The MEMSYS code represent a powerful set of ME algorithms developed by Gull and Skilling (Gull and Skilling 1991). The MEMSYS algorithms probably represent the best commercial software package available for image reconstruction. The MEMSYS reconstruction was performed by Nick Weir of Caltech, a recognized MEMSYS expert, and were supplemented with his multi-channel correlation method which has been shown to enhance the quality of MEMSYS reconstructions (Weir 1991; Weir 1994). The true, noise-free, unblurred image presented in the top row is constructed from a broad, low-level elliptical Gaussian (i.e. a 2-dimensional Gaussian with different FWHMs in perpendicular directions), and 2 additional narrow, radially symmetric Gaussians. One of these narrow Gaussians is added as a peak on top of the low-level Gaussian. The other is subtracted to make a hole. To produce the input image, the true image was convolved with a Gaussian PSF of FWHM=6 pixels, then combined with a Gaussian noise realization. The resulting input image is displayed in the top row. The signal-to-noise ratio on the narrow Gaussian spike is roughly 30. The signal-to-noise on the peak of the low level Gaussian is about 20. The signal-to-noise at the bottom of the Gaussian "hole" is 12.

As can be seen, the Fractal Pixon Based (FPB) reconstruction is superior to the multi-channel MEMSYS result. The FPB reconstruction is free of the low-level spurious sources evident in the MEMSYS 5 reconstruction. These false sources are due to the presence of unconstrained degrees of freedom in the MEMSYS 5 reconstruction and are superimposed over the entire image, not just in the low signal to noise portions of the image. Furthermore, the FPB reconstruction's residuals show no spatially correlated structure, while the MEMSYS 5 reconstruction systematically under-estimates the signal, resulting in biased photometry.

4.5.2. *The Gravitational Lens of FSC 10214+4724*

We next discuss the application of pixon-based methods to image restoration of Keck K-band ($2.3\mu m$) imaging of the gravitational lens system FSC 10214+7424 (data presented in Graham and Liu 1995). This object was initially thought to be the most luminous object in the universe with a luminosity approaching 10^{15} solar luminosities. However, it is now recognized as a gravitational lens with a magnification of roughly 100 (Broadhurst and Lehar 1995; Close et al. 1995; Graham and Liu 1995). Hence the luminosity is about 100 times fainter than initially estimated.

In order to make sure that our image description language would not be biased against faint Einstein rings, and was able to describe inherently thin structures, we decided it would be wise to use a set of elliptical pixon shape function as our language basis. Our elliptical pixon language is displayed in Figure 9 and contains 196 elliptical pixon shape

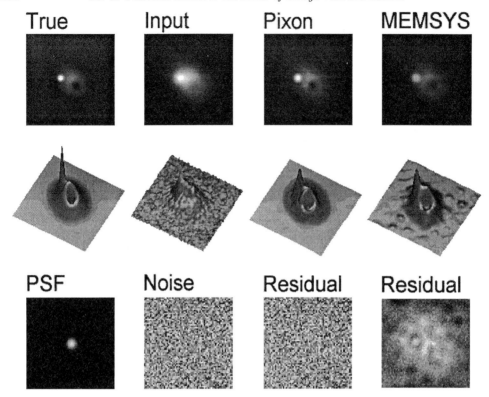

FIGURE 8. Comparison of pixon-based image restoration with Maximum Entropy restoration for a mock data set. Shown is the PSF (Gaussian with FWHM=6 pixels), the true, underlying image, and the Gaussian noise. The restored image and residuals for radially symmetric, truncated parabolic fuzzy pixons are shown. Also shown are the results and residuals for the multi-channel MEMSYS 5 method. Reproduced from Puetter 1994.

functions with 6 different ellipticities, having major to minor axis ratios logarithmically spaced between 1 and 7.25. For each ellipse there are a number of major axis orientation angles. The ellipses of high ellipticity have a larger number of orientations.

Our elliptical pixon restoration of the Keck K-band image is shown in Figure 10. Also displayed in this figure is the Maximum Entropy reconstruction of Graham and Liu (Graham and Liu 1995). As can be seen from this figure, our reconstruction reveals the entire Einstein ring. This low-level ring was not detectable by the Maximum Entropy methods employed by Graham and Liu because of the large number and strength of the spurious sources produced by that method. These sources are obvious in the Maximum Entropy result displayed in Figure 10. We have also plotted in Figure 11 our reconstruction overlaid with the local pixon shape functions. In this image one can easily see how the elliptical pixons follow the the image's spatial structure (information content). Smaller pixons are used in locations of detailed structure. Larger pixons are used in regions in which the image is smooth, and the ellipses align with the structures present so that the largest pixon can be used to describe the image at each location. As argued in Chapters 2 and 3, relative to simpler languages, e.g. radially symmetric pixons, this richer elliptical pixon language should be able to describe the image using fewer degrees-of-freedom. We also have performed a reconstruction for this data with radially symmetric pixons and find that the elliptical pixon restoration uses half as many pixons.

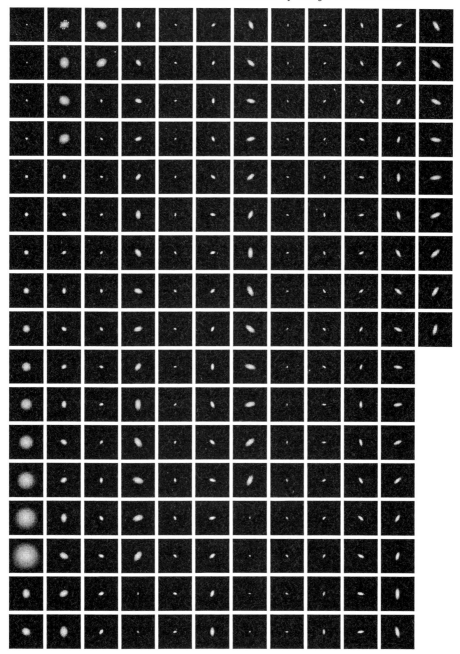

FIGURE 9. The elliptical language used to reconstruct the image of FSC 10214+7424.

The science that our better reconstruction allows for this data is quite interesting. First, Graham and Liu concluded on the basis of their inability to see an entire Einstein ring, that the size of the emitting source (which has a redshift of 2.3) is 0.25 arcsec or less. They point out that given the lens geometry, if a complete ring was seen, then the size of the emitting region would have to be at least 0.5 arcsec. Hence our first conclusion is that the emitting region of FSC 10214+7424 has significant emission out to a size of 0.5 arcsec, roughly twice that deduced by Graham and Liu. Our second result is related

Raw, Flat-Fielded Data

The Einstein Ring of FSC 10214+4724

Keck K-band imaging data of Graham and Liu (1995)

Elliptical Pixon Reconst.

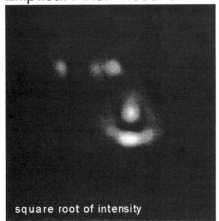

Max Entropy Reconst.

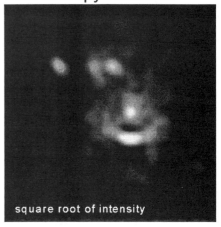

FIGURE 10. Comparison of image reconstruction techniques for the Einstein ring of FSC 10214+4724. The elliptical pixon reconstruction used 196 different elliptical pixon kernel functions. Relative to radially symmetric reconstructions, the elliptical pixon reconstruction used roughly 50% fewer DOFs in the image model. Reproduced from Puetter 1996.

to the nature of the gravitational lens. Recent HST images of the Einstein ring at far-UV rest-frame wavelengths ($0.25\mu m$) were obtained by Eisenhardt et al. 1996. From the narrow nature of the arc, these authors conclude that the size of the UV source is of order 0.005 arcsec (40 pc), and that the best fit to the shape of the arc requires a contribution to the gravitational lensing potential from the next brightest elliptical galaxy near the main lensing galaxy. Our restored image directly confirms the gravitational contribution of this object to the lensing potential. The K-band Einstein ring is clearly warped in the direction of this source, directly confirming the speculations of Eisenhadt et al. Further discussion of these results will appear in a paper which is in preparation.

4.6. Conclusions

This lecture has discussed the effects of random noise on detectable structure in a restoration, the effects of discrete sampling, and the nature and effects of spatially correlated

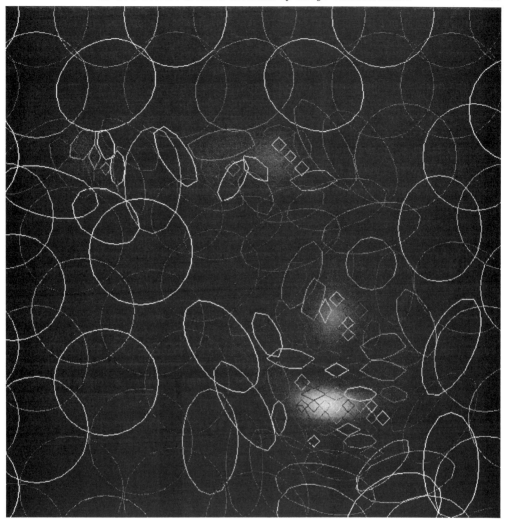

FIGURE 11. Reconstruction of the Einstein ring of FSC 10214+7424 with overlaid pixon kernel functions. Note that for a fuzzy pixon reconstruction a pixon kernel shape is assigned to each pseudo-pixel in the image. However plotting this many kernel functions would be confusing to the eye. Thus for the purposes of illustration, we have plotted a new pixon kernel function shape such that the area overlap is 25%. Pixon kernel contours are plotted at 50% of maximum.

noise caused by data frame co-adding. We have outlined a rule of thumb to estimate the level at which random noise limits spatial resolution and discussed how image dithering can greatly improve our ability to detect sub-pixel scale structure. We have also described the nature of the spatial correlation resulting from data frame coadding and have indicated how these effects can be incorporated into pixon-based methods. Finally, we have presented 2 sample restorations. The first was of a mock data set and directly demonstrated the advantages of pixon-based methods over the best Maximum Entropy methods. The second was an elliptical pixon-based reconstruction of K-band imaging data of the gravitational lens system FSC 10214+7424. This example also confirmed the greater sensitivity afforded by pixon-based methods relative to Maximum Entropy methods. Furthermore we illustrated how these benefits translate directly into science.

4.7. *Acknowledgements*

The author would like to thank the many people who have contributed over the years toward making pixon-based methods a reality. In this regard I would like to especially thank Robert Piña, the co-inventor of the pixon. I would also like to thank James Graham for providing his imaging data for FSC 10214+7424 and Amos Yahil for valuable discussions during the development of the elliptical pixon-based code.

5. Reconstruction of Complexly Encoded Images

5.1. *Introduction*

This lecture concentrates on image reconstruction, i.e. deducing the spatial structure and intensity distribution of an underlying source from a data set in which the source structure and brightness are encoded in a complex manner. Examples of complexly encoded data include (1) radio interferometry in which the source brightness is encoded in a sparsely sampled interferogram, (2) medical CAT scans and ultrasound images in which the spatial structure of a patient's brain, heart, or other organs are studied by scanning a source of radiation, and recording the transmission or echo of this radiation, and (3) LIDAR, RADAR, and SONAR imaging in which reflected light or sound is used to encode the spatial structure of a field of objects in front of broadcasting antennae. These are only a few examples of the many important applications of reconstruction of complexly encoded images.

An especially important aspect of image reconstruction in the case of complexly encoded images is the high degree of ambiguity that may be encountered in solution space. In the astronomical radio interferometry imaging problem, for example, the spatial Fourier components of the image are very incompletely sampled. This means a direct inversion of the problem is impossible. Too many Fourier components are absent. Thus in order to find a solution, one is forced to provide some sort of "regularization" for the problem, e.g. making a plausible limited physical model and estimating its parameters. It is not surprising, therefore, that successful methods of radio interferometry are characterized by approaches such as CLEAN, in which the radio image is successively approximated in terms of point sources, and Maximum Entropy (ME) methods, in which the regularization is provided by insisting on a very smooth underlying source distribution.

Because of the great ambiguity in the solution of such problems, Bayesian methods are a natural choice for a solution method. They do not attempt to perform an exact inversion of the problem (which is generally impossible), but use all of the available information to constrain the range of solutions and then select the most likely solution given what is known. Indeed, the complexly encoded image problem is an excellent showcase for the power of Bayesian priors. In such cases simple Goodness-of-FIT (GOF) or Maximum Likelihood (ML) methods usually fail miserably—such methods avoid priors. The GOF and ML criteria used simply do not adequately constrain the solution. Consequently these methods often select unphysical or non-sensical solutions since such solutions represent the typical element of solution space.

To find valid solutions to the complexly encoded image problem one must apply realistic and powerful constraints. As mentioned above, Bayesian methods are often employed for this purpose since they do not attempt to directly invert the problem, yet they make use of all of the information available to constrain the solution. Even so, the range of possible solutions can be much too large without further constraints. ME methods, for example, add the additional constraint that the solution have "maximal smoothness".

The CLEAN algorithm insists on building radio images from collections of point sources. The ME method is generally too naive, while the CLEAN algorithm is much too restrictive, and suffers the typical problems associated with parametric fitting when the model is inappropriate. (A parametric fitting procedure assumes a fixed model which is characterized by a small set of parameters. The data is then used to fit the parameters.) What we require is a powerful and flexible non-parametric fitting procedure. (Non-parametric fitting attempts to model an underlying function without reference to a particular functional form. This really amounts to modeling the underlying function by using a large number of parameters—see, for example, Heden 1975; Sprent 1981; and Hardle 1990.)

Pixon-based image reconstruction is a non-parametric modeling approach for the image reconstruction problem. The language used to model the image (e.g. the pixon shape functions in the case of multiresolution pixons) is the large space of possible parameters out of which the model for the underlying image will be constructed. The GOF criterion and minimization of the Algorithmic Information Content (AIC) of the image model are the criteria by which the specific non-parametric model is selected. As argued in Chapters 2 and 3, minimization of the AIC is an ideal criterion for selecting a model to constraint a reconstruction, and use of a multiresolution language should provide a concise expression for the AIC with a resulting highly optimized Bayesian prior.

While the above arguments indicate that Bayesian methods in general, and pixon-based methods in particular, provide an excellent approach for "phrasing" or formulating the problem of reconstruction, the question of how to numerically find the solutions is still an open issue. In the case of image restoration, standard multi-dimensional optimization methods seem to work fine. This is largely because the data and the image are very closely related. Indeed, often the data is used directly as the image value, i.e. often image restoration is not even performed. In any event, if image restoration is desired, in this case the data provides an excellent first guess for the image value, and only modest extrapolations are required from this first guess to find a statistically acceptable solution. Things are much different in the case of complexly encoded images. Here the data does not provide a good first guess for the image. Indeed, it may be impossible to easily obtain a good first guess for the image. Furthermore, because of the nature of the encoding process, or incomplete sampling, or other loss of information, there is often great ambiguity in solution space. This presents an exceedingly difficult challenge for numerical methods. Thus before presenting sample reconstruction for the complexly encoded image case, it is appropriate to discuss numerical methods.

5.2. *Robust Numerical Methods for Inverse Problem Solution*

Large ambiguity in the solution to an optimization problem essentially means that there are many local extrema for the problem and that solution space is very "convoluted". Hence simple optimization approaches, such as the method of steepest decent (multi-dimensional Newton-Ralpson iteration), have an extremely small radius of convergence. (Press *et al.* 1994 presents an excellent discussion of the "standard" algorithms and their general convergence properties). An example of the type of solution space "terrain" that we have in mind appears in Figure 12.

As can be seen in Figure 12, the sample solution space we have selected is filled with many local extrema (we shall assume from here on that we are attempting to solve a minimization problem). Our goal is to find the *global* minimum. However, if the method of steepest descent is used as the solution algorithm, the resulting answer depends strongly on the starting point. As illustrated in the figure, proceeding down the gradient from the marked starting point results in finding a local minimum which is separated by many "hills" and "valleys" from the true global minimum.

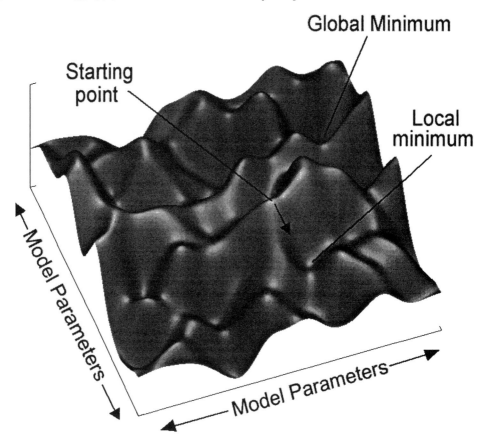

FIGURE 12. Sample solution space for the complexly encoded image reconstruction problem. There are many local extreme (illustrated here as minima) to examine before finding the globally optimal solution. Simple methods, such as the method of steepest decent, get caught in local extrema before finding the global extrema. Shown is the location of the global minimum, a sample starting point for a solution algorithm, and the direction of the gradient (direction of steepest descent).

While one can immediately see the problems with steepest descent methods, other methods, which are usually more robust, also fail if the solution space terrain is sufficiently complex. These techniques include conjugate gradient methods (which currently is our standard minimization technique for pixon-based reconstruction) and variable metric methods. Conjugate gradient methods use information on the local gradient to search for a solution, but insist at each point that the search proceeds in a direction perpendicular to the last search direction. Variable metric methods have a similar strategy, i.e. using past search directions to speed the solution. However both methods effectively assume that the function to be minimized is well approximated by a quadratic form. This, of course, will be true for almost all problems sufficiently close to the solution. (This is the case in Figure 12, too.) The practical difficulty is that one may have no way in which to get *sufficiently close* to the solution in a first guess. The starting point indicated in Figure 12 is clearly not *sufficiently close*.

As mentioned above, some image reconstruction methods try to circumvent the problem of small radii of convergence by adding mathematical constraints which *regularize* the problem, i.e. make the problem *convex* by increasing the range over which the function is well approximated by a quadratic form. Adding the ME prior often performs this feat for χ^2 minimization problems, and ME is often said to regularize the image reconstruction problem. The pixon-prior also performs this function very well in most cases. However, it is not uncommon to see the ME prior fail in the complexly encoded image case, and we have seen at least one situation (and expect many others) in which the pixon prior was totally inadequate for increasing the radius of convergence of the conjugate gradient method in a practical way—see the example of the OSSE Virgo Survey data below.

In order to effectively search for the global extrema in complex solution spaces, a different approach is required. Recently, great success has been achieved for such problems by using stochastic search methods. Such methods include simulated annealing (Cerny 1982; Kirkpatrick et al. 1983; Kirkpatrick 1984; German and German 1984), genetic algorithms (Fraser 1957a; Fraser 1957b; Fraser 1968; Holland 1969; Holland 1975), and evolutionary programing (Conrad 1974; Kampfner and Conrad 1983; Conrad 1984; Conrad 1988). (An excellent review of genetic and evolutionary algorithms can be found in Fogel 1995.) Powerful mathematical theorems have been proven for these methods that show that provided that the annealing schedule is slow enough, or the evolutionary process is carried out long enough in the presence of random mutation, the resulting answer will be the globally optimal solution. While a globally optimal solution is guaranteed with these methods provided that sufficient care is taken, it is by no means guaranteed that the solution will be obtained in a computationally acceptable time, and it is how such algorithms can be made fast and efficient where significant current research effort is being applied.

In the examples presented below, we have used both standard multi-dimensional optimization methods (conjugate gradient) as well stochastic approaches (simulated annealing) to solve the image reconstruction problem. Even for the more standard methods, however, we have introduced a partial annealing process, i.e. we start with large pixon scales and gradually introduce smaller scales as the iterations proceed. Furthermore, this annealing process was found to be essential for developing a robust algorithm—see Chapter 3.

5.3. *Sample Reconstructions of Complexly Encoded Images*

This section presents three examples of pixon-based reconstruction of complexly encoded images. Each of these examples have been presented before (see Puetter 1995). The first example is from the $60\mu m$ survey scans of the IRAS satellite and reconstructs the image of the interacting galaxy pair M51. This example first appeared in print in Puetter and Piña 1993a. The second example is from coded aperture X-ray imaging made with the HXT instrument on the Yohkoh satellite (see Metcalf et al. 1996). The final example is of OSSE Virgo Survey data (see Dixon et al. 1996).

5.3.1. *60 Micron IRAS Survey Scans of M51*

In Figure 13 we present a reconstructed image from $60\mu m$ IRAS survey scans of the interacting galaxy pair M51. This particular data set is especially useful for comparing the performance of various different image reconstruction techniques since it was chosen as the basis of an image reconstruction contest at the 1990 MaxEnt Workshop (Bontekoe 1991). Furthermore, the IRAS data for this object is particularly strenuous for image reconstruction methods. This is because all the interesting structure is on "sub-pixel scales" (IRAS employed relatively large, discrete detectors—1.5 *arcmin* by 4.75 *arcmin* at $60\mu m$) and the position of M51 in the sky caused all scan directions to be nearly parallel. This means that reconstructions in the cross-scan direction (i.e. the

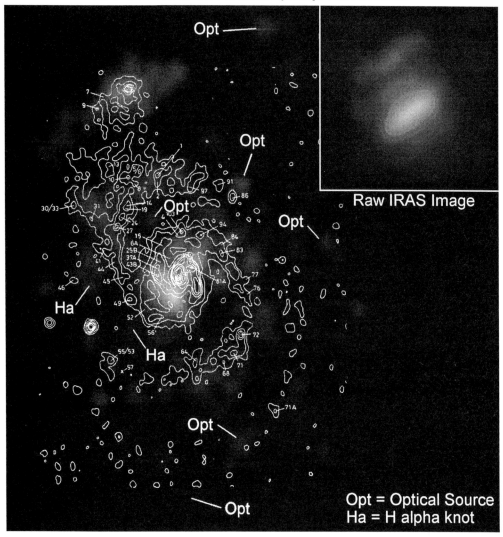

FIGURE 13.   Pixon-Based reconstruction of $60\mu m$ survey scans of M51.   Grey-scale image: The pixon-based reconstruction. The faintest sources visible have a formal SNR of roughly 30. Contours: The 5 GHz radio flux contours of van de Hulst 1988. Noted are several of the stronger features in the reconstruction that can be identified with optical sources (labeled "Opt") and $H\alpha$ knots (labeled "$H\alpha$"). Figure reproduced from Puetter and Piña 1993b.

4.75 *arcmin* direction along the detector length) should be significantly more difficult than in the scan direction. In addition, the point source response of the 15 IRAS $60\mu m$ detectors (pixel angular response) is known only to roughly 10% accuracy, and finally, the data is irregularly sampled.

In Figure 14, our FPB reconstruction is compared to the results from other methods, including the Lucy-Richardson, the Maximum Correlation Method (MCM) (see Rice 1993, and MEMSYS 3 reconstructions Bontekoe et al. 1991). The winning entry to the MaxEnt 90 image reconstruction contest was produced by Nick Weir of Caltech and is not presented here since quantitative information concerning this solution has not been published—however, see Bontekoe 1991 for a gray-scale image of this reconstruction. Nonetheless, Weir's solution is qualitatively similar to Bontekoe's solution. Both

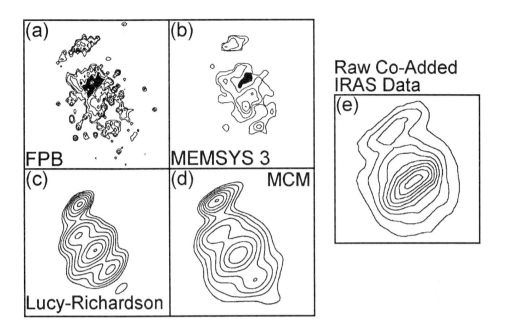

FIGURE 14. Comparison of the performance of various image reconstruction methods for the M51 IRAS survey data. Shown are the results for our pixon-based method (fuzzy, radially symmetric pixons), the Lucy-Richardson method, the "in-house" IPAC Maximum Correlation Method (MCM), and the Maximum Entropy Method (MEMSYS 3). Panel (b) is reproduced from Bontekoe 1991b by permission of the author, and panels (c) and (d) are reproduced from Rice 1993 by permission of the author. This figure first appeared in Puetter and Piña 1993a.

were made with MEMSYS 3. Weir's solution, however, used a single correlation length channel in the reconstruction. This constrained the minimum correlation length of features in the reconstruction, preventing break-up of the image on smaller size scales. This is probably what resulted in the "winning-edge" for Weir's reconstruction in the MaxEnt 90 contest (Weir 1993).

As can be seen from Figure 14, our FPB-based reconstruction is superior to those produced by other methods. The Lucy-Richardson and MCM reconstructions fail to significantly reduce image spread in the cross-scan direction, i.e. the rectangular signature of the 1.5 by 4.75 *arcmin* detectors is still clearly evident, and fail to reconstruct even gross features such as the "hole" (black region) in the emission north of the nucleus—this hole is clearly evident in optical images of M51. The MEMSYS 3 reconstruction by Bontekoe is significantly better. This image clearly recovers the emission "hole" and resolves the north-east and south-west arms of the galaxy into discrete sources. Nonetheless, the level of detail present in the FPB reconstruction is clearly absent, e.g. the weak source centered in the emission hole (again, this feature corresponds to a known optical source).

To assess the significance of the faint sources present in our FPB reconstruction, our reconstruction in Figure 13 is overlaid with the 5 GHz radio contours of van der Hulst et al. (1988). The radio contours are expected to have significant, although imperfect, correlation with the far infrared emission seen by IRAS. Hence a comparison of the two maps should provide an excellent test of the reality of structures found in our reconstruction. Also identified in Figure 13 are several prominent optical sources and (hydrogen Balmer line emission) knots. As can be seen, the reconstruction indicates excellent correlation with the radio. The central region of the main galaxy and its two brightest arms align remarkably well, and the alignment of the radio emission from the north-east companion and the IRAS emission is excellent. Furthermore, for the most part, whenever there is a source in the reconstruction which is not identifiable with a radio source, it can be identified with either optical or Hα knots. An excellent example is the optical source in the "hole" of emission to the north-east of the nucleus of the primary galaxy or the bright optical source to the north-west of the nucleus (both labeled "Opt" in Figure 13). Because of the excellent correlation with the radio, optical, and H line images, we are quite confident that all of the features present in our reconstruction are real. (The faintest sources visible in Figure 13 have a formal SNR of 30.)

5.3.2. *HXT Solar Flare Imaging*

We next present a sample image reconstruction of coded mask X-ray data from the Hard X-ray Telescope (HXT) instrument on board the Yohkoh spacecraft (Kosugi et al. 1991). Figure 15 shows hard X-ray (23-33 keV) images of a solar flare which occurred on 20 August 1992. Since there is no effective method of manufacturing optics with which to focus hard X-ray light, the HXT instrument takes X-ray images by taking pictures through a series of coded masks. The series of coded images is then inverted to yield the underlying source structure. Figure 15 shows a time series of 3 images. Each row of panels shows three different reconstructions. All inversions were performed by Tom Metcalf (U. Hawaii). As can be seen, direct linear inversion produces an enormous amount of spurious structure. In addition to hiding low contrast features, the presence of such spurious structure is particularly worrisome since the flux conserving nature of the algorithm requires that flux placed in spurious sources must come from the true sources thereby grossly affecting photometry. In this regard, the ME and FPB reconstructions can be seen to be a great improvement over the direct linear inversion. Relative to the pixon-based reconstruction, however, the ME inversion still produces a wealth of spurious emission, resulting in poor photometry and often over resolves real features. (We know the resolution of the ME image is too high since the quality of the pixon fit is just as good—in fact slightly better for these images—and uses a lower resolution. Hence the ME resolution is unjustified.)

5.3.3. *OSSE Gamma-Ray Imaging Toward Virgo*

As a final example of pixon-based image reconstruction, we present a comparison of two image reconstruction techniques using 50 to 150 keV data from the Oriented Scintillation Spectrometer Experiment (OSSE) aboard the Compton Gamma-Ray Observatory (GRO). The OSSE instrument consists of four shielded detectors with a field of view of $3.8° \times 11.4°$ (FWHM). Each detector is mounted on an independent single-axis pointing system which allows for sub-stepping the detector field of view. Figure 16 presents a comparison of a pixon-based reconstruction of the OSSE survey data with the Non-Negative Least-Squares (NNLS) method developed at the University of California Riverside (UCR). Both the NNLS and pixon-based reconstruction presented in Figure 16 were performed by D. Dixon of UCR (Dixon et al. 1996). Details of the UCR implementation of the pixon-based algorithm are given below. The dark area of the figure represents all of the points for which there is significant exposure time during the scanned observation. Each of the pixels in the reconstructed images in Figure 16 has an angular size of $2° \times 2°$.

As can be seen from Figure 16, the NNLS reconstruction produces an image which has the appearance of a random noise field. From this image, it is unclear whether or not there are any real detections of sources. On the other hand, the pixon-based reconstruction clearly finds the two bright sources expected to be seen in this data, i.e.

HXT: 1992 August 20, 23-33 keV

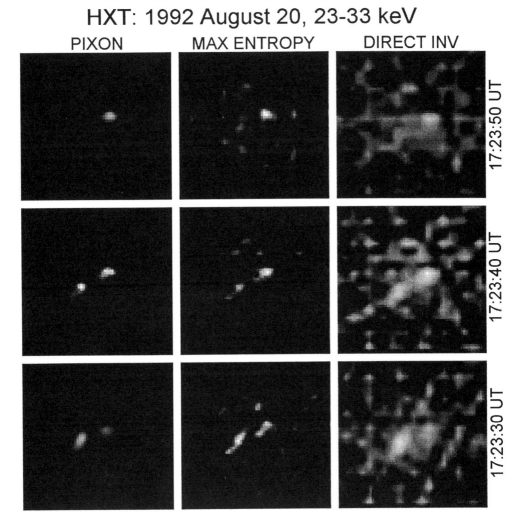

FIGURE 15. Hard X-ray (23-33 keV) images of the 20 August 1992 solar flare. The observations are from the hard X-ray Telescope (HXT) on board the Yohkoh satellite. Each column shows a time series of three image reconstructions. The background patterns which show up in the direct inversion and the ME inversion are artifacts resulting from the sparse UV-plane coverage in the HXT instrument. The pixon-based reconstruction essentially eliminates all of these artifacts. The reconstruction were performed by Metcalf of the University of Hawaii. Figure reproduced from Puetter 1995.

3C273 and NGC 4388 (the active galaxy M87 is also in the pixel occupied by NGC 4388, but is not believed to contribute significant 50-150 keV emission), and it can be seen that the pixon-based method was very successful in suppressing spurious sources in the reconstruction.

The OSSE data, and γ-ray data in general, present an especially difficult challenge for image reconstruction methods. This is because the signal-to-noise of the collected data is very low. At γ-ray energies there are only very few photons to count. Consequently the standard pixon-based methods developed at UCSD were found to be inadequate—the conjugate gradient method could not find the global minimum. It was for this reason that D. Dixon (U.C., Riverside) decided to use a simulated annealing approach to finding the

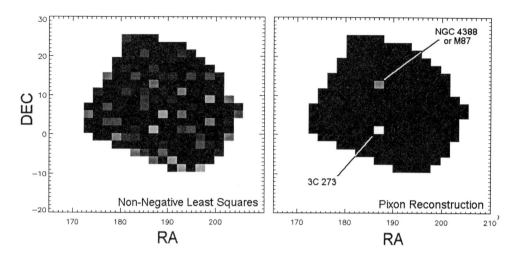

FIGURE 16. Non-Negative Least Squares (NNLS) and pixon-based image reconstructions from the 50-150 keV data of the OSSE Virgo Survey. The reconstructions were performed by Dixon of the University of California, Riverside. While the NNLS method gives a result that looks like random noise, Dixon's simulated annealing, pixon-based method clearly reveals the two strong sources expected in these data, i.e. NGC 4388 and 3C 273. It also detects a mild gradient in the γ-ray background. Figure reproduced from Puetter (1995).

optimal pixon reconstruction. As previously discussed, simulated annealing is well known for its robust ability to find shallow global minima in the presence of numerous local minima. However, this ability has significant computational costs. To speed the method, Dixon adopted a mean field approach and used only two pixon scales, one appropriate for point sources and one appropriate for the diffuse background. Given the expected nature of the image, this is quite a suitable assumption and gives rise to the excellent reconstruction presented in Figure 16. It is clear from the success of this method, that pixon-based approaches such as that of Dixon et al. (1996) have dramatically increased the scientific capability of the OSSE instrument. Current plans at UC, Riverside are to apply these techniques to COMPTEL data as well, and to explore simulated annealing pixon-based approaches with multiple pixon scales.

5.4. *Conclusions*

This lecture has presented a number of applications of pixon-based methods to the reconstruction of complexly encoded images. Comparisons of the performance of pixon-based techniques to other methods demonstrate that pixon-based methods provide consistently superior solutions with greater spatial resolution and sensitivity to faint sources. In addition, unlike other methods, pixon-based techniques robustly reject spurious sources. We have also discussed the special problems associated with complexly encoded images, i.e. a highly complex solution space with many local minima. Such problems typically have very small radii of convergence for standard multi-dimensional optimization methods (e.g. methods that approximate the optimization criterion as a quadratic form). Alternative, stochastic methods for solving the inverse problem were discussed, and a practical example of a simulated annealing, pixon-based reconstruction was given, i.e. the reconstruction of OSSE Virgo Survey data.

5.5. *Acknowlegements*

The author would like to thank a number of people for their valuable contributions to this work. Special thanks go to T. Metcalf (U. of Hawaii) and D. Dixon (U.C., Riverside) for extending the application of pixon-based methods to Yohkoh X-ray and GRO γ-ray data.

REFERENCES

ABRAMS, M. C., DAVIS, S. P., RAO, M. L. P., ENGLEMAN, R., AND OTHERS 1995, High ResolutionFourier Transform Spectroscopy of the Meinel System of OH *Ap. J. (Suppl.)***93**, 351-395.

AKAIKE, H. 1973, Information Theory and an Extension of the Maximum Likelihood Principle, *Proc. Second International Symp. on Inf. Sci.*, eds. B. N. Petrov and F. Csáki, pp. 267-281. Akadémia Kiadó.

AKAIKE, H. 1973, On Entropy Maximization Principle, *Proc. Symp. on Appl. of Statistics*, ed. P. R. Krishnaiah, p. 267. North-Holland.

BONTEKOE, T. R. 1991, The Image Reconstruction Contest, in *Maximum Entropy and Bayesian Methods*, eds. W. T. Grady, Jr., and L. H. Schick, Kluwer Academic.

BONTEKOE, T. R., KESTER, D. J. M., PRICE, S. D., DE JONGE, A. R. W., AND WESSELIEUS, P. R. 1991, Image Reconstruction from the IRAS Survey, *Astron. and Astrophys.*, **248**, pp. 328-336. eds. W. T. Grady, Jr., and L. H. Schick, Kluwer Academic.

BROADHURST, T.; LEHAR, J. 1995, A Gravitational Lens Solution for the IRAS Galaxy FSC 10214+4724, *Ap. J. (Letters)*, **450**, pp. L41-4.

BRYAN, R. K., AND SKILLING, J. 1980, *M.N.R.A.S.***191**, 69.

CERNY, V. 1982, A Thermodynamical Approach to the Traveling Salesman Problem: An Efficient Simulation Algorithm, preprint, Inst. Phys. and Biophys., Comenius Univ., Bratislava.

CHAITIN, G. J. 1966, *J. Ass. Comput. Mach.* **13**, p. 547.

CHAITEN, G. J. 1982, Algorithmic Information Science, *Encyclopedia of Statistical Science* **1**, pp. 38-41. Wiley.

CHERRY, C. 1978, *On Human Communication,* 3rd edition. MIT Press.

CLOSE, L.M.; HALL, P.B.; LIU, C.T.; HEGE, E.K. 1995, Spectroscopic and Morphological Evidence that IRAS FSC 10214+4724 is a Gravitational Lens, *Ap. J. (Letters)*, **452**, pp. L9-12.

COHEN, J. G. 1991, In *Maximum Entropy and Bayesian Methods*, ed.s W. T. Grady, Jr. and L. H. Schick. Kluwer Academic Publishers

CONRAD, M. 1974, Evolutionary Learning Circuits, *J. Theo. Bio.*, **46**, pp. 167-188.

CONRAD, M. 1984, Microscopic-Macroscopic Interface in Biological Information Processing, *BioSystems*, **13**, pp. 303-320.

CONRAD, M. 1988, Prolegomena to Evolutionary Programming, in *Advances in Cognitive Science: Steps toward Convergence*, eds. M. Kochen and H. M. Hastings, AAAS Selected Symposium, pp. 150-168.

CORNWELL, T. J., AND PERLEY, R. A., EDS. 1991, *Radio Interferometry: Theory, Techniques, and Applications*, (San Francisco: Astronomical Society of the Pacific)

DIPLAS, A., BEAVER, E. A., BLANCO, P. R., PIÑA, R. K., AND PUETTER, R. C. 1993, Application of Pixon Based Restoration to HST Spectra and Comparison to the Richardson-Lucy and Jansson Algorithms, *The Restoration of HST Images and Spectra–II*, ed.s R. J. Hanisch and R. L. White, pp. 272-276. Space Telescope Science Institute.

DIXON, D. D., TUMER, O. T., KURFESS, J. D., PURCELL, W. R., WHEATON, W. A., PIÑA, R. K., AND PUETTER, R. C. 1996, Pixon-Based Image Reconstruction from OSSE Data, *Ap. J.*, submitted.

EISEN HADRT, P. R., ARMUS, L., HOGG, D. W., SOIFER, B. T., NEUGEBAUER, G., AND WERNER, M. W. 1996, HST Observations of the Luminous IRAS Source FSC 10214+7424:

A Gravitaionally Lensed Infrared Quasar, *Ap. J.*, in press.

FEINSTEIN, A. 1958, *Foundations of Information Theory*. McGraw-Hill.

FOGEL, D. B. 1995, *Evolutionary Computation*, IEEE Press.

FRASER, A. S. 1957a, Simulation of Genetic Systems by Automatic Digital Computers. I. Introduction, *Australian J. of Biol. Sci.*, **10**, pp. 484-491.

FRASER, A. S. 1957b, Simulation of Genetic Systems by Automatic Digital Computers. I. Effects of Linkage on Rates of Advance Under Selection, *Australian J. of Biol. Sci.*, **10**, pp. 492-499.

FRASER, A. S. 1968, The Evolution of Purposive Behavior, in *Purposive Systems*, eds. H. von Foerster, J. D. White, L. J. Peterson, and J. K. Russell, pp. 15-23, Spartan Books.

GERMAN, S., AND GERMAN, D. 1984, Stochastic Relaxation, Gibbs Distributions, and the Bayesian Restoration of Images, *IEEE Trans. Pat. Anal. and Mach. Intel.*, **PAMI-6(6)**, pp. 721-741.

GHEZ, A. M., NEUGEBAUER, G., AND MATHEWS, K. 1993, The Multiplicity of T Tauri Stars in the Star Forming Regions Taurus-Auriga and Ophiychus-Scorpius: A 2.2 Micron Speckle Imaging Survey *A. J.***106**, 2005-2023.

GRAHAM, J.R.; LIU, M.C. 1995, High-Resolution Infrared Imaging of FSC 10214+4724: Evidence for Gravitational Lensing, *Ap. J. (Letters)*, **449**, pp. L29-32.

GULL, S. F., AND SKILLING, J. 1991, *MemSys5 Quantified Maximum Entropy User's Manual*.

HALL, P. 1990, Akaike's Information Criterion and Kullback-Leibler Loss for Histogram Estimation, *Prob. Th. and Rel. Fields* **85**, pp. 449-467.

HARDLE, W. 1990, *Applied Non-Parametric Regression*, Econometric Society Monographs, Cambridge University Press.

HEDEN, H. 1975, *Non-Parametric Emperical Bayes Estimation*, Arboga.

HOBSON, A. 1971, *Concepts in Statistical Mechanics*. Gordon and Breach.

HOLLAND, J. H. 1969, Adaptive Plans Optimal for Payoff-Only Environments, in *Proc. of the Second Hawaii Intern. Conf. on Sys. Sciences*, pp. 917-920.

HOLLAND, J. H. 1975, *Adaptation in Natural and Artificial Systems*, University of Michigan Press.

JAYNES, E. T. 1957, Information Theory and Statistical Mechanics, *Phys. Rev.***106**, pp. 171-190.

JANES, E. T. 1963, Information Theory and Statistical Mechanics, In *Statistical Physics*, ed. K. W. Ford, pp. 181-218. W. A. Benjamin, Inc.

JONES, R. 1983, *Holographic and speckle Interferometry: A Discussion of the Theory, Practice and Application of the Techniques*. Cambridge University Press.

KAMPFNER, R. P., AND CONRAD, M. 1983, Computational Modeling of Evolutionary Learning Processes in the Brain, *Bull. Math. Bio.*, **45**, pp. 931-968.

KHINCHIN, A. I. 1957, *Mathematical Foundations of Information Theory*. Dover.

KIKUCHI, R., AND SOFFER, B. H. 1976, In *Image Analysis and Evaluation*, Society of Photographic Scientists and Engineers, Toronto, Canada, July 1976, 95.

KIRKPATRICK, S., GELATT, C. D., AND VECCHI, M. P. 1983, *Science*, **220**, pp. 671-680.

KIRKPATRICK, S. 1984, *J. Stat. Phys.*, **34**, pp. 975-986.

KITTEL, C. 1969, *Thermal Physics*. Wiley.

KOLMOGOROV, A. N. 1965, *Inf. Transmission* 1, p. 3.

KORESKO, C. D., HARVEY, P. M., CURRAN, D., AND PUETTER, R. C. 1995, Pixon Deconvolution of Far-Infrared Images from the UT Multichannel Photometer, Proc. Airborne Astronomy Symp. on the Galactic Ecosystem: From Gas to Stars to Dust, *A.S.P. Conference Series*, **73**, pp. 275-278.

KOSUGI, T., MAKISHIMA, K., MURAKAMI, T., SAKAO, T., AND OTHERS 1991, The Hard X-Ray Telescope (HXT) for the Solar-A Mission, *Solar Phys.*, **136**, pp. 17-36.

MACKAY, D. J. C. 1994 in *Models of Neural Networks III*, eds. E. Domany, J. L. van Hemmen, and K. Schutten, Chapter 6, Springer-Verlag.

METCALF, T. R., HUDSON, H. S., KOSUGI, T., PUETTER, R. C., AND PIÑA, R. K. 1996,

Pixon-Based Multiresolution Image Reconstruction for Yohkoh's Hard X-Ray Telescope, *Ap. J.* in press.

NARAYAN, R. AND NITYANANDA, R. 1986, Maximum Entropy Image Restoration in Astronomy, *Ann. Rev. Astron. & Astrophys.* **24**, 127.

PIÑA, R.K., AND PUETTER, R.C. 1993, Bayesian Image Reconstruction: The Pixon and Optimal Image Modeling *P.A.S.P.* **105**, pp. 630-637.

PRASAD, C. V. V., AND BERNATH, P. F. 1994, Fourier Transform Spactroscopy of the Swan (D(3)PI(G)-A(3)PI(U)) System of the Jet-Cooled C2 Molecule, *A. J.* **426**, 812-821.

PRESS, W. H., TEUKOLSKY, S. A., VETTERLING, W. T., AND FLANNERY, B. P. 1994, *Numerical Recipes in C: The Art of Scientific Computing, Second Edition.* Cambridge University Press.

PUETTER, R.C., AND PIÑA, R.K. 1993a, The Pixon and Bayesian Image Reconstruction, *Proc. S.P.I.E.* **1946**, pp. 405-416.

PUETTER, R.C., AND PIÑA, R.K. 1993b, Pixon-Based Image Reconstruction, *Proc. MaxEnt '93*, in press.

PUETTER, R.C., AND PIÑA, R.K. 1994, Beyond Maximum Entropy: Pixon-Based Image Reconstruction, *Science with High Spatial Resolution Far-IR Data*, ed/s S. Terebey and J. Mazzarella, JPL Pub. 94-5, pp. 61-68.

PUETTER, R.C. 1994, Pixons and Bayesian Image Reconstruction, *Proc. S.P.I.E.* **2302**, pp. 112-131.

PUETTER, R. C., 1995, Pixon-Based Multiresolution Image Reconstruction and the Quantification of Picture Information, *Int. J. of Imaging Sys. and Tech.*, **6**, pp. 314-331.

PUETTER, R.C. 1995, Pixon-Based Multiresolution Image Reconstruction and the Quantification of Picture Information Content, *Int. J. Image Sys. & Tech.*, in press.

PUETTER, R. C. 1996, Pixon-Based Multiresolution Image Reconstruction and Quantification of Image Information Content, *Proc. MaxEnt '95*, in press.

RABBANI, M., AND JONES, P. W. 1991, *Digital Image Compression Techniques.* SPIE Optical Engineering Press.

RICE, W. 1993, An Atlas of High-Resolution IRAS Maps of Nearby Galaxies, *A. J.*, **105**, pp. 67-96.

RIDGWAY, S. T., AND BRAULT, J. W. 1984, Astronomical Fourier Transform Spectroscopy Revisited, *Ann. Rev. Astron. & Astrophys.* **22**, 291-317.

SHANNON, C. E. 1948, A Mathematical Theory of Communication, *Bell Systems Tecnical Journal* **27**, 379.

SHANNON, C. E., AND WEAVER, W. 1949, *The Mathematical Theory of Communication.* University of Illinois Press.

SERABYN, E., AND WEISSTEIN, E. W. 1995, Fourier Transform Spectroscopy of the Orion Molecular Cloud Core, *Ap. J.* **451**, 238-251.

SIBISIS, S. 1990 in *Maximum Entropy and Bayesian Methods*, ed. J. Skilling. Kluwer Academic Publishers.

SKILLING, J. 1989, Classic Maximum Entropy, In *Maximum Entropy and Bayesian Methods*, ed. J. Skilling, p. 45. Kluwer Academic Publishers.

SMITH, C. H., AITKEN, D. K., MOORE, T. J. T., ROCHE, P. F., PUETTER, R. C., AND PIÑA, R. K. 1994, Mid-infrared Studies of η Carinae–I: Sub-Arcsecond Imaging at 12.5 and 17μm, *M.N.R.A.S.*, **273**, pp. 354-358.

SOLOMONOFF, R. 1964, *Inf. Control* **7**, p. 1.

SPRENT, P. 1981, *Quick Statistics: An Introduction to Non-Parametric Methods*, Penguin Books.

THOMPSON, A. R., MORAN, J. M., AND SWENSON, G. W. JR. 1986, *Interferometry and Synthesis in Radio Astronomy.* Wiley.

VAN DE HULST, J. M., KENNICUTT, R. C., CRANE, P. C., AND ROTS, A. H. 1988, Radio Properties and Extinction of the H II Regions in M51, *Astron. and Astrophys.*, **195**, pp.

38-52.

VERES, S. M. 1990, Relations Between Information Criteria for Model-Structure Selection. Part3. Strong Consistency of the Predictive Leaast Squares Criterion, *Int. J. Control* **52**, p. 737.

WEIR, N. 1991, Applications of Maximum Entropy Techniques to HST Data, in Proc. of the ESO/ST-ECF Data Analysis Workshop, April 1991, ed.s P. Grosbo and R. H. Warmels, p. 115. ESO

WEIR, N. 1993, private communication.

WEIR, N. 1994, A Maximum Entropy-Based Model for Reconstructing Distributions with Correlations at Multiple Scales, *J. Opt. Soc. Am.*, in press.

WEIR, N. AND DJORGOVSKY, S. 1990, MEM: New Techniques, Applications, and Photometry, in *The Restoration of HST Images and Spectra*, Proc. of a Workshop at the Space Telescope Science Institute, 20-21 August 1990, Baltimore, ML, ed.s R. L. White and R. J. Allen, pp. 31-38.

WICKEN, J. 1987, Entropy and Information: Suggestions for a Common Language, *Philos. Sci.* **54**, pp. 176-193.

WOHLLEBEN, R., MATTES, H., AND KRICHBAUM, TH. 1991, *Interferometry in Radio Astronomy and Radar tecniques*, (Dordrecht; Boston: Kluwer Academic Publishers).

Spectroscopic Techniques for Large Optical/IR Telescopes

By KEITH TAYLOR[1],

[1]Anglo-Australian Observatory, PO Box 296, Epping, NSW 2121, Australia

This series of five lectures, given at the December, 1995, IAC Winter School, will cover the following topics in the subject of ground-based, optical/NIR, astronomical spectroscopy:

◇ **BASIC PRINCIPLES of ASTRONOMICAL SPECTROSCOPY**

◇ **DIFFRACTION GRATINGS**

◇ **MULTI-OBJECT SPECTROSCOPY**

◇ **2dF - A MULTI-FIBRE CASE STUDY**

◇ **FABRY-PEROTS AND FOURIER TRANSFORM SPECTROMETERS**

◇ **INTEGRAL FIELD AND HYBRID SYSTEMS**

---

1. Basic Principles Of Astronomical Spectroscopy

1.1. *Fundamental Principles*

Figure 1 represents an entirely generalized astronomical spectrograph receiving light from a telescope and dispersing it onto an area detector. The symbols are defined as follows:

- $\mathcal{A} \equiv$ telescope collecting area ($\mathcal{D} \equiv$ diameter)
- $\Omega \equiv$ solid angle subtended by telescope aperture (angle $\equiv \phi$)
- $a \equiv$ beam area of collimator ($d \equiv$ diameter)
- $\Omega_1 \equiv$ acceptance solid angle of spectrograph (angle $\equiv \beta$)
- $f \equiv$ focal-ratio of telescope
- $f_1 \equiv$ focal-ratio of spectrograph camera
- $dp \equiv$ pixel-size of detector (linear dimensions: l_x-by-l_y)

The wavelength resolving power ($\mathcal{R}$) of an astronomical spectrograph can be defined as:

$$\mathcal{R} = \frac{\lambda}{d\lambda} \tag{1.1}$$

Furthermore, the *information* flux, or $\mathcal{E}$tendue, through any optical system (e.g., telescope through spectrograph) is conserved and given by:

$$\mathcal{A}\Omega = a\Omega_1 \quad (or \ \mathcal{D}\phi = d\beta) \tag{1.2}$$

If the source has a surface brightness B (ergs^{-1}s^{-1}cm^{-2}sterad^{-1}) then the flux gathered by the spectrograph is given by $\epsilon B A \Omega$ (ergs s^{-1})) where ϵ = spectrograph efficiency.

- i.e.: $A\Omega$ = "**Information Throughput**"
- and: $\mathcal{L}$**uminosité** = $\epsilon A \Omega$ $(= \mathcal{L})$
- and: "**Information Content**" = $\mathcal{L}$uminosity $\times$ $\mathcal{R}$esolving Power
- or: "$\mathcal{L}.\mathcal{R}$-**product** = $\epsilon A \Omega \mathcal{R}$ (a general *figure-of-merit*)

Now, the scale at the input aperture is ϕ/fD while at the detector it is given by $\phi/f_1 D$. The magnification $= f_1/f$ = the camera-collimator ratio (f_1/f determines the pixel scale).

Note that $A\Omega$ implies circular apertures.

The advent of area detectors permits spectrograph slit apertures giving, simultaneously, 1D of spatial information and 1D of spectral information.

Now a given pixel-size, dp, is given by:

$$dp = df_1 \, d\beta = D f_1 \phi \tag{1.3}$$

While spectral ; spatial multiplexes are given by:

$$\mathcal{M}_x = l_x/(2dp) \quad ; \quad \mathcal{M}_y = l_x/(2dp) \tag{1.4}$$

Thus **Information Capacity** (or *Figure-of-Merit*) is given by:

$$\epsilon A \Omega \mathcal{R} \mathcal{M}_x \mathcal{M}_y \tag{1.5}$$

where:
- $A\,\Omega$ = $\mathcal{E}$ntendue
- $\epsilon A\,\Omega$ = $\mathcal{L}$uminosité
- $\epsilon A\,\Omega \mathcal{R}$ = $\mathcal{L}\mathcal{R}$-product

Thus:

$$(\mathcal{L}\mathcal{R})\mathcal{M}_x \mathcal{M}_y = \epsilon A \Omega \mathcal{R} \mathcal{M}_x \mathcal{M}_y \tag{1.6}$$

Now substituting equations (1.1) to (1.4) in (1.6) gives:

$$(\mathcal{L}\mathcal{R})\mathcal{M}_x \mathcal{M}_y = \epsilon \lambda (\frac{\pi}{4})^2 D\phi \frac{d\beta}{d\lambda} d \mathcal{M}_x \mathcal{M}_y \tag{1.7}$$

$$= \epsilon \lambda (\frac{\pi}{4})^2 D\phi \frac{dp}{d\lambda} \frac{1}{f_1} \mathcal{M}_x \mathcal{M}_y \tag{1.8}$$

$$= \epsilon (\frac{\pi}{4})^2 \mathcal{R} \frac{l_x l_y}{4} \frac{1}{f_1^2} \tag{1.9}$$

Note however that there is no dependency of *Information Capacity* on D or on d! Instead it depends purely on the camera f-ratio, f_1, and the size of the detector, $l_x l_y$.

So ... why build large telescopes ? or large spectrographs ?

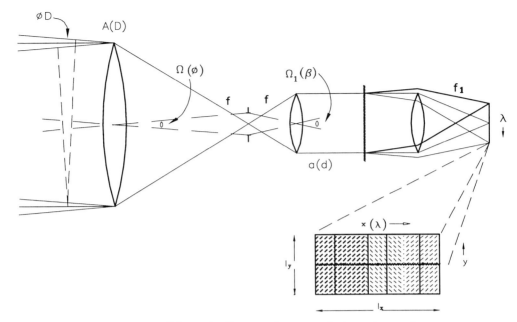

FIGURE 1. Schematic diagram of astronomical spectrograph

1.2. *Practical Constraints*

The following are the *general* rule for modern astronomical spectrographs:
  (*a*) Input Aperture: $\phi_x \leq 1\overset{\wedge}{}$ (generally);
  (*b*) Pixel-size: $dp \sim 20\mu$m;
  (*c*) Camera f-ratio: $f_1 \geq 1$
  Constraints (a)&(b) combined with equation (1.3) immediately imply $f_1 \mathcal{D} \sim 8$ metres (assuming *Nyquist* – 2-pixel – sampling).

Large telescopes are required to give spatial resolution
NOT $\mathcal{L}$uminosité !

1.3. *The Large Telescope Game*

Once $\mathcal{D} > 4$ metres, we need either (or both) $f_1 < 2$ which is increasingly difficult and $\phi < 1\overset{\wedge}{}$ which implies adaptive optics

Dispersing devices supply an intrinsic angular dispersion, $d\beta/d\lambda$; hence if $\mathcal{D}$ increases, from equation (1.2), d increases proportionately. But, from equation (1.3), to keep the same pixel matching requires that f_1 decreases proportionately. Thus larger telescopes imply larger spectrographs and/or faster cameras; hence instrumentation costs and complexity goes non-linear!

2. Diffraction Gratings

The simplest realization of a reflection grating is shown in a Littrow configuration in Figure 2, where:
  • the incident angle is normal to the grating facets;
  • the reflected rays leave the grating at an angle, $d\beta$, from the facet normal;

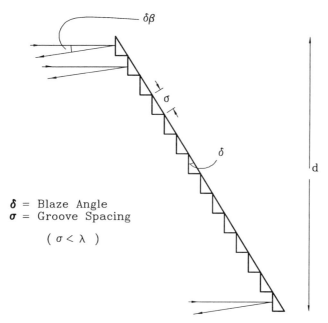

FIGURE 2. Reflection grating in a Littrow configuration

- δ is the blaze angle;
- σ is the groove spacing (where: $\sigma < \lambda$)
- d is *depth* of grating collimated beam.

Now for constructive interference:

$$\text{if } d\beta = 0 \ : \ then \ \ 2\sin\delta = \frac{m\lambda}{\sigma} \tag{2.10}$$

Differentiating equation (2.10) gives the angular dispersion:

$$\frac{d\beta}{d\lambda} = \frac{m}{\sigma\cos\delta} \tag{2.11}$$

where m is the order of interference.

Now, from equation (2.11):

$$\mathcal{R} = \frac{\lambda}{d\lambda} = \frac{\lambda m}{\sigma\cos\delta\, d\beta} \tag{2.12}$$

which from equation (1.3) gives:

$$\mathcal{R} = \frac{2d\tan\delta}{\phi D} = \frac{\text{"Delay" across grating}}{\text{"Delay" across incident wavefront}} \tag{2.13}$$

Diffraction limited resolution is obtained when $\phi D = \lambda$ and hence:

$$\mathcal{R}_{\mathrm{D}} = \frac{2d\tan\delta}{\lambda} \tag{2.14}$$

which from equation (1.2) gives:

$$\mathcal{R}_{\mathrm{D}} = \frac{md}{\sigma \cos \delta} \tag{2.15}$$

But: $d/(\sigma \cos \beta) = \mathcal{N} = $ No. of "recombining beams"; thus:

$$\mathcal{R}_{\mathrm{D}} = m\mathcal{N} \tag{2.16}$$

(A general *interferometric* result - *cf:* FPs and FTSs.)

2.1. *Practical Considerations*

Practical constraints usually require:
- the spectrograph to be **slit**-limited (i.e.: equation (2.13) not (2.16) applies);
- separate camera and collimator f-ratios for slit and pixel matching (i.e.: non-Littrow) which further requires the control of slit magnification (i.e.: $f_1 \neq f$) and the need to extract light beam from grating.

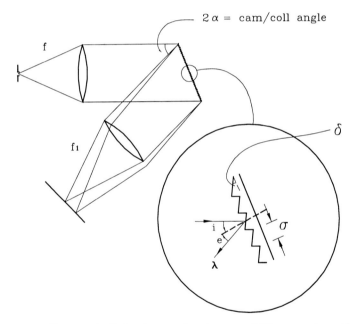

FIGURE 3. Reflection grating in a non-Littrow configuration

The schematic geometry for a non-Littrow reflection grating spectrograph is shown in Figure 3 where:
- α is the semi-angle between the collimator and the camera;
- i and e are, respectively, the incident and emergent angles as measured from the grating normal (with the sign sense preserved).

Now the general equation for constructive interference becomes:

$$\sigma(\sin i + \sin e) = m\lambda \tag{2.17}$$

This is generally referred to as the **GRATING EQUATION**.

So if $i = \delta = e$ and $m = 1$ then the *Blaze wavelength*, λ_0, is given by:

$$\lambda_0 = 2\sigma \sin \delta \qquad (2.18)$$

(NB: the sign of i and e are always in reference to the grating normal with the sign sense preserved.)

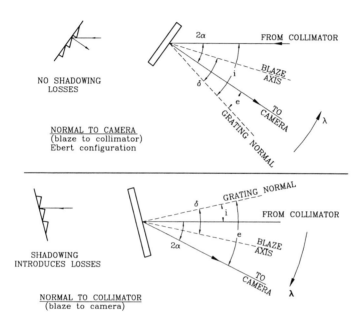

FIGURE 4. Ebert and non-Ebert grating configurations

Note that there are 2 cases to consider; blaze to camera and blaze to collimator. The two configurations are shown diagrammatically in Figure 4, where α is the semi-angle between the collimator and camera axes. For both configurations, the *grating equation*, (2.17), becomes;

$$\frac{m\lambda}{\sigma} = 2\sin(\frac{i+e}{2})\cos(\frac{i-e}{2}) \qquad (2.19)$$

On the camera axis, $|i - e| = 2\alpha$ and hence at the blaze wavelength (λ_c), $(i + e) = 2\delta$.

Therefore:

$$\frac{m\lambda_c}{\sigma} = 2\sin \delta \cos \alpha \qquad (2.20)$$

or, from equation (2.18):

$$\lambda_c = \lambda_0 \cos \alpha \qquad (2.21)$$

(i.e.: The blaze wavelength is always blueward of the Littrow condition.)

Now, differentiating the *grating equation*, (2.17), dispersion at blaze wavelength becomes:

$$\frac{de}{d\lambda} = \frac{\sin i + \sin e}{\lambda \cos e} = \frac{2 \sin \delta \cos \alpha}{\lambda_c \cos e} \qquad (2.22)$$

Angular dispersion increases with e and hence is greater for the non-Ebert condition. However the *realized* spectral resolution $(dp/d\lambda)$ is not necessarily smaller.

2.2. Grating Anamorphism

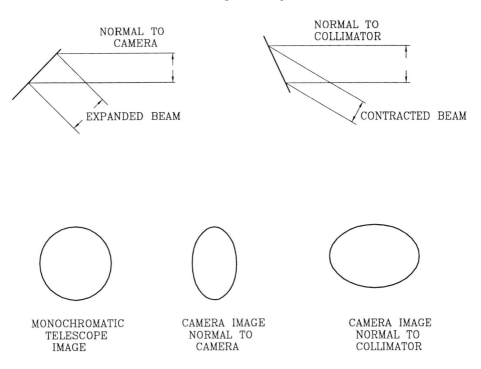

FIGURE 5. Anamorphism for non-Littrow configurations

As is shown in Figure 5, in the Ebert (normal to camera) configuration the incident beam size is expanded by reflection off the grating while in the non-Ebert (normal to collimator) configuration the beam is contracted.

Thus from the grating equation (2.17):

$$-\frac{di}{de} = +\frac{\cos e}{\cos i} = anamorphic\ factor \qquad (2.23)$$

With the grating in the Ebert configuration, we have $e < i$ so the anamorphism is >1, a condition which also maximizes wavelength coverage. Although the dispersion is less the resolution is maintained since the projected slit width is narrower. However the Ebert configuration may give under sampled spectra and, if the camera optics cannot accept the full, expanded, beam there also may be some light loss.

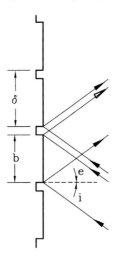

FIGURE 6. Schematic diagram of the blaze effect

2.3. *The Blaze Function*

How much light gets distributed into the various orders m of a grating fundamentally determines spectrograph efficiency. An exact treatment of this effect has to resort to Maxwell's equations however we will simplify matters here by using a *scalar* approximation which neglects the important polarization effects.

Using the formalism defined with reference to Figure 6, simple diffraction theory gives:

$$\mathcal{I} = (\frac{\sin^2 \mathcal{N}\gamma'}{\sin^2 \gamma'})(\frac{\sin^2 \gamma}{\gamma^2}) \qquad (2.24)$$

or

$$\mathcal{I} = (\mathcal{IF} : \mathcal{I}\text{nterference } \mathcal{F}\text{unction})(\mathcal{BF} : \mathcal{B}\text{laze } \mathcal{F}\text{unction}) \qquad (2.25)$$

where:
- γ' = phase difference between facets;
- γ = phase difference across facets;
- $\mathcal{N}$ = No. of recombining beams.

Thus:

$$\gamma' = \frac{\pi \sigma}{\lambda}(\sin i + \sin e) \qquad (2.26)$$

and:

$$\gamma = \frac{\pi b}{\lambda}(\sin i + \sin e) \qquad (2.27)$$

with equation (2.26) being the *grating equation* with $\gamma' = m\pi$.

Note that the blaze function ($\mathcal{BF}$) is maximum when $i = -e$ (or $m = 0$) – equivalent to specular reflection.

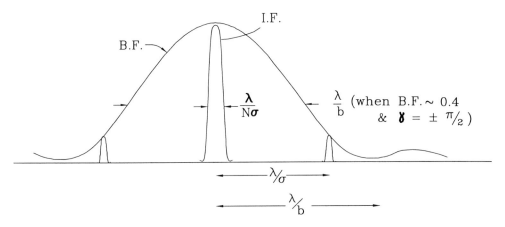

FIGURE 7. Intensity distribution through grating orders

2.4. *Grating Instrumentation Profile*

With reference to Figure 7 and equations (2.26) and (2.27):
- FSR, the free-spectral-range $= \lambda/\sigma$
- $\mathcal{F}$inesse $= \mathcal{N} = d\lambda$

This formalism is identical to that of the Fabry-Perot, but with most of the light being generated at the $m = 0$, zeroth order, where, of course, it is undispersed and hence useless.

To be of any use, therefore, the blaze function has to be shifted so that it peaks at an integer $m \neq 0$. This requires modifying the phase delay, *gamma* by tilting the blaze facets and thus producing a **blaze** grating as shown in Figure 8 where:
- θ is the angle between the incident ray and the facet normal (FN);
- the incident (i) and emergent (e) rays are, as usual, defined with respect to the grating normal (GN)

Here:

$$\gamma = \frac{\pi\sigma\cos\delta}{\lambda}[\sin(i - \delta) + \sin(e - \delta)] \tag{2.28}$$

where: $\gamma = 0$ when $2\delta = (i + e)$

Hence from the *grating equation*, (2.17):

$$m\lambda_0 = 2\sigma\sin\delta\cos\theta \tag{2.29}$$

where λ_0 is the **blaze** wavelength.

2.5. *Shape of the Blaze Function*

Figure 9 demonstrates the changing blaze structure with order, m, as a function of both angle $(i + e)$ and wavelength, λ. If ϵ defines a small change in the *emergent* angle, e_0, at blaze, as shown in Figure 9, then:

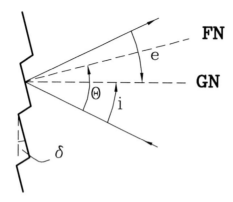

FIGURE 8. Tilting the facets to create a blazed grating.

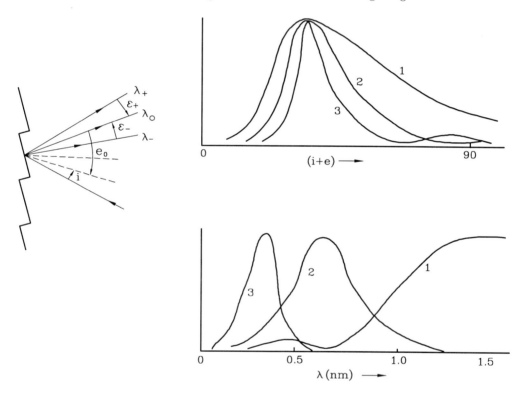

FIGURE 9. The blaze facets and corresponding blaze function

$$m\lambda_\pm = \sigma \sin i + \sigma \sin(e_0 \pm \epsilon_\pm) \qquad (2.30)$$

If we now choose $\epsilon_\pm$ such that the blaze function, $\mathcal{BF} \sim 0.4$, then:

$$\epsilon_\pm = \frac{\lambda_\pm}{2\sigma \cos \delta} \qquad (2.31)$$

GRISM

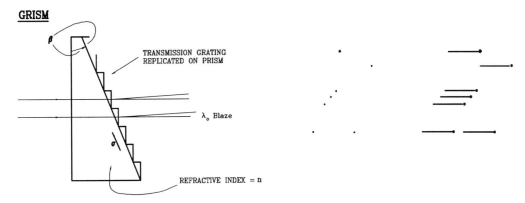

FIGURE 10. A transmission grism showing the undeviated path of the blaze wavelength. A field of objects (stars) as imaged by the focal-reducer is transformed into spectra by the inclusion of the grism in the collimated beam.

and:

$$\lambda_{\pm} \sim \frac{m\lambda_0}{\left(m \mp \frac{1}{2}\right)} \tag{2.32}$$

2.6. *Problems with non-Littrow Reflection Gratings*

There are two problems which exacerbate spectrograph camera design in non-Littrow configurations:

(*a*) Beam dilation: the Camera-Collimator angle (2α) creates variable beam dilation, which implies that the camera cannot be optimized for all grating angles.

(*b*) Pupil imagery: since the grating represents the last surface of diffraction it forms its own system pupil outside the physical limits of the camera. This requires that the camera be substantially *faster* than its nominal f-ratio.

In special circumstances, these can be alleviated. Two *partial* solutions may be adopted.

2.7. *Grisms*

Grisms, as their name suggests, are combinations of gratings and prisms such that a transmission grating is replicated onto the prism hypotenuse (prism angle, β), generally, so that the blaze wavelength (λ_0) is undeviated. The advantage of the grism is that it can be used as a dispersive element in a simple focal-reducer camera, where the camera-collimator angle $\alpha = 0$, and gives spectra anywhere in its field of view (FoV).

In reference to Figure 10 and assuming the refractive index of the replication resin, n, is the same as that of the prism, constructive interference requires:

$$m\lambda = \sigma(n-1)\sin\beta \tag{2.33}$$

It is readily shown also that the angular dispersion is given by:

$$\frac{d\beta}{d\lambda} = \frac{m}{\sigma\cos\beta} \tag{2.34}$$

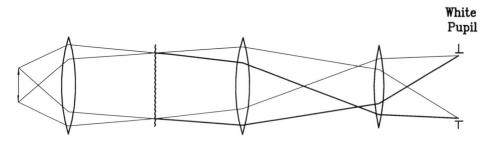

FIGURE 11. A schematic, fully transmitting, Baranne optical train, showing the re-imagery of the pupil formed by the grating. This *white-pupil* is located within the final camera optics which relay the intermediate spectral image onto the detector.

and comparing equation (2.34) with equation (2.11) gives the following relationship:

$$\frac{\beta_{GRISM}}{\beta_{GRATING}} = \frac{2}{(n-1)} \sim 4 \qquad (2.35)$$

thus demonstrating that a given grating ruling will give much higher dispersions in reflection than its equivalent transmission grism counterpart. (Note: For $\beta \sim 45°; \to \sigma \sim 1.4$ – i.e.: 700 groves/mm.)

2.8. *White-Pupil Spectrograph*

In order to avoid both the beam dilation effects of the non-Littrow configuration while alleviating the camera design problems caused by the external pupil formed by the grating, a *white-pupil* (Baranne 1988) design can be invoked. In a white-pupil design, an intermediate spectral image is formed, which is subsequently relayed by auxiliary optics to a final detected image (see Figure 11). This has a number of benefits:

(a) Geometrical constraints no longer demand severely non-Littrow camera-collimator angles, thus alleviating beam dilation effects;

(b) The relay optics can be designed to re-image the grating (hence *white*) pupil to a position within the final imaging camera to permit optimal aberration balance;

(c) the size of the final, *white*, pupil can be chosen arbitrarily to ease camera design and, in the case of a catadioptric systems, to minimize vignetting.

Of course, the over-riding disadvantage of the *white-pupil* design is that it entails more optics and hence greater absorption/reflection losses than a standard reflection grating configuration - it also is more prone to scattered light. Nevertheless a thorough evaluation of the relevant trade-offs in throughput, imagery and versatility are required before this fundamental design decision can be made.

3. Multi-object Spectroscopy

Spectra are 1-dimensional whereas detectors are now generally 2-dimensional. This bald fact, combined with

• the ever increasing ambitions of astronomers to gather information at the faintest limits, and,

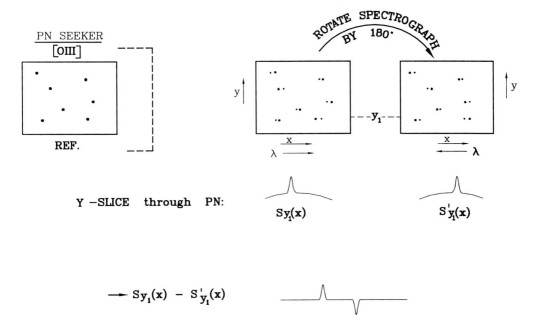

FIGURE 12. On the left is shown, schematically, a narrow-band image. The pair of dispersed images $(S(x[\lambda], y)$ & $S'(x[-\lambda], y))$ with the grating rotated by 180° are shown on the right. Note the change in wavelength (λ) direction. The pair of Y-profiles $(S_{y_1}(x)$ & $S'_{y_1}(x))$ through a PN location are displayed below the respective images. Differencing these two profiles gives the location (X, Y) flux and relative velocity $(\Delta\lambda[\Delta v])$ of the PN.

- the FoV of telescopes being generally much larger than the spectrograph slits that are being fed by the telescope,
  drives spectrograph usage (and hence design) towards **multi-object** spectroscopy.

While individual spectra of astronomical objects can be used for a multitude of *analytic* purposes (determination of distance, composition, physical conditions, dynamics, mass, etc.) the drive for *statistical* spectroscopic information, to interface with imaging surveys, both ground- and satellite-based, in a large variety of wave-bands, requires that astronomical spectroscopy evolves efficient multi-object strategies.

3.1. *Slitless Spectroscopy*

In principle, point-like (stars, compact galactic and extra-galactic) objects can be observed without the use of a spectrograph slit to define their input aperture. In this case, the wavelength resolution, $d\lambda$, is defined by the size of the object as projected onto the detector. Nevertheless, a dispersive element within an imaging system can, in principle, give spectra of **all** objects within the FoV and hence represents, under special circumstances, an enormously powerful multi-object spectroscopic facility.

A classic example of slit-less spectroscopy can be found on the UK Schmidt Telescope, whereby a full aperture, objective prism is placed in front of the telescope aperture. In this way spectra of all objects (perhaps $10^5 - 10^6$ above the detected flux limit) can be obtained within the 6° FoV of the telescope. There are, however, severe limitations to this type of technique.

(*a*) while the light from point-like objects is dispersed, the light from the back-ground sky is unchanged. This effect critically decreases the contrast between sky and dispersed object and places severe constraints on sky-limited observations. This feature emphasise the values of using slits, which minimize the superimposed sky flux in the sky-background limited regime.

(*b*) given that the detector is finite and has to cover both spectral and imaging information, there is always a trade off between FoV and length of spectra. In the case of the UK Schmidt Objective Prism surveys, the dispersion is very low (<1000Å/mm) and hence only spectral classification, and crude red-shift determination, can be achieved.

(*c*) spectra can (and do) overlap in the wavelength direction. Generally, as the magnitude limit increases the surface density of targets increases giving rise to serious spectral confusion. In principle this can be alleviated by rotating the objective prism by 90°.

3.2. *Planetary Nebula in* $\mathcal{E}$-*Galaxies - a special case*

While slitless spectroscopy is not a technique in the mainstream of observational astronomy and is limited generally to obtaining ultra-low dispersion spectra of relatively bright objects, it is a technique which can, under special circumstances, be used to great effect. A particular example is in the detection and kinematic mapping of Planetary Nebulæ (PNe) in elliptical galaxies. The motivation behind this PN work is two-fold: firstly PNe act as very effective *standard candles* permitting distances of the host ellipticals to be obtained; secondly the PNe act as dynamical probes which can be used to trace the mass and shape of the elliptical out to unprecedented distances from the nucleus.

What distinguishes these PNe from the diffuse back-ground light of their parent elliptical galaxy is:
• their point-like morphology, and
• their characteristically pure emission-line spectrum

Traditionally, the emission-line spectrum of PNe has been used to distinguish them from the background sky and stellar continuum. The [OIII], $\lambda5007$ line is generally the strongest and has a wavelength which is well placed for CCD observations. The strategy has been to take on- and off-band images through interference filters to look for candidate PNe. This method suffers from a number of problems, however:
• Obtaining reliable fluxes from narrow-band images is fraught with difficulties. In principle the temperature dependence of the interference filter and the detailed shape of its bandpass as a function of field position, and the radial velocity of the detected PNe, must be known before accurate photometry can be achieved. In practice, simple flat fielding is used in the expectation that these effects are not dominant. For similar reasons, continuum subtraction of the off-band image is also unreliable;
• Variations in seeing and source position between the on- and off-band images degrade the sensitivity of the technique;
• For optimum contrast against the sky and galaxy background continuum, the on-band interference filter bandwidth should be matched to the velocity dispersion of the galaxy (~10–15Å). For safety's sake however the bandpass of the filter is usually made large enough (~30–40Å) that variations in radial velocity of the PNe do not induce significant throughput variations within the filter;
• Follow-up fibre spectroscopy is needed to determine the PNe kinematics, and this demands very accurate astrometry. The logistics of locating and then using the PNe for dynamical studies may take two seasons of observations.

There is, however, a slitless spectroscopic method which can yield the position, flux

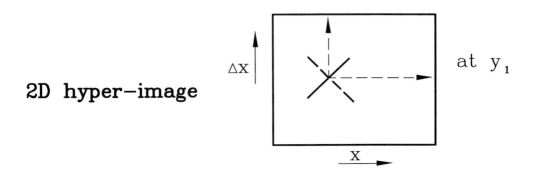

FIGURE 13. The 2D *hyper*-image, $\chi(x, \Delta x)$, at y_1 showing the positive/negative-going *cross* which defines the (x, y, v) location of the detected PN.

and velocity of the PNe in a single experiment. Given the same on-band interference filter and instrumental efficiency, the discrimination between continuum background and emission-line source is the same as with direct narrow-band imaging. The PNe simply appears as a *seeing-disk limited* emission spike above a dispersed continuum. The position of the spike is a function of both location in the field and radial velocity about the galaxy and hence some other method has to be employed to separate these two parameters. The method used has been to combine the slitless image with a direct image (sometimes taken through the spectrograph by replacing the grating with a mirror). This, however, still has many of the problems of direct imaging noted earlier.

A simple solution to this problem is to take *two* slitless images with the spectrograph rotated through 180° (in principle any angle will do). Given a *double* detection, one is then able to recover both the location and velocity of each PNe (Taylor & Douglas (1995)). Since the continuum is expected to vary slowly with wavelength compared with the emission lines, detection should be straightforward. Indeed, by differencing the two images the background continuum will, to a certain approximation, be nulled. In principle the detection sensitivity should be limited only by the shot-noise in the background, systematic effects being reduced to a minimum.

The technique is also very robust in regard to fluxing, since stars in the observed field readily supply the bandpass response of the filter enabling the flux of each PNe to be calibrated without the uncertainties inherent in on- and off-band imaging. This allows us to match the bandpass of the interference filter to a nominal maximum galaxy velocity dispersion, reducing background continuum contamination by a large factor.

The procedures for detecting PNe in the image pairs ($\mathcal{S}_{y_1}$ & $\mathcal{S}'_{y_1}$); (see Figure 12) rely on pattern recognition algorithms.

The function $[\mathcal{S}_{y_1}(x) - \mathcal{S}'_{y_1}(x)]$ has a characteristic *P-Cygni*-like signature in a typically back-ground noise dominated image. Now, for each slice at $Y = y_1$ a 2-dimensional *hyper*-image, $\chi(x, \Delta x)$ can be formed such that:

$$\chi(x, \Delta x) = \mathcal{S}_{y_1}(x - \Delta x) - \mathcal{S}'_{y_1}(x + \Delta x) \ ... \ \text{for all } y_1 \qquad (3.36)$$

As demonstrated schematically in Figure 13, each detected PN will give a characteristic orthogonal, positive/negative-going *cross* centred on its x_1-location and at a Δx which

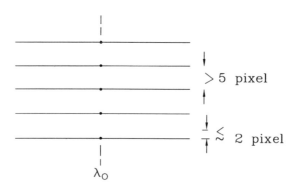

FIGURE 14. A schematic presentation of multi-fibre spectra

uniquely defines its velocity relative to systemic. Hence a 3-dimensional *hyper*-cube can be created from the (S, S') image pairs to give the x, y, v and flux of each detected PN.

3.3. *Multi-slit spectroscopy*

In sky-noise dominated conditions, where sky-subtraction is an imperative, the use of slits is generally recognised to be essential, both:

(a) to reduce sky back-ground to a minimum, given the necessity to also accept light from the object itself;

(b) to optimize sky-subtraction by recording localized sky on adjacent regions of the detector - this also minimizes systematic instrumental effects.

Faint object, spectroscopic sky surveys clearly require some form of wide-field imaging spectrograph in order to observe significant numbers of objects simultaneously, although at the faintness level (B≤23) achievable on 4-metre class telescope at low dispersion (~100Å/mm), the surface density of suitable targets can be quite large (typically ≥100) and hence only moderately wide fields (~5′) may be thought adequate. Over the last decade, or so, many simple, fully transmitting, focal-reducer, imaging spectrographs have been deployed (CRYOCAM (KPNO); LDSS (AAT); LDSS-2 (WHT)) and we are seeing now a new generation of the same basic principles being applied to the 8-10m class telescope (GMOS (Gemini); DEIMOS (Keck-II); FORS (VLT)).

For this type of system, the ability to achieve direct imaging by removing the dispersive element (usually a grism) from the collimated beam gives a powerful combination whereby sky-limited imaging and spectroscopy is facilitated through the same instrument. It is, of course, necessary to manufacture multi-slit masks (through laser or NC milling machines) for each field configuration and this can be organised *on-line* through use of the direct image. Experience with this type of technique has, however, often argued for *off-line* manufacture (photo-chemical etching, for example) given the complexity of the target selection process and hence the advantages of the *on-line* mask manufacturing capabilities can be illusory.

3.4. *Multi-fibre spectroscopy*

Fibres are ideally information re-formatting devices. In principle, a source, anywhere within the field of an optical telescope, can be relayed via a fibre to a point on the input

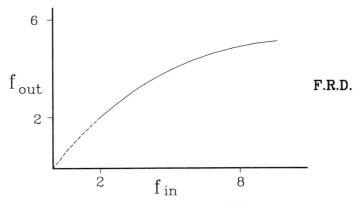

FIGURE 15. A typical fibre FRD curve

slit of a spectrograph and, of course, this can be achieved simultaneously for as many fibres as can be accommodated by such a slit. The process of manufacturing multi-mode, step index fibres used in astronomy is now sufficiently advanced that very low attenuation is achieved for metre-length fibres over a broad wavelength region from the UV to the NIR, while coupling losses, at fast f-ratios are typically $\leq 20\%$.

Furthermore, as is graphically represented in Figure 14, in contrast to the multi-slit case, the multi-fibre spectra can be organized on the detector so that the same wavelength range is achieved in all cases.

However, in order to utilize fibres over a large field of view, one needs a technology to re-position fibres in the focal-plane of the telescope to astrometric accuracy and it is here that difficulties arise. Through the last decade, or so, various schemes have been tried.

(*a*) Plug-plate (Gray (1986)) – this is the simplest solution, whereby holes to locate the fibres are drilled in a field-plate which is then mounted at the telescope focus. This requires a new field-plate per exposure and manual installation of the fibres.

(*b*) Fisherman-around-a-pond – here each fibre is located on its own positioning arm which is independently controlled (in x, y, θ) into position in the focal-plane of the tele-scope.

(*c*) AutoFib (Ellis & Parry (1987)) – taking its name from the original AAT robotic system; this technique uses a pick-and-place robot to configure fibre buttons which are held magnetically to a ferro-magnetic field-plate in the focal-plane of the telescope.

The advantage of a) is that it is cheap, manual and easy to install, however it can never be regarded as a long-term solution. It is, however, the technology chosen by the SLOAN Digital Sky-Survey project. The advantage of b) is that it can address fibre placements simultaneously and hence fibre configuration times are minimized. However it suffers severely in crowded field conditions and can never be contemplated for multiplexes beyond a few 10s of objects. There is little doubt that the AutoFib technique (c), first developed by AAO and later adopted by NOAO, the WHT and finally the AAT's new 2dF facility, is the most flexible scheme so far envisaged. While the re-configuration times can be severe, it does permit robotic, automated, control and can be used over a very wide field and in highly clustered environments. We will return, in Section 4 to the question of how the 2dF facility alleviates the problem of configuration speed.

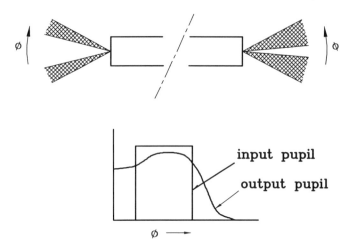

FIGURE 16. FRD in-filling of the telescope central obstruction by a fibre.

3.5. *Fibre characteristics*

For the fibre to act as an optical wave-guide, total-internal reflection has to be maintained throughout the passage of light along the fibre. This is achieved by constructing the fibre in two layers of differing refractive indices. The fibre *core* has a refractive index of n_f while the *cladding* has a refractive index of n_c. Total internal reflection then requires that:

$$\sin \theta_{max} = \frac{\sqrt{n_f^2 - n_c^2}}{n_0} \tag{3.37}$$

where n_0 is the refractive index of the external medium.

Generally, $\theta_{max} \le 40°$ equivalent to an input f-ratio of $\sim f/2.2$. However because of imperfections within the fibre, the output f-ratio is always somewhat faster than the input compromising, to some degree, the $A\Omega$-product of the system. This effect is know as focal-ratio-degradation (or FRD).

Figure 15 demonstrates the FRD effect, showing how fibres retain more of their $A\Omega$-product at faster f-ratios.

However spectrographs are often designed with camera or collimator central obstructions which are designed to match those of the telescope itself. In this case, as is shown in Figure 16, the central obstruction is always at a relatively slow f-ratio so that in-filling is always severe. This fact alone requires fibre spectrographs to be custom designed for optimum performance.

3.6. *Sky-subtraction with fibres*

It is generally believed that sky-subtraction with a multi-slit system is more accurate than with a multi-fibre system, and so to do spectroscopy at the faintest possible levels one should avoid using fibres because they do not work as well. However several workers have made very deep observations with fibres and have achieved sky-limited spectra close to the Poisson limit given from the photon shot-noise (e.g.: Cuby & Mignoli 1994).

So why is there such scepticism about using fibres for ultra-faint spectroscopy ? Basically it is because with fibres one has to take much greater care when making the observations and doing the data-reduction in order to eliminate systematic errors. To achieve good sky-subtraction one has to derive a sky spectrum that very accurately estimates the sky component of the object-plus-sky spectrum. For a multi-slit system the sky spectrum and the object-plus-sky spectrum are adjacent both on the sky and on the detector and the systematic errors in both are practically the same and therefore cancel out in the sky-subtraction process. Interpolation along the slit so that sky regions either side of the object are used helps here. However, for a fibre system the two spectra to be subtracted from one another are typically neither adjacent on the detector nor on the sky and so the systematic errors associated with each often do not cancel completely if a simplistic data reduction algorithm is used. The most dominant systematic errors are due to a lack of adjacency at the detector and include wavelength sampling, scattered light and spectrograph vignetting.

However, with care these systematic effects can be successfully dealt with to give results as good as those of a multi-slit system. When one considers that fibres also offer higher spectral resolution and bigger field sizes than multi-slits it is clear that using fibres is well worth the extra effort.

4. 2dF - A Multi-fibre Case Study

The AAT now has a new Prime-focus corrector capable of producing aberration-corrected images over a 2° field of view. This is at the heart of a new and powerful 400-fibre spectroscopic survey facility known as the **2dF** (Taylor & Douglas (1995)).

4.1. *The Fibre Positioner*

The 2dF positioner is designed to accommodate 400 fibres which are fed to two separate spectrographs (200 fibres each). The basic technology used to position and place fibres is based on the original AutoFib concept; that of robotically positioning fibre *buttons* magnetically held to a *field-plate* placed at the focal surface of the telescope. However major changes are required in order to optimize such a system for the 2dF. Given the increased positional accuracy requirements imposed by the finer plate-scale at Prime-focus and the fact that we now have a factor of 8 more fibres than AutoFib, we are faced with the design goal that fibre configurations should be interchangeable within times significantly shorter than a nominal *short* integration to minimize duty cycle losses. Indeed the necessity to deal with differential atmospheric refraction at the edges of the 2° field places a hard limit of ∼60 minutes on the **maximum** integration time before re-configuration. Clearly re-configuration dead-times of ∼30 minutes, scaled simply from a typical AutoFib speed, are unacceptable. The adopted solution to this basic problem is to *double-buffer* the 2dF positioner so that data from the *active* field is being acquired while the fibres for the next field are being configured. This approach allows a relaxation of the re-configuration time facilitating an increase in positional accuracy while ensuring negligible duty-cycle losses. It also means we need a total of 800 fibres for the full system.

The *double-buffer* philosophy has been realized with the *tumbler* design, shown in the upper section of Figure 17. The fibre re-configuration takes place on the upper surface using a robotic gripper head mounted on an X-Y Gantry. A TV viewing system is used to fine-tune the fibre positioning to an accuracy of $\leq 0.15''$. Meanwhile the *active* field-plate is located at the lower surface of the tumbler in the flat focal-plane of the corrector, the fibres feeding their collected light to the two spectrographs. A similar X-Y gantry,

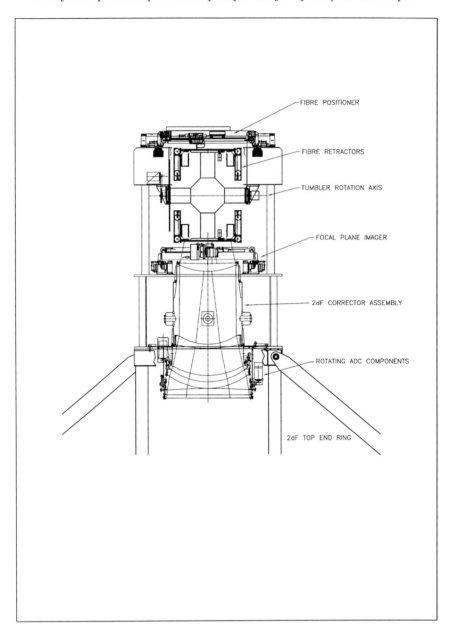

FIBRE POSITIONER

FIBRE RETRACTORS

TUMBLER ROTATION AXIS

FOCAL PLANE IMAGER

2dF CORRECTOR ASSEMBLY

ROTATING ADC COMPONENTS

2dF TOP END RING

FIGURE 17. 2dF Prime-focus top-end showing 4 optical elements of the corrector, with ADC; the *double-buffered* tumbling fibre positioner; and the focal-plane imager. The fibres are routed down the veins of the 2dF top-end to the two spectrographs which are located, out of the view, on the outer ring.

holding a *focal-plane imaging CCD*, is located at the lower surface of the tumbler. This will be used to map field distortions and assist in field acquisition and guiding. The positioner system uses linear motors and encoders rather than the older AutoFib lead-screw technology. Impressive speed and accuracy performance give us confidence that our design goals (10μm *rms* positioning accuracy in ~4 sec. over the full field) can be achieved.

A fundamental requirement in positioning fibres over the 2dF's field of view is to be able to successfully acquire fundamental reference stars at all orientations of the telescope. In order to achieve this, one needs to calculate the combined effects of a number of coordinate transformations, representing the atmosphere, telescope, 2dF corrector, positioner and CCD camera. All but the last are non-linear transformations which have to be evaluated and/or calibrated.

4.2. *The Fibre Spectrographs*

The design of a dedicated fibre spectrograph to service ∼400 fibres, supplying 1st order spectral resolving power, R≤2500, over a wavelength range matched to the 2dF corrector, is predicated on whether one adopts a model whereby all 400 fibres are fed to a single spectrograph or to several similar, but smaller, spectrograph units. Given the ready availability of 1024^2 CCDs at the time, two 200 fibre spectrograph system were specified.

Fibre FRD has the effect of smearing out the far-field pupil profile, incident on the fibre input, so as to both speed up the output beam and fill in any central hole thus requiring a fibre spectrograph to have a faster collimator and smaller central obstruction than would be the case in an equivalent slit-spectrograph.

The simple corollary here is that transmission cameras and collimators are to be preferred over catadioptric systems which generally have some degree of central obstruction. However, in order to match detector pixels with fibre diameters requires camera f-ratios faster than ∼f/1.3 so as to maximize spectral coverage and fibre-multiplex. Such fast f-ratios are very difficult to achieve over a broad wavelength range and wide field with transmission systems. It was for this reason that we decided on an off-axis Maksutov collimator design which allows the input fibre slits and attendant filters and shutters to be conveniently located to the side of the spectrograph and out of the beam thus avoiding any vignetting. The 150mm diameter collimated beam is then fed to one of a set of standard ruled reflection gratings and thence, at a collimator/camera angle of 40°, to a fast, wide-field, Schmidt camera. The difficulty with the camera is that pixel matching requirements force it to have an f-ratio as fast as f/1.2 [f/1.0] in the fibre [spectral] direction while retaining its wide ±5.5° field. The hybrid Schmidt camera design uses a single, severely aspheric, plate which has good aberration performance over the full wavelength range and field angle required to image onto a 1024^2 Tektronix CCD. The spectrograph design is shown diagrammatically in Figure 18.

For typical spectrograph configurations, numerical modelling suggests that camera losses amount to ∼10% for realistic FRD values. However, when spectral coverage is not at a premium, the gratings can be used *blaze-to-camera* leading to anamorphic factors which will reduce the central-obstruction losses while not impacting on the spectral resolution. There seems no doubt that minimizing fibre length will facilitate improved system efficiency, but at the cost of having to accommodate gravitational and temperature stability effects. However there is now sufficient experience in designing instruments for the Cassegrain focus that the light losses implicit in the employment of long ∼30 metre fibres argues most forcibly for mounting the spectrographs at Prime. This decision is bolstered by our 2-spectrograph approach; it is unlikely that a *monolithic* spectrograph, catering to 400 fibres, could have been accommodated on the 2dF top-end, however, space for at least two smaller spectrographs can readily be found around the the the ring's circumference.

4.3. *2dF Data Reduction System*

The scientific productivity of the 2dF is heavily dependant on the ease, or otherwise, of the data reduction process. It is clear that rapid processing of object CCD frames

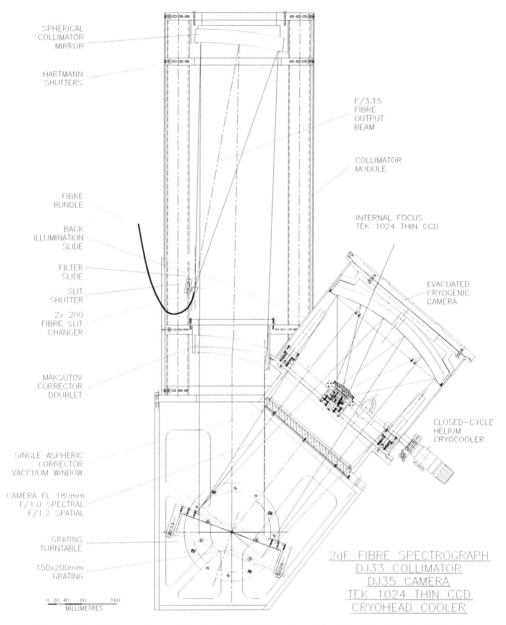

SPHERICAL
COLLIMATOR
MIRROR

HARTMANN
SHUTTERS

F/3.15
FIBRE
OUTPUT
BEAM

COLLIMATOR
MODULE

FIBRE
BUNDLE

INTERNAL FOCUS
TEK 1024 THIN CCD

BACK
ILLUMINATION
SLIDE

FILTER
SLIDE

EVACUATED
CRYOGENIC
CAMERA

SLIT
SHUTTER

2× 200
FIBRE SLIT
CHANGER

MAKSUTOV
CORRECTOR
DOUBLET

CLOSED−CYCLE
HELIUM
CRYOCOOLER

SINGLE ASPHERIC
CORRECTOR
VACCUUM WINDOW

CAMERA FL 180mm
F/1.0 SPECTRAL
F/1.2 SPATIAL

GRATING
TURNTABLE

150x200mm
GRATING

2dF FIBRE SPECTROGRAPH
DJ33 COLLIMATOR
DJ35 CAMERA
TEK 1024 THIN CCD
CRYOHEAD COOLER

0 20 40 80   160
MILLIMETRES

FIGURE 18. 2dF spectrographs showing the fibre input feeding the off-axis Maksutov collimator. The grating feeds a hybrid f/1 Schmidt cryogenic camera with CCD at its internal focus.

into a set of fully calibrated spectra requires a minimum of user interaction. The goal of achieving fully automated, and thereby largely non-interactive, 2dF data reduction within the time frame of the *next* exposure (say, 30 minutes) is made easier because the spectrographs are illuminated by fibre slits whose geometry is fixed and definable and use dedicated fixed format CCDs. The data reduction procedure has to be designed so as to allow the data reduction process as much system information as possible while

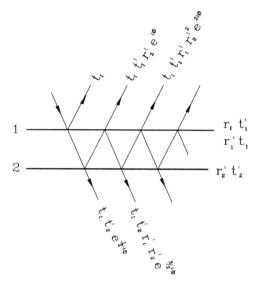

FIGURE 19. The Fabry-Perot is formed from two parallel, highly reflective surfaces, whose reflectance, r and transmittance, t are given suffixes 1 & 2 to distinguish them. The two surfaces are separated a distance, l in a medium of refractive index, μ. Reflectance within, and transmission from, this medium are designated by a superscript, '. With ϕ as the phase delay on double reflection within the medium, the complex amplitudes of the first few emergent rays, both in transmission and reflection are given.

permitting storage and retrieval of configurational information relevant to the current data reduction process.

These two requirements are driven by the recognition that it is wasteful to re-evaluate system parameters from scratch when they possess only slow (or low wavelength) variability - a statement which is intuitively obvious, from the design point of view, but that is at odds with most of the presently available systems for single- and multi-object spectroscopy. A new set of data reduction algorithms were specified with such design goals in mind.

5. Fabry-perots And Fourier Transform Spectrometers

A Fabry-Perot is formed essentially by two parallel reflective surfaces separated by distance, l in a medium of refractive index, μ. Light of wavelength, λ, incident at an angle θ to the normal is partially absorbed, reflected and transmitted at each surface with the resultant wave-trains interfering. The underlying symmetry of the physics, as apparent in Figure 19, is broken by the direction of the incoming photon.

Defining ϕ as the phase delay on double reflection within the medium:

$$\phi = 2\pi \frac{2\mu l \cos\theta}{\lambda} \tag{5.38}$$

Hence, for the mth ray, the transmitted amplitude, $\mathcal{U}_{tm}$ is given by:

$$\mathcal{U}_{tm} = \frac{t_1 t_2'(1 - r_1'^m r_2'^m e^{im\phi})}{1 - r_1' r_2' e^{i\phi}} \tag{5.39}$$

and for $m \to \infty$:

$$\mathcal{U}_t = \frac{\mathcal{T}}{1 - \Upsilon e^{i\Psi}}$$ (5.40)

where:
- $\mathcal{T} = |t_1 t_2'|$
- $\Upsilon = |r_1' r_2'|$
- $\Psi = \phi + \chi(\lambda)$; $\chi(\lambda)$ represent the wavelength dependant term in ϕ

From this we can obtain the Airy Function, $\mathcal{A}$, such that:

$$\mathcal{A}(\Psi) = \frac{\mathcal{T}^2}{(1 - 2\Upsilon \cos \Psi + \Upsilon^2)}$$ (5.41)

from which we can obtain:

$$\mathcal{A}(\Psi) = (1 - \frac{a}{1 - \Upsilon})^2 (1 + \frac{4\Upsilon}{(1 - \Upsilon)^2} \sin^2 \frac{1}{2} \Psi)^{-1}$$ (5.42)

where:
- $a = 1 - \Upsilon - \mathcal{T} (\to 0$, if using high reflectivity dielectric stacks).

Now from equation (5.42):

$$d\Psi \sim \frac{2(1 - \Upsilon)}{\sqrt{\Upsilon}}$$ (5.43)

the FP $\mathcal{F}$inesse is given by:

$$\mathcal{N} = \frac{\Delta \Psi}{d\Psi} = \frac{\pi \sqrt{\Upsilon}}{(1 - \Upsilon)}$$ (5.44)

[i.e.: $\mathcal{N} \sim 30$; for $\Upsilon \sim 0.9$]

the Peak Transmission $\mathcal{T}_{peak}$ by:

$$\mathcal{T}_{peak} \sim \frac{\mathcal{T}^2}{(1 - \Upsilon)^2} = (1 - \frac{a}{(1 - \Upsilon)})^2$$ (5.45)

the Contrast by:

$$(\frac{MAX}{MIN}) = (\frac{1 + \Upsilon}{1 - \Upsilon})^2$$ (5.46)

and the "purity", $\mathcal{P}$, by:

$$\mathcal{P} = \frac{\int_0^{d\Psi} \mathcal{A}(\Psi) d\Psi}{\int_0^{\pi} \mathcal{A}(\Psi) d\Psi}$$ (5.47)

or:

$$\mathcal{P} = \frac{2}{\pi} \tan^{-1} [\frac{2\mathcal{N}}{\pi} \tan \frac{\pi}{\mathcal{N}}]$$ (5.48)

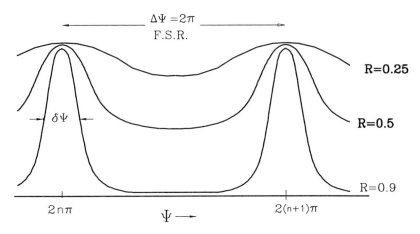

FIGURE 20. FP instrumental profiles for a range of reflectivities, Υ.

and as can be seen from inspection of Figure 20, Fabry-Perots are "dirty"; half of the energy transmitted by its instrumental profile lies outside the FWHM of its peak. This makes the use of FPs in general highly non-optimal for absorption-line work and is why they have been employed traditionally for observations of the kinematics of strong emission, weak continuum sources such as gaseous nebulæ and late-type galaxies.

The "Airy" Function of the FP (equation 5.42) can now be re-written as:

$$A(\Psi) = (1 + \frac{4\mathcal{N}}{\pi^2} \sin^2[2\pi\mu l \cos\theta/\lambda])^{-1} \qquad (5.49)$$

from which all else is derived!

The condition for constructive interference then becomes:

$$m\lambda = 2\mu l \cos\theta \qquad (5.50)$$

where m is the order of interference.

Note: Any change is $\mu; l; \theta$ or λ will change A.

5.1. *FP (or Interference Filter) at image plane*

Figure 21 shows the FP at the telescope input image plane subsequently relayed at an arbitrary f-ratio onto the detector surface. The FP acts as a *periodic* monochromator but since its interference rings are at infinity with respect to the FP plates, no rings are imaged onto the detector itself.

Now from equation (5.50):

$$\mathcal{R} = \frac{\lambda}{d\lambda} \sim \frac{2\mu^2}{\theta^2} \sim \frac{2\pi\mu^2}{\Omega} \qquad (5.51)$$

Note: from equation (5.51) the standard formula for predicting interference band-pass variations with angle, θ, is obtained.

F.P. (or I.F.) at image plane

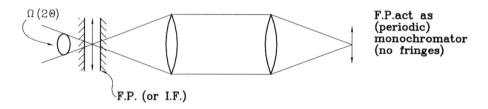

FIGURE 21. FP placed, non-classically, in the focal surface of the telescope.

For air gap FPs, we take $\mu=1$ and obtain:

$$\mathcal{R}\Omega = 2\pi \tag{5.52}$$

$\mathcal{R}\Omega$ is generally referred to as the **Jacquinot Spot** and is always significantly greater than the equivalent $\mathcal{L}.\mathcal{R}$-product for a grating instrument, basically because FP systems and their FTS counterparts (see Section 5.7) are on-axis, axially symmetric, systems.
Also the free-spectral-range (FSR) of the FP is given by:

$$\Delta\lambda = \frac{\lambda^2}{2\mu l} \tag{5.53}$$

and hence:

$$\mathcal{R} = m\mathcal{N} \tag{5.54}$$

(c.f.: equation (2.16) derived from grating theory).

Now from equation (5.44), $\mathcal{N}$ is purely a function of $\mathcal{R}$, however two other effects have to be considered.

(*a*) Aperture Finesse, $\mathcal{N}_A$:

This represents the instrumental broadening due to the variation of θ at each point in the focal-plane. While an equivalent effect is also of importance when the FP is mounted in the collimated beam the use of area detectors makes it obsolete in this context (see Section 5.2).
In the non-classical, focal-plane positioning of the FP we have:

$$\mathcal{N}_A = \frac{2\pi}{m\Omega} = 8f^2\mu^2/m \tag{5.55}$$

Thus, for $\mathcal{N} > 30$ and $m \sim 100 \to f \geq 20$. However, unlike the classical case, the FoV is limited by the physical size, d, of the FP.

CLASSICAL CONFIGURATION

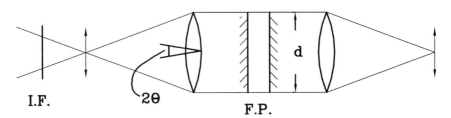

F.P. FRINGES FORMED AT INFINITY

FIGURE 22. FP placed, classically, at the pupil-plane of a focal reducer. For completeness, I.F. designates an order-sorting interference filter.

(*b*) Defect Finesse, $\mathcal{N}_D$:

FP plates are never perfectly parallel nor are they perfectly flat, hence across the FP aperture. If we simply characterize the resultant gap variations by an r.m.s. measure of micro-irregularities, dl, then:

$$\mathcal{N}_{\mathrm{D}} = \frac{\lambda}{2dl} \qquad (5.56)$$

Note: $dl < \lambda/100$ for $\mathcal{N} \sim 50$.

The *Effective Finesse* is then given by:

$$\frac{1}{\mathcal{N}_{\mathrm{E}}^2} = \frac{1}{\mathcal{N}_{\mathrm{R}}^2} + \frac{1}{\mathcal{N}_{\mathrm{A}}^2} + \frac{1}{\mathcal{N}_{\mathrm{D}}^2} \qquad (5.57)$$

However, it should be noted that $\mathcal{N}_D$ is irrelevant in the case where the FP is located at the focal-plane since each recorded spatial resolution element *sees* a unique part of the FP; it is only the peak wavelength which is modulated by dl. This is not the situation in the classical mounting, to be described in the next section, since the FP is now in the pupil and each spatial resolution element *sees* the full FP aperture, implying a net efficiency loss.

5.2. FP at pupil plane - the classical case

Figure 22 shows a simple focal reducer with the FP located in the re-imaged pupil of the telescope. The detector sees a direct image of the sky, within the wave-band of the interference filter, I.F., modulated by the FP interference rings.

As is demonstrated in Figure 23, a diffuse monochromatic source gives rise to a regular ring pattern as predicted from inspection of the Airy Function, equation (5.49). When illuminated by a diffuse, and yet non-monochromatic, source (such as a late-type galaxy, where its wavelength variations are a direct measure of its internal kinematics) then

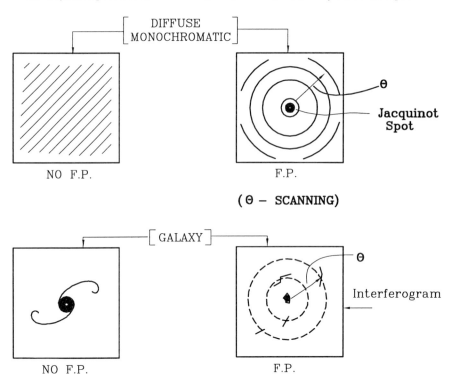

FIGURE 23. Two pairs of images are shown; the first pair represents the response of the system to a diffuse monochromatic input while the second pair is a cartoon representing an emission-line galaxy input.

deviations from circularity of the FP rings in the *interferogram* are a direct measure of the wavelength variations from point to point in the image.

Unlike a grating spectrograph, the spectral dispersion is not constant but varies radially from the centre of the FP rings, such that:

$$\frac{d\theta}{d\lambda} = -\frac{1}{\lambda_0 \sin \theta} \tag{5.58}$$

When used with an area detector whose spatial resolution is limited to, let us say, the seeing disk, ϕ, the usable field, as derived from equation (5.58) is now vastly greater than the central spot Jacquinot limit (as given in equation (5.52)) and is thus given by:

$$\phi_F = 2(\frac{d}{D})^2 \frac{1}{\mathcal{R}\alpha} \tag{5.59}$$

From the previous description of the FP interferogram it is natural to assume that emission-line wavelength shifts across an image can be deduced from a single interferogram; and indeed this technique has been used in the past to obtain velocity field information from galaxies and diffuse nebulæ, however there is a *fatal* ambiguity to the

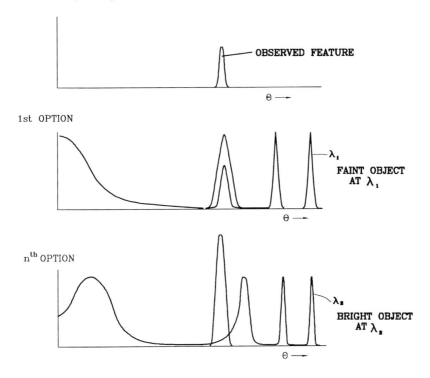

FIGURE 24. The intrinsic ambiguity inherent in FP interferograms. Each graph represents a cut through a point source in a 2D image. The first is without an FP, showing the location of the source; the second and third (options) demonstrate the effect of in peak wavelength (λ_1 to λ_2) of the FP when in the optical path.

technique which gives rise to serious errors in the analysis and which severely limits the use of FP interferograms for this type of work.

As can be seen from inspection of the Airy Function, equation (5.49), a given wavelength, λ, only uniquely defines a set of θs at peak transmission: The fact is that all θs are transmitted at some level and hence a feature in the image can either be:

• a relatively faint source detected at the peak of the Airy Function - hence the wavelength derivation given in equation (5.50) will be correct.

• or a relatively bright source (not necessarily even an emission-line source) seen in the wings of the Airy Function - any wavelength derivation will hence be invalid.

Figure 24 gives a schematic representation of the ambiguity. Given a single FP interferogram, one cannot distinguish between these two possibilities and hence, in principle at least, interferograms contain no wavelength information - however, this technique has been used (relatively) successfully in the past! It can only be used *safely* in conditions where the source is highly diffuse and has no strong flux contrasts. While this is hardly ever the case, before the advent of fast readout area detectors (such as the IPCS) or very high DQE detectors (such as CCDs), the technique, although *dangerous*, was nevertheless seen to be highly effective in obtaining velocity field information for diffuse emission-line sources provided suitable caveats were acknowledged (Tully (1974)).

5.3. *The TAURUS technique*

Of course it is also true that a single FP interferogram only contains $(1/\mathcal{N})^{th}$ of the spatial information subtended at the image surface. A natural way to recover **all** of the spatial information in the image and, as an outcome, to resolve the basic ambiguity of the interferogram, is to scan some parameter of the system. If we refer back to the Airy Function of equation (5.49) it can be seen that we have one of three choices:

(*a*) Scan θ: - by tilting the FP (or tracking the telescope!);
(*b*) Scan μ: - by changing the pressure of gas in the FP cavity;
(*c*) Scan l: - by moving the FP plates apart.

While all these technique have been used in practical systems, in the early 80s, two technological strands came together in the TAURUS instrument which revolutionized FP usage on large telescopes:

(a) The highly accurate and reliable technique of **capacitance micrometry** had been developed by a group at Imperial College, London (later to evolve into Queensgate Instruments) which facilitated the controlled gap (*l*) scanning. Recall (from equation (5.53)) that the FP gap has to be actively maintained both in parallelism and absolute separation by δl where:

$$\delta l \sim \frac{\lambda}{2\mathcal{N}} \ ... \ (or \geq \lambda/200 \ !) \tag{5.60}$$

and the FSR is given by:

$$\Delta l = \lambda/2 \tag{5.61}$$

(b) The Image Photon Counting System (IPCS) was developed by a group in University College, London. This, at the time, revolutionary new detector permitted the output image of relatively high DQE (prior to CCDs) image tubes to be relayed to a TV camera in a way which facilitated true photon counting, giving one the facility to read images at TV frame rates in photon shot-noise limited conditions.

... and it took a hick from the home counties to bring these two London-based forces together! (Taylor (1980))

The combination of capacitance micrometry, which could sequence the FP gap accurately and repeatedly at kHz rates, and the IPCS, which could sequence images at TV frame rates, facilitated the development of a scanning FP imaging spectrometer (TAURUS) which could cycle through a series of $2\mathcal{N}$ etalon gap settings sampled at $\sim \lambda/4\mathcal{N}$ over a full, $\lambda/2$, FSR such that, for the first time, a true data cube (x,y,z[v]) could be acquired in real time. The ability to record and update the cube as the scans progressed permitted *rapid-scanning* of the cube to permit averaging out of atmospheric transparency and seeing changes.

5.4. *The TAURUS data-cube*

In Figure 25 we see two spectral emission-lines within the data cube whose spatial (x, y) axes are orthogonal to the z (or wavelength) axis. The spectra are the result of a convolution of the source function, $\mathcal{I}_{xy}(\lambda)$ and the Airy Function, $\mathcal{A}_{xy}(\lambda)$. In reality this z-axis is a *time* axis since each 2D spatial slice of the cube is obtained sequentially, however since TAURUS is a FP gap-scanning device, z is linearly encoded in l which is,

TAURUS CUBE

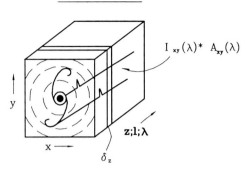

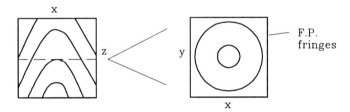

FIGURE 25. A cartoon of a TAURUS data cube, showing a 3D spectral image of a galaxy. Below are 2D views of a TAURUS data cube taken with a diffuse monochromatic source. On the left is shown an x, z-slice through the cube, while the right hand view is of a spatial x, y-slice.

through equation (5.50), related to λ. Given the nature of the Airy Function, the l (or z) to λ transformation is unfortunately rather complex.

As shown in Figure 25 the lines of constant wavelength, λ, as seen in the (x, z)-slice are in fact a set of nested parabolæ equally spaced in z. Any (x, y)-slice within the cube cuts through these nested paraboloids to give the familiar FP ring pattern. Wavelength calibration thus requires converting $z \rightarrow \lambda$ where:

$$l(z) = l(0) + az \tag{5.62}$$

a is a constant of proportionality.

Constructive interference on the axial position, (x_0, y_0) gives:

$$az_0 = \frac{n\lambda_0}{2\mu} - l(0) \tag{5.63}$$

but an off-axis (x, y) location transmits the same, λ_0, wavelength at $(z_0 + p_{xy})$ where:

$$ap_{xy} = l(z_0)(\sec \theta_{xy} - 1) \tag{5.64}$$

Hence, as stated earlier, the surface of constant wavelength is a set of nested paraboloids, as shown in Figure 26.

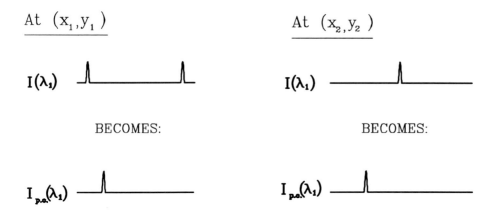

FIGURE 26. A demonstration of TAURUS phase-correction.

Now from equation (5.50) a 2D *Phase-Map*, $p(x, y)$, can be defined such that:

$$p(x, y) = n\Delta z_0(\sec \theta - 1) \tag{5.65}$$

The phase-map, $p(x, y)$, can be obtained by illuminating the FP system with a diffuse monochromatic source of wavelength λ'.

5.5. *Phase-correction - or how to tame your TAURUS cube !*

The Phase-Map is so called since it can be used to transform the raw TAURUS cube, with its strange multi-paraboloidal iso-wavelength contours into a well-mannered (or *flat*) data cube where all (x, y)-slices are now at constant wavelengths. The process, called Phase-correction is simply a matter of "rolling" each $\mathcal{I}_{xy}(\lambda)$ spectrum in the data cube by the phase-map value, $p(x, y)$.

[The term "rolling" rather than "sliding" is used here since each spectrum, $\mathcal{I}_{xy}(\lambda)$, is periodic in l (or z) and hence appears periodic in λ so the process has to be thought of as cyclical in that as regions of the phase-corrected spectra are shifted beyond the z-limits of the cube they are simply folded back to the beginning of the cube by one FSR (Δz).]

In this manner the TAURUS cube is "aligned" so that surfaces of constant wavelength are represented simply by (x, y)-planes in the phase-corrected cube. Figure 26 refers to this alignment.

It will be noted that the phase-map, as defined in equation (5.65) is independent of wavelength and hence, in principle, any calibration line can be used for all observed wavelengths. Also from the Airy Function, equation (5.49), the phase-map as expressed in wavelength-space, $\delta\lambda_{xy}$, is given by:

$$\delta\lambda_{xy} = \lambda_0(1 - \cos \theta) \tag{5.66}$$

and hence is independent of gap (l) and hence applicable to any etalon aligned to the same optical axis.

While these features of the phase-map are formally worthy of note, in practice, it is only the wavelength independence which is routinely used, and even then it is safest, for wavelength calibration purposes, to use calibration wavelengths which are not too distant from those actually observed.

5.6. *TAURUS Wavelength Calibration*

Some would regard wavelength calibration of TAURUS data as a *black art* however, while it is not as straight forward as it is for diffraction systems it is tractable provided one uses at least two comparison calibration wavelengths that are suitably separated.

Using two calibration wavelengths, λ_1 and λ_2 which, on-axis, peak at z_1 and z_2 in order m_1 and m_2, respectively; then from equations (5.49) and (5.62) we obtain:

$$m_2 - m_1 = \frac{\lambda_1}{\lambda_2}\frac{z_2 - z_1}{\Delta z_1} - \frac{m_1}{\lambda_2}(\lambda_2 - \lambda_1) \qquad (5.67)$$

If m_1 can be estimated from the etalon manufacturers *nominal* gap using equation (5.50) then equation (5.67) can be used to search for a solution whereby m_2 is an integer. Clearly the further apart λ_1 is from λ_2, the more accurate an estimate of the interference order is obtained and there is sometimes a need for more than two wavelengths to be used in extreme circumstances.

Once the interference order for a known calibration line has been identified then wavelength calibration is trivially achieved through the following relationship:

$$\lambda = \lambda_0 \left[\frac{(z - z_0)}{m_0 \Delta z_0} + 1\right] \qquad (5.68)$$

5.7. *Imaging Fourier Transform Spectrographs - IFTS*

Figure 27 is a schematic representation of an IFTS. Light from an image subtending a solid angle, Ω, is collimated and relayed via a beam-splitter to the two arms of the Michelson Interferometer. The light is then recombined to form interference fringes at the detector surface, D. One arm is adjustable to give path length variations to facilitate Michelson scanning. The intensity at D is determined by the path difference, Δx, between the two arms of the interferometer.

Given that frequency, $\nu = 1/\lambda$, the phase difference is thus $2\pi\nu\Delta x$ and hence the recorded intensity, $\mathcal{I}$, is:

$$\mathcal{I}(\nu, \frac{\Delta x}{2}) = \frac{1}{2}[1 + \cos(2\pi\nu\Delta x)] \qquad (5.69)$$

Now if we vary x in the range: $-\infty \leftarrow x/2 \rightarrow \infty$ continuously then:

$$\mathcal{I}(x) = \int_{-\infty}^{\infty} \mathcal{B}(\nu)(1 + \cos 2\pi\nu x)\,d\nu \qquad (5.70)$$

$$\mathcal{B}(\nu) = \int_{-\infty}^{\infty} \mathcal{I}(x)(1 + \cos 2\pi\nu x)\,dx \qquad (5.71)$$

Equations (5.70) and (5.71) represent Fourier Transform pairs and hence spectrum $\mathcal{B}(\nu)$ is obtained from the *cosine* transform of the interferogram $\mathcal{I}(x)$.

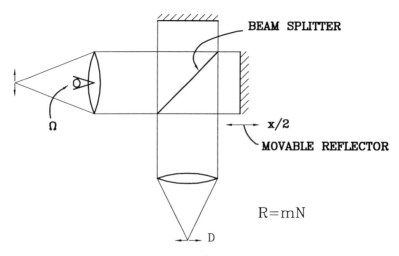

2–BEAM INTERFEROMETER   (SCANNING MICHELSON)

FIGURE 27. The basic principles underlying the double beam Fourier Transform Spectrograph.

In reality, of course, x goes from $0 \rightarrow \Delta x(max)$ which limits the spectral resolving power to:

$$\mathcal{R}_0 = \frac{\lambda}{\delta\lambda_0} = \frac{2\Delta x(max)}{\lambda} \tag{5.72}$$

For example: if $\Delta x(max) = 10$cm and $\lambda = 500$nm then: $\rightarrow \mathcal{R} \sim 4.10^5$.

Since we are talking here about an imaging FTS then consideration of FoV is relevant and the circular symmetry and the IFTS gives an angular dependance which is the same as for the FP. Hence, as with the FP:

$$2\mu l \cos\theta = m\lambda \tag{5.73}$$

and again, in analogy with the FP, we have $\Omega\mathcal{R} \gg 2\pi$ which is only limited by the wavelength variation, $\delta\lambda$, across each pixel of the detector. However the consequences of such *field-widening* imply that an analogue to TAURUS phase-correction is required in order to correct for path difference variations over the image surface.

The IFTS has many advantageous features for 3D imaging spectroscopy (Maillard 1995).
- Arbitrarily variable wavelength resolution to the $\mathcal{R}$ limit set by $\Delta x(max)$;
- A large 2D FoV;
- A very clean, sinc function, instrumental profile (cf: FP's Airy profile!)
- A finesse, $\mathcal{N} = 2\Delta\lambda/\delta\lambda$ which can have values higher than 10^3

The disadvantages of the IFTS are:

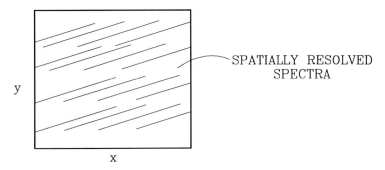

FIGURE 28. A cartoon representing a TIGER 3D spectrum. The spectra are angled to avoid overlap.

• Sequential scanning - like the FP. However the effective integration time of each interferogram image can be monitored through a separate *complementary* channel, if required;
• Especially in the optical domain, very accurate control of the scanned phase delay is required.

Because of the technical challenge, the IFTS is a technique which is only just beginning to be used effectively for 3D imaging spectroscopy (mainly in the NIR) and I expect the technology to percolate downwards into the optical domain within the next few years.

6. Integral Field And Hybrid Systems

The FP and IFTS are classical 3D imaging spectrographs in that they facilitate the sequential detection of images while either wavelength (FP) or phase delay (IFTS) scanning to produce the 3rd dimension. There are, however, techniques which use a 2D area detector to sample 2D spatial information with spectral information simultaneously. These we refer to as **hybrid** systems.

6.1. TIGER

This is a technique (Courtès et al. 1987) which re-images the telescope focal plane onto a micro-lens array which is located at the entrance aperture of a standard focal-reducer, grism system. The micro-lens array acts to form multiple telescope pupils which are re-imaged through the grism onto the detector. Without the grism, each pupil would be recorded as a grid of points matching the micro-lens distribution in the image plane, however the grism act to disperse each pupil image independently to give a spectrum for each individual pupil image.

Of course in general the pupil spectra will overlap on the detector, however by selecting the wavelength region and angling the grism appropriately each pupil spectrum can be recorded separately. Figure 28 demonstrates the principle schematically. While the number of spatial samples and their resolution on the sky are highly constrained by the dimensions of the micro-lens and the requirement for non-overlapping spectra, nevertheless *TIGER* represents an extremely effective technique for true 3D spectroscopy. It supplies low dispersion (grism) integral field spectroscopy; it separates spectral and

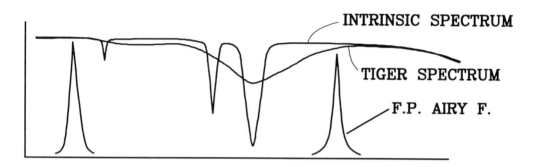

FIGURE 29. Being a low-dispersion device, TIGER smears out the *intrinsic* high resolution spectrum; the FP Airy profile of PYTHEAS, however, permits a reconstruction of the original high-dispersion spectrum through FP scanning

spatial information; uses area detectors efficiently and is very good for diffuse sources with complex spectra. This makes it very complementary to FP devices which have a far greater spatial multiplex but which are generally limited to strong emission-line regions at relatively high dispersion.

6.2. *PYTHEAS*

PYTHEAS (Georgelin & Comte (1995)) is a very interesting extension of the TIGER concept to give it a high dispersion capability. By placing a scanning FP at the focal surface, just in front of the micro-lens array, the FP act as a multi-order pre-filter to the grism. Each setting of the FP transmits a periodic series of wavelengths, corresponding to its interference orders throughout the wavelength range defined by the grism, pre-filter combination. Thus Figure 28 is transformed from a series of parallel spectra to a set of FP *dots* along each spectral locus. The spectral consequences are shown in Figure 29.

While PYTHEAS represents an imaginative and effective use of detector real-estate, it is a highly specialized device, useful for high-dispersion, spatial mapping of relatively bright targets. Problems of photometric registration of the FP scans and basic data reduction complexities are, however, non-negligible.

6.3. *Fibre Image Slicers*

Optical fibre technology offers the astronomical spectrograph designer vast opportunities which are only just beginning to be realized. As I hope to have already demonstrated, the art of recording spatial and spectral information simultaneously onto a 2D detector is basically one of subtly re-formatting information in one way or another. If we were able to arbitrarily define the geometry of our detectors (even to make them 3D!) then none of the sophisticated optical design would be necessary. Fibres however are perfectly matched to the demands of information re-formatting, as has already been demonstrated in Section 3.4, and they give us a supreme opportunity to perform integral field spectroscopy.

The simplest approach is to feed a 2D array of fibres at a telescope focal-plane. This

FIBRE IMAGE SLICERS

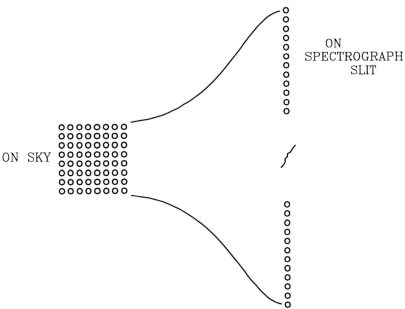

FIGURE 30. Simple fibre re-formatting from sky to spectrograph slit.

array can be any shape but can be re-formatted onto the slit of any spectrograph, as demonstrated in Figure 30. This represents a trivial, but powerful, technique.

- It captures light over a full seeing disk (and more) without degrading the intrinsic resolving power of the spectrograph;
  - it facilitates spatially resolved spectroscopy;
  - there is no requirement to centre a point object on the slit;
  - there is no requirement to match slit-width to the seeing;
  - it effectively detaches spectral and spatial information;
  - it facilitates spatially integrated spectroscopy;
  - and supplies robust spectrophotometry.

Examples of such systems are:
(a) F.I.S. on AAT (1981) - 100 fibres
(b) SILFID on CFHT (1988) - 400 fibres
(c) HEXAFLEX on WHT (1991) - 61 fibres
(d) 2D-FIS on WHT (1994) - 125 fibres

Clearly for a fibre of diameter, ϕ_{fibre}, each individual fibre aperture (α_{fibre}) on the sky is given by:

$$\alpha_{fibre} = \frac{\phi_{fibre}}{D\mathcal{F}_{fibre}} \tag{6.74}$$

where $\mathcal{F}_{fibre}$ is the input focal ratio of the fibre.
(As an example, take $\alpha_{fibre} = 0.5''$; $D = 8\text{m}$ and $\mathcal{F}_{fibre} = 5 \rightarrow \phi_{fibre} \sim 80\mu$.)

Generally this type of fibre integral field unit (IFU) is retro-fitted to existing long-slit spectrographs, however, Cassegrain f-ratios are generally not well matched to fibres, due to their intrinsic focal-ratio-degradation (FRD), thus requiring rather lossy f-ratio conversions at both input and output of the fibres. Furthermore spatial information is lost in the inter-fibre regions. A conceptually simple solution to both input problems is to use hexagonally packed micro-lenses to feed the fibre at the focal-plane of the telescope. Such schemes are now being proposed for several large telescope instrument projects, however they suffer several disadvantages:

- the micro-lenses are very small (typically $\sim 150\mu m$);
- the micro-lens/fibre alignment is highly critical;
- the output from the fibres have generally to be micro-lensed as well, in order to match the f-ratio of the long-slit spectrograph;
- the input feed to the fibres is non-telecentric.

This last problem can be illustrated as follows:

Given a micro-lens placed at the focal-plane of a telescope of f-ratio, $\mathcal{F}_T$, the telescope's pupil is then imaged onto a fibre of diameter ϕ_{fibre} at an f-ratio $\mathcal{F}_{fibre}$. Equation (6.74) represents the on-axis relationship between fibre diameter, ϕ_{fibre}, and fibre aperture, α_{fibre}, on the sky, however off-axis rays impact the fibre at steeper angles implying that larger fibres (and hence lower spectrograph $\mathcal{R}$s) are achievable. In the general case equation (6.74) becomes:

$$\alpha_{fibre} = \frac{\phi_{fibre}}{D.\mathcal{F}_{fibre}}(\mathcal{F}_T - \mathcal{F}_{fibre}) \tag{6.75}$$

In other words, by placing the micro-lens at the bare Cassegrain (or Nasmyth) focus of a large telescope, losses in the potential $A\Omega$ product, $(\mathcal{F}_{fibre}/\mathcal{F}_T)^2$ can easily be as large as $\sim 50\%$.

6.4. SPIRAL

The solution to this and other problems to do with micro-lens arrays is in the recognition of the fact that equation (6.74) holds for an physical size of lenslet. In other words it is independent of $\mathcal{F}_T$. Thus, given an image relay to any arbitrary input f-ratio, there is complete freedom to choose lenslet dimensions to suit optimal manufacturability. Furthermore, by building a dedicated fibre spectrograph whose collimator f-ratio is matched to that of the fibre, $\mathcal{F}_{fibre}$, fibre output micro-lenses can be avoided.

Such a system has been devised for the AAT. It is called SPIRAL (**S**egmented **P**upil and **I**mage **R**elay **A**rray **L**ens) and is described in Parry, Kenworthy and Taylor (1996). The system preserves $A\Omega$ product, permits both image and pupil segmentation, allows for variable telescope scales (ie: variable $\mathcal{F}_T$) and feeds a dedicated, fibre-optimized spectrograph. Its most powerful feature, however, is that the SPIRAL fibre unit act as an ideal interface between telescope and spectrograph so that the same spectrograph can be optimized for any size of telescope simply by making a trivial change to the fore-optics which relay the telescope focal-plane (or pupil) onto the lenslet array.

SO ... WHO NEEDS SLITS?
Go Forth and Multiplex

7. Acknowledgements

Almost all of the developments described in these chapters are not my own. Even those with my name attatched, tenously or otherwise, are the result of input from a huge variety of sources and influences too numerous to mention. I do however wish to acknowledge the great companionship of my fellow lecturers during the Winter School which was much apprecaited and to the warm hospitality we were all shown by the staff of the IAC. I am especially grateful for Prof. Rodríguez Espinosa who demonstrated enormous patience with me in the somewhat protracted process of transcribing these lecturers into their present form. Thank you, Jose.

REFERENCES

BARANNE, A. 1988. White Pupil Story or Evolution of a Spectrograph Mounting. In *ESO Conf. on Very Large Telescopes and their Instrumentation* (ed. H.-H. Ulrich), vol. II, pp. 1195–1206. ESO.

COURTÈS G., GEORGELIN, Y., BACON, R., MONNET, G. & BOULESTEIX, J. 1987. A New Device for Faint Object High Resolution Imagery and Bidimensional Spectroscopy (First Observational Results with "TIGER" at the CFHT 3.6-Meter Telescope). In *Instruments for Ground-Based Optical Astronomy: Present and Future* (ed. L. B. Robinson), pp. 266–274. Springer-Verlag.

CUBY, J-G. & MIGNOLI, M. 1994. Sky subtraction with Fibres. In *Instrumentation in Astronomy VIII* (ed. D. L. Crawford) Proc. of S.P.I.E. vol. 2198, pp. 98–107.

ELLIS, R. S. & PARRY, I. R. 1987. Multiple Object Spectroscopy. In *Instruments for Ground-Based Optical Astronomy: Present and Future* (ed. L. B. Robinson), pp. 192–208. Springer-Verlag.

GEORGELIN, Y. & COMTE, G. 1995. The PYTHEAS concept and applications. In *Tridimensional Optical Spectroscopic Methods in Astrophysics* (ed. G. Comte & M. Marcelin) Proc. of IAU Coll. No. 149. ASP. Conf. Ser., vol. 71, pp. 300-307.

GRAY, P. M. 1986. Anglo-Australian Observatory Fibre System. In *Instrumentation in Astronomy VI* (ed. D. L. Crawford) Proc. of S.P.I.E. vol. 627, pp. 96–104.

MAILLARD, J. P. 1995. 3D-Spectroscopy with a Fourier Transform Spectrometer. In *Tridimensional Optical Spectroscopic Methods in Astrophysics* (ed. G. Comte & M. Marcelin) Proc. of IAU Coll. No. 149. ASP. Conf. Ser., vol. 71, pp. 316–327.

PARRY, I. R., KENWORTHY, M. & TAYLOR, K. 1996. SPIRAL Phase A: A Proto-type Integral Field Spectrograph for the AAT. In *Optical Telescopes of Today and Tomorrow* (Joint S.P.I.E., ESO, Lund Observatory, Symposium.) *In Press.*

TAYLOR, K. 1980. Seeing-limited radial velocity field mapping of extended emission line sources using a new imaging Fabry-Perot system. *M.N.R.A.S.* **191**, 675–684.

TAYLOR, K. 1995. The Anglo-Australian Telescope's 2dF Facility: Progress and Predicted Performance. In *Wide Field Spectroscopy and the Distant Universe* (ed. S. J. Maddox & A. Aragón-Salamanca) Proc. of 35th Herstmonceux Conference, pp 15–24.

TAYLOR, K. & DOUGLAS, N. G. 1995. A New Technique for Characterising Planetary Nebulae in External Galaxies. In *Tridimensional Optical Spectroscopic Methods in Astrophysics* (ed. G. Comte & M. Marcelin) Proc. of IAU Coll. No. 149. ASP. Conf. Ser., vol. 71, pp. 33–37.

TULLY, R. B. 1974. The Kinematics and Dynamics of M51. I. The Observations. *ApJ. Suppl. Ser.*, **27**, 415–448

High Resolution Spectroscopy

By D A V I D   F .   G R A Y

Department of Astronomy, University of Western Ontario,London, Ontario N6A 3K7, Canada;
dfgray@uwo.ca

1. Here We Go

An alternative title of this material could be "The Data Everyone Would Like to Get for their Research!" The first thing we seem to do in astronomy is 'see' something, be it simply looking in the sky, using a big telescope, or helping ourselves with sophisticated adaptive optics or space probes. But the very next thing we want to do is get that light into a spectrograph! We might get spectral information from colors, energy distributions, modest resolution or real honest high resolution spectroscopy, but we desperately need such information. Why? Well, because that's where most of the physical information is, and higher spectral resolution means access to more and better information. *High resolution* implies actually resolving the structure of the spectrum. Naturally we want to do this as precisely as possible, not only pushing toward good spectral resolution and high signal-to-noise, but also by understanding how the equipment has modified the true spectrum and by weeding out problems and undesirable characteristics. The main focus here will be on the machinery of spectroscopy, but oriented toward optical spectrographs and the spectral lines they are best suited to analyze. I do not concentrate on the specific instruments, but rather on the techniques and thought patterns we need. These are the fundamental things you can take with you and apply to any spectroscopic work you do. Of course, you will always have to fill in specific details for the particular machinery and tools you use.

2. What is a Spectrograph?

From a theoretician's point of view, this should be basically a black box into which light from the telescope goes, connected to a nice computer on the output end, and from the computer emanates an A&A manuscript. But then, we aren't theoreticians here are we? Still, let's start with the simple. The basic point of a spectrograph is 1) to image the entrance slit on the light detector, and 2) provide dispersion, i.e., separation of light into its colors. The conceptual setup is illustrated in Fig. 1, a spectrograph seen from the 'top', meaning that the dispersion is in the plane of the diagram. The dispersers I will talk about are reflection diffraction gratings, and frequently the collimator and the camera(s) are mirrors. But in the figures, I indicate a transmission grating and lenses instead. Lenses have incoming light on one side and outgoing light on the other; mirrors place both sets on the same side, making it harder to see what is happening. The ray diagrams for a lens can be turned into ray diagrams for a mirror by simply folding the diagram across the central plane of the lens. The same basically holds for the transmission versus reflection grating.

So here, in Fig. 1, we have the basics: entrance slit at which the telescope seeing disk is focused, divergent beam travelling the focal length of the collimator before entering the collimator and being converted into parallel rays, through the grating where now different colors are sent in different directions, and finally the focusing of these beams by the camera onto the light detector. Obviously we end up with an image of the entrance slit for each color in the incident light.

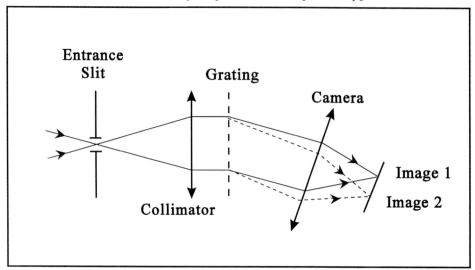

FIGURE 1. Here is a schematic diagram of a spectrograph. The light from the telescope comes from the left. Image positions for two different wavelengths are shown on the right.

Notice the *plane* grating. Experience has shown plane gratings to be easier to make, easier to use, and to have higher efficiency. All astronomical spectrographs employ plane gratings. Let's spend a moment looking at the basics of plane gratings.

3. Diffraction Gratings

A grating introduces a periodic disturbance into the beam. The 'disturbances' in our case are grooves ruled into the surface of a mirror, or in our conceptual diagrams, they are long parallel slits in an otherwise opaque screen. But more generally, they could be standing wave in a gas or regular ripples on a pain of glass. Light coming out of the grating has a phase shift that is a linear function of position along the face of the grating perpendicular to the grooves. This is illustrated in the diagram shown in Fig. 2. The individual slits are spaced a distance d apart, they each have a width b, and the grating plane is assumed to extend to large distances in both directions. You can see the arbitrary ray coming in at an angle α and leaving at an angle β. The expression for the combined light leaving the grating can be written as the sum of the outgoing waves with the path differences converted to phase shifts,

$$g(\beta) = \int F_0 \exp[2\pi i(x/\lambda)\sin\alpha]G(x)\exp[2\pi i(x/\lambda)\sin\beta]dx \qquad (3.1)$$

where F_0 is a measure of the strength of the incoming light, $G(x)$ is the grating function, and x is the linear dimension along the grating. If we set $\theta = \sin\alpha + \sin\beta$, and if we express x in units of wavelength, then this becomes

$$g(\beta) = F_0 \int G(x)\exp[2\pi i x\theta]dx. \qquad (3.2)$$

So we come to the important result that the diffraction pattern, $g(\beta)$, produced by the grating is the Fourier transform of the grating function, $G(x)$.

In our case, $G(x)$ is a series of 'boxes' of width b, equally spaced d apart, and extending over the much larger width of the whole grating, call it W. In other words,

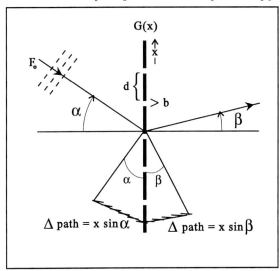

FIGURE 2. The parameters associated with a diffraction grating are shown here. Light comes to the grating at an angle α to the normal to the grating (horizontal line), and leaves after diffraction at some arbitrary angle β. The slits are spaced d apart along the face of the grating, and they each have the same width of opening, b, through which light can pass. We imagine the grating to have very long slits (in and out of the page). The path difference for plane waves striking the grating is given by x sinα, where x is a linear coordinate perpendicular to the slits, as indicated. On the output side, there is also a path difference given by x sinβ.

$$G(x) = B_1(x) * III(x) \bullet B_2(x), \tag{3.3}$$

in which $B_1(x)$ is the box for the individual slits, $B_2(x)$ is the box representing the whole width of the grating, and $III(x)$ is the shah function composed of an infinite set of equally-spaced delta functions. The star in this expression indicates a convolution and the dot stands for normal multiplication. Therefore,

$$g(\theta) = \frac{b \sin \pi \theta b}{\pi \theta b} III(\theta) * \frac{W \sin \pi \theta W}{\pi \theta W} \tag{3.4}$$

or if we do the convolution,

$$g(\theta) = \sum \frac{W \sin \pi(\theta - n/d)}{\pi(\theta - n/d)W} \frac{b \sin \pi \theta b}{\pi \theta b} \tag{3.5}$$

In other words, we have the product of some 'sinc' functions, as shown in Fig. 3. This is the situation for one wavelength. The individual interference peaks appear at the positions given by

$$\theta = n\lambda/d \tag{3.6}$$

or

$$\sin \alpha + \sin \beta = n\lambda/d, \tag{3.7}$$

which you recognize as the grating equation.

Now in Fig 3, we see the situation for two wavelengths. The pattern for the longer wavelength is simply scaled up in the θ coordinate in proportion to the wavelength. But

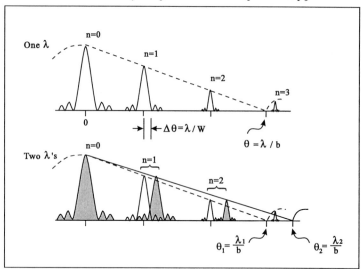

FIGURE 3. The diffraction pattern of the grating can be expressed as a function of $\theta = \sin \alpha + \sin \beta$. The upper diagram shows the pattern for a single wavelength. It is composed of a diffraction envelope (dashed line) under which there are interference peaks with orders indicated (n). The θ coordinate is proportional to the wavelength. The patterns for two wavelengths are shown in the lower portion of the figure. The collection of peaks for each order comprises a spectrum.

from this simple diagram, we immediately see that the collection of $n = 1$ peaks is the 1st-order spectrum; of $n = 2$, the 2nd order, etc. And because the θ scale is proportional to the wavelength, the dispersion increases with the order number. That is,

$$d\theta/d\lambda = n/d \tag{3.8}$$

or

$$d\beta/d\lambda = (n/d)/\cos \beta. \tag{3.9}$$

This is the angular dispersion, and it tells us how well the grating is performing its basic function in life (here bigger is better). The finest wavelength separation that can be discerned is the width of the individual interference peaks,

$$\Delta\theta = \lambda/W. \tag{3.10}$$

That means the resolving power of the grating is

$$\lambda/\Delta\lambda = nW/d, \tag{3.11}$$

and we might also notice that W/d is the total number of grooves. Be careful here. The resolving power we actually have to work with is for the whole spectrograph, and that is typically an order of magnitude poorer. Now if you take a moment to digest Fig. 3 a little more, you will notice a serious problem: the main maximum appears at zero order where there is no dispersion. This problem is cured by blazing the grating. That amounts to shifting the diffraction envelope relative to the interference maxima, i.e., the spectrum. The shift is accomplished in a plane reflection grating by slanting the reflecting grooves of the grating. The center of the blaze or envelope, of course, is the specular reflection from the grooves (see *Photospheres*, Chapter 3. I will refer to Gray 1992 by this title, and Gray 1988 I will call *Lectures*). Gratings are said to be blazed at such-and-such a wavelength in some given order. For example, 10000 Å in 1st order. Then, because the blaze is always in the same direction (i.e., α and β the same for the blaze), the grating equation tells us that $n\lambda$ is a constant, and we then know that our grating is also blazed at 5000 Å in second order, 3333 Å in 3rd order, and so on. This also implies that each order has a width $\approx \lambda/n$. Consider what this means when n is $\approx$ 1-10 for classical gratings versus $\approx$ 100-200 for echelles.

4. The Basic Spectrograph

Find the width of the entrance slit image on the detector. Incoming rays, as seen from the collimator, span an angle $\Delta\alpha = W'/f_{coll}$. Using the grating equation with $\lambda =$ constant to map the rays through the grating, we deduce that

$$d\beta = -(\cos\alpha/\cos\beta)d\alpha, \tag{4.12}$$

and since the scale of the image of W', call it w, is given by the focal length of the camera, namely $w = \Delta\beta f_{cam}$, we have

$$w = \frac{\cos\alpha}{\cos\beta}\frac{f_{cam}}{f_{coll}}W'. \tag{4.13}$$

The ratio of focal lengths is just what you would expect from simple considerations of lenses. The ratio of cosines you might not expect, and it can be significant when you consider that blaze angles can range up to 60-70°.

Now, one more basic point. The focal length of the camera is a parameter of convenience. We choose it to get the proper scale on the light detector, namely w should match 2-3 pixels. Less will loose spectral resolution, more will increase exposure times. Choosing the camera focal length fixes the linear dispersion, i.e., expresses the linear dispersion as a product of the (inverse) angular dispersion and the camera focal-plane scale,

$$d\lambda/dx = (d\lambda/d\beta)(d\beta/dx). \tag{4.14}$$

Using the expression noted above for the first term, and recognizing the second term as the reciprocal of the camera's focal length, we can also write

$$\frac{d\lambda}{dx} = \frac{w(d\cos\beta)}{n}\frac{1}{f_{cam}} \tag{4.15}$$

When we ask now about the resolution of the spectrograph, we are asking about the wavelength increment that corresponds to w. Now that we have the above expression for the linear dispersion, we can write

$$\Delta\lambda = w(d\lambda/dx) \tag{4.16}$$

or

$$\delta\lambda = -\cos\alpha\frac{W'}{f_{coll}}\frac{d}{n}. \tag{4.17}$$

This is the basic resolution equation for a spectrograph. It depends on the angular size of the entrance slit as seen from the collimator, and it depends on the grating parameters,

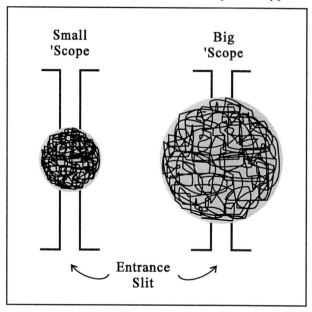

FIGURE 4. The size of the seeing disk is proportional to the focal length of the telescope. Seeing disks for a 4-meter and for an 8-meter telescope are shown as they might appear on the entrance slit of a spectrograph.

n and d. The grating blaze will also largely determine α. Notice in particular that there is no reference to the camera parameters; f_{cam} cancelled out. Bear in mind, we want $\Delta\lambda$ small for high-resolution spectroscopy, and usually the smaller, the better from the scientific point of view. We now turn our attention to how we get $\Delta\lambda$ small enough to do the science we want to do and the compromises we have to endure.

5. The Fundamental Reasoning

The science we wish to do dictates the maximum value of $\Delta\lambda$. The available gratings limit the values of d and n (and hence α) we can work with. The size of the seeing disk tells us roughly how small we dare make W' and still get an acceptable amount of light into the spectrograph. So what's left? Only the focal length of the collimator. It therefore has to be as large as possible.

There are two main paths along which we can progress: 1) observe in relatively low diffraction orders, say $n \leq 20$, or 2) observe in high orders, say $n \approx 100\text{-}200$. As we shall see, the first path gives us the maximum accuracy, the second the maximum wavelength coverage. Let us proceed along Path 1; Path 2 will surface later on its own.

When we select a grating to work in low orders, f_{coll} typically comes out to be 5-10 meters. Such huge dimensions naturally lead to using the coude fixed focus. The spectrograph is just too big to hang on the telescope! Losses incurred with the extra mirrors to bring the light to the coude focus are more than gained back in the wider entrance slit. By using super-reflecting coating on the relay mirrors, the losses can be kept to a few per cent. And there are significant additional gains in going to a coude spectrograph. These include mechanical stability, thermal stability, convenience and precision of alignment, and convenience of use with such things as calibration lamps and liquid nitrogen dewars. Several of these translate into greater accuracy.

But the focal length of the collimator cannot be increased without limit because after the telescope focus, the beam is diverging. Therefore the size of the beam and all the optics needed in the spectrograph increase in proportion to the collimator's focal length. We can build huge mirrors, but we can't build huge diffraction gratings; they are the limiting element.

6. Telescope-Spectrograph Coupling

We all believe that bigger telescopes are better, don't we? But consider the following. Real-world limits on the f-ratio of large telescopes means that larger telescopes come with longer focal lengths. The scale in the focal plane of the telescope therefore goes up as the size of the telescope. And that, in turn, says that the seeing disk will grow with the size of the telescope. The situation is illustrated in Fig. 4. In the seeing disk, the surface brightness is the same for the two cases since that depends on the f-ratio, which is constant. More light can get through the slit along the slit, but the slit width can't be opened more without destroying the spectral resolution. The extra scaled-up width of the seeing disk spills light onto the slit jaws, and it is lost. Conclusion: light into a spectrograph increases linearly with the diameter of the telescope, not with the area of collection.

So, what options do we have left. Well, we can further increase the focal length of the collimator, building mosaic gratings to accommodate even larger beam size. Mosaics were first used at the Dominion Astrophysical Observatory in Victoria, but with 4-meter class telescopes we are already using four-grating mosaics. Is it practical to go to 16-element mosaics for 8-meter telescopes?

Another option is to go to high diffraction orders. This allows the entrance slit to be opened while keeping the spectral resolution under control. It also turns out to require cross dispersion since without it we could use only one order covering at most a few Angstroms. Fortunately, CCD developments have led to rectangular and square detector arrays well suited to handle 'stacked' spectral orders. This is the option selected by almost everyone building instruments for the 8-10 meter telescopes.

There are other options as well. We can place the telescope at a better site where the seeing is better. That too will get more light through the limited width of the entrance slit. Of course that is exactly what we are doing with our big telescopes, although most people get confused and think we are after good seeing because of the improvements in direct imaging. (Was that a political statement?)

The final option, a very practical one, is to use image slicers, to which we now turn our attention.

7. Image Slicers – Another Way Out

Although there are in fact several types of image slicers, I will talk only about the one invented by Harvey Richardson of the Dominion Astrophysical Observatory in Victoria. This slicer has the most advantages for almost all types of spectroscopy, as we shall see. The basic layout is shown in Fig. 5, where we see two concave spherical mirrors facing each other and separated by their common radius of curvature, that is, twice their focal length. Each mirror is split in the middle across a diameter, but the two splits differ 90° in orientation. The opening in the first mirror we call the slot; the opening in the second mirror is the actual entrance slit into the spectrograph. Preceding the mirrors is a cylindrical lens with its axis parallel to the opening in the first mirror.

The bottom part of Fig. 5 shows how the beam from the telescope is focused into a line image at the slot. The length of the line image is the diameter of the telescope beam at

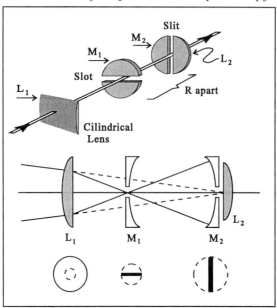

FIGURE 5. The workings of the Richardson image slicer are shown. On top we see a 3D sketch of the optical elements. Lens L_2 is hidden behind the slit mirror M_2. The lower diagram shows the telescope beam entering the slicer from the left. Below the optical elements we see a cross section of the beam at these locations: a circular beam before the cylindrical lens, a horizontal line image inside the slot, and a vertical line image on the entrance slit.

the position of the slot. The height of the line image is the seeing. Therefore, the height of the slot is made large enough to let through any reasonable size seeing. After the light passes the plane of the slot, it diverges in the vertical direction, but is still converging in the horizontal direction to the original focus position of the telescope. At the slit mirror, another line image is formed. This time the long dimension of the line is along the slit and its actual length depends on the focal length of the cylindrical lens and the distance between the two mirrors. Its width is the seeing. This second line image might appear as in Fig. 6, which shows the reflecting surface of the slit mirror. That portion of the light falling on the slit opening passes into the spectrograph where we want it. Since the slit width is always narrower than the seeing for high resolution spectroscopy, we see light to the left and right of the slit that did not make it into the spectrograph. This is the light that would be wasted without a slicer, and the light that the slicer has to re-cycle and get into the spectrograph.

Consider what happens to the reflected light on the right side of the slit. Since the mirror is concave, and since the light comes from the line image located in the slot of the first mirror located two focal lengths away, we expect the reflected light to form a horizontal line image again back in the slot. Of course if this were true, then the light would pass through the slot and back out the telescope to be lost in space. Instead, the one side of the slit mirror is tilted slightly so that its axis is not on the optical axis of the whole slicer, but rather passes through a point located half a slot width above the center of the slot, making the new image lie completely on the upper slot mirror, as shown in the figure. The upper slot mirror would then return the light back to the slit mirror where it would fall to the left of the slit, but as you will now have guessed, the upper slot mirror is also twisted slightly such that its axis passes through the right-hand edge

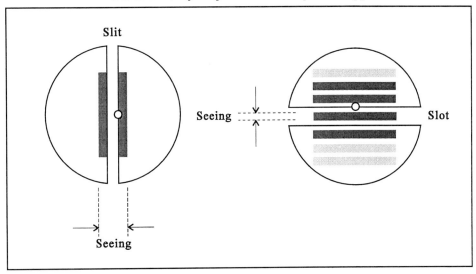

FIGURE 6. The reflecting surfaces of the slicer mirrors are seen here. The shaded portions indicate the line images formed by the slicer. The small circle on the top edge of the slot is where the axis of the left slit mirror is pointed, and the small circle on the left side of the slit shows where the axis of the upper slot mirror is pointed.

of the slit instead of 'on axis' through the middle of the slit. So, the actual new image is not formed to the left of the slit, but goes right through the slit, just where we want it!

The other pair of half mirrors work independently of the first, and serve to move light falling initially off the slit on the left systematically into the slit. If the first cycle of reflections within the device is not enough to span the seeing dimension, the remaining light will cycle again, moving over one slit width each time. For each cycle, another line image is added to the stack on the reflecting surface of the slot mirror, and this is shown in Fig. 6. Typically four such images are used; one that is from the original beam, going straight through and having no reflections, two that have cycled once, and one that has cycled twice. Each cycle causes some reflection loss within the slicer, but naturally, the mirror have super-reflection coatings on them, and the losses are not large. The final element of the slicer is a field lens located at the back of the slit mirror. Its purpose is to image the surface of the slot mirror onto the collimator (and the grating, since the light past the collimator is collimated). Therefore, on the collimator and on the grating we have horizontal lines of light. The slicer is designed to have the slices fill the width of the grating. The height of the lines is either the projected dimension of the seeing or the width of the slot, which ever is smaller.

There are some small details in the actual design that I have not covered (see Richardson 1966), but as astronomers, we can leave such thing to the engineers and opticians. The slicers I use at Western Ontario cover an area of sky equal to 4.5" × 6.8" and typically results in a throughput gain of a factor of two. Among the advantages of the Richardson slicer are a) the height of the spectrum is constant, essentially independent of seeing, and it can be specified in the design of the slicer, b) the design can incorporate a central gap in the illumination of the grating, so that central obstructions like camera pick-off mirrors will not obstruct light nor cause field errors, c) the 'beam' of the telescope is what is imaged on the entrance slit, so seeing noise and spectrum 'wander' is eliminated, d) the collimated beam fits and fills the grating, unlike the usual annular

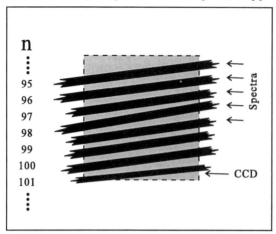

FIGURE 7. Several orders formed by a cross-dispersed echelle spectrograph spill across the CCD detector at an angle. The spectral lines are vertical however. Characteristic order numbers are shown on the left.

beam that often has to overfill the grating to maximize throughput. On the disadvantage side, one cannot do slit-image spectroscopy.

Obviously we stand to gain significantly by using a slicer. This is true in geographical locations having poor seeing, but particularly so for large telescopes where the struggle to get light into a high-resolution spectrograph is very difficult.

8. High-Order Solution: Echelle Spectrographs

Another option in struggling to make the spectral resolution high and at the same time make the spectrograph entrance slit wide is to go to high orders. If n is ≈ 100, then the collimator focal length can even be small enough for Cassegrain installation. Along with this high-order option comes the serious problem of overlapping orders. The free spectral range of a typical order is $\approx \lambda/n$, say 60 Å at 6000 Å, for example. To reach reasonable wavelength coverage, cross dispersion is forced upon us. Cross dispersion is accomplished with a low dispersion grating orthogonally oriented to the main grating, or with a prism. The orders become 'stacked,' as illustrated in Fig. 7. The typical CCD rectangular format facilitates this arrangement. But bear in mind that the images of the entrance slit, that is the spectral lines, are not perpendicular to the dispersion, that each order is slanted across the detector, and that the brightness of the light changes drastically across each order since it is designed to encompass a substantial fraction of the blaze.

In an attempt to mimic the stability of a coude spectrograph, some 'Cassegrain' echelle spectrographs are placed on a bench and the light brought to them with an optical fiber.

When wide wavelength coverage is essential to the science you wish to do, then the cross dispersed echelle spectrograph may be what you need. As we shall see below, this advantage is paid for in generally lower accuracy in the final data, a result of problems such as larger scattered light and the much steeper curvature of the blazed light distribution compared to lower-order configurations.

9. Comments on Detectors

Undoubtedly you all know that we need linearity in a detector; output signal should be proportional to input light. This allows one to work near the bottom or the top of the dynamic range and still get the correct shapes for line profiles. It also means that we do not have to calibrate nor use a non-linear mapping function to get the results we need. Quantum efficiency tells us what fraction of the input light we can expect to register as a detected signal. When one talks about a thick CCD having a quantum efficiency of 40%, or a thinned CCD having a quantum efficiency of 80%, remember that this is the peak value. The actual value at the wavelength you want to work at can be very much less. Check the curves carefully before any observing run you go on. The exposures may be substantially longer than you might at first guess. Readout noise is often negligible compared to photon noise on modern CCD's, at least for high signal-to-noise spectroscopy, and high signal-to-noise goes with high resolving power in most cases. With infrared detectors though, we still have to include the readout noise in the total noise. Likewise, you should know the full well capacity of the detector you will be working with. There are only so may electrons in a tiny pixel, and that's the maximum number of photons the pixel can handle. We need to detect large numbers of photons to get low noise, and we can fairly easily push the capacity of the well. Be aware of the pixel size and the field and format arrangements. The pixel should span no more than half the spectral resolution element ($\Delta\lambda$ above), but it should also not span less than one third of it either, or the exposure times will be larger than necessary (and we all know that telescope time is a most precious commodity!). There may be a preferred section of the detector, and if you are working with a single order, you might be able to produce better data be positioning the spectrum on this better section.

10. Some Examples of High Resolution Spectrographs

Although there are many spectrographs we could talk about, I will choose only a few. Be warned that the words 'high resolution' mean rather different things to stellar and extra-galactic people. The first group usually thinks in terms of 50000 to 100000 or more, while the second group may well consider 10000 to be high resolution. I am definitely in the first camp.

First let me direct your attention to the coude spectrograph at the University of Western Ontario (since that's 'my' machine). See also Gray (1968). The entrance slit is 200 microns wide, and the slicer turns 4.5" × 6.0" of sky into four strips on the grating with a fifth light-less one in the center giving space for the pick-off mirror without light loss. The collimator focal length is 6.96 m. The grating has 316 l/mm, is 154 mm high by 306 mm wide, blazed at 63° 26' so that it works in the 7th through the 14th orders across the visible window. The spectral resolving power is approximately 100000.

There are two cameras. One has a 559 mm focal length and is of the Schmidt type with a typical corrector lens (610 mm diameter) and spherical mirror. It has been used continuously since 1983 with a Reticon self-scanned diode array with 1872 pixels 0.75 mm high and 15 microns wide. The second camera is on the other side of the collimator beam, and it has a focal length of 2080 mm. That's so long that a Schmidt corrector lines is not needed. Currently mounted at this focus is a CCD having 200 × 4096 pixels 15 microns square.

In the orange spectral region, the field is some 70 Å, and light outside 300 Åcentered on this band is prevented from entering the spectrograph by an interference order-sorting filter. This gives the most protection against scattered light, as I will discuss later.

Thermal inertia is built into this coude room by having large amounts of concrete in the floor, ceiling, and walls surrounded by insulation and a double metal cladding allowing convection to move heat from the daytime sunshine out of the building's skin. Temperature stability varies with the season, but can be as good as one degree over several days. Such thermal stability minimizes focus changes and drifts in alignment.

The new coude spectrograph at the Canada-France-Hawaii telescope is an improved version of the one at Western Ontario. It uses four of the same gratings mounted in a mosaic. Images slicers are used, and the collimator focal length is 6000 mm. Instead of a Schmidt corrector, which is prohibitively expensive at this size, a three-element corrector lens and field flattener is used near the focus. This very practical design was worked out by Harvey Richardson. Interference filters can be used to remove all but the needed order, giving up to $\approx$ 90% transmission in the red. In the blue, where the efficiency of interference filters is less good, grisms can be used instead.

This CFHT coude installation yields a resolving power of 120000. The focus and orientation of the detector is fully computer controlled thanks to the effort of John Glaspy and his staff. Alignment of the four gratings in the mosaic is also remotely controlled, and this offers the valuable option of placing the spectrum generated by each of the gratings one above the other on the CCD. In effect, one is taking four simultaneous exposures of the same spectrum. Our natural inclination is to try and superimpose the four spectral images, and if one were ever looking at a source so faint that the readout noise is significant compared to the photon noise, then this is certainly the way to do it. But most of the time we are working to much larger signal-to-noise ratios, and then it is definitely better to use the 'expanded mode' of observing. The reasons it is better stems from the fact that the gratings are not identical and cannot be mounted perfectly. They will have slightly different dispersions and alignments. In the expanded mode, these differences are explicitly accounted for, but if we overlap the four beams, we get an average of the imperfectly matched spectra resulting in somewhat degraded resolution.

The CFHT installation has two sets of collimator and camera mirrors for the blue or the red spectral regions, and either set can be rotated easily into place. The HIRES spectrograph at the Keck telescope is a cross-dispersed echelle system having a resolving power of 67000 (Vogt 1992, Epps and Vogt 1993, Vogt et al. 1994). The instrument is located in a room at the Nasmyth focus, not terribly different from a coude focus, with an f/13.7 beam. The collimator focal length is 4180 mm. The three-element mosaic is mounted on a granite slab and has a total ruled dimension of 1200 mm. These gratings have 52.68 l/mm, and parts of 20 to 40 orders are recorded simultaneously. The whole visible region can be covered in two exposures. The camera is a very fast one, working at f/1.0 and has a 6.7' field.

A number of higher-resolution spectrographs are in the design or construction phases. Most are associated with the 8-10 meter telescopes. Among these are the coude spectrograph at the Hobby-Eberly telescope at the McDonald Observatory, the ESO UVES spectrograph for the VLT, and the HROS instrument planned for the Gemini telescopes (please see Pilachowski et al. 1995). Each of these instruments has innovative design characteristics worthy of study by observing astronomers. Consider, for example, the 'white pupil' approach to the UVES, the simultaneously-run dual beam system for the Texas system, and the so-called 'immersed' gratings of the Gemini HROS spectrograph.

Although several of these spectrographs have a truly high resolution option $\approx$ 100000 or more, they are mainly optimized to run in cross dispersed mode at lower resolving powers $\approx$ 40000.

11. Using Spectrographs, or So That's the Way it Works!

11.1. *Some Basic Practical Details*

There is a basic rule that I recommend be followed by each observer: get to the facility well ahead of your observing run, learn how to run things, and practice.

Just who is responsible for what varies from one observatory to the next, but as an observer you certainly should test a few basic things to help assure the integrity of your data. There should be a flat field lamp of some sort. Since each pixel of the detector has a slightly different electronic gain and/or quantum efficiency, the source spectrum has to be divided by the exposure of the flat field lamp. This division removes the 'structure' of the array. It may also be necessary to first subtract a zero-level pattern or bias. At Western Ontario, I use a ribbon filament lamp for generating a flat continuum spectrum. A simple lens images the uniform filament onto the entrance slit of the spectrograph. The whole device fits on a small optical bench permanently mounted in front of the entrance slit. When the lamp exposures have been completed, the lamp is slipped off the bench and out of the way of the telescope beam.

Do some simple linearity checks. If you have a stable lamp, try taking a few exposures of increasingly long times. A plot of signal strength versus exposure time should be linear. Or, take a strong and a weak exposure of the same line and divide them. The result should be flat and show no variation across the position of the line. While you're at it, check out the saturation level. All CCD-type detectors will saturate if enough light is put on them. Simply increase the strength of the exposure (time × light level) until the output shows a leveling off or a 'blooming' from charge spilling across barriers separating pixels. This will not permanently damage CCD-type detectors, but it might make them noisy for several subsequent exposures, so do this early on in your testing and well before your plan to take serious data. Be careful with intensified systems that use photocathodes of any sort. Over exposure for them could mean serious damage even destruction.

Take a look at the level of the readout noise by taking some dark exposures. You may not be able to translate the numbers directly into equivalent photons unless you have the conversion factor from the documentation, but at least you will gain a feel for the count level where readout noise is important. Likewise try dividing a couple of equal flat-field lamp exposures. The RMS deviations divided by $2^{1/2}$ give you a measure of the noise at that level of exposure.

Now let's turn our attention to the spectrograph itself. The wavelength is selected by rotating the grating. Don't expect this to be fool proof. Verify ahead of time that you have the right wavelength region. You will need a map of a known source so that you can recognize the line pattern in the exposure. This might be done with an emission-line lamp, the daytime sky, or with a bright star during twilight (if you have no other option). Nothing is more embarrassing or frustrating than to head into your first night and not be able to find the spectral region you intend to measure!

Unwanted orders in the spectrograph are removed with color glass or interference filters, or the orders may be spread out with a cross disperser. It is usually a good idea to make sure the proper filters are actually available; you may have to bring your own.

11.2. *Focus that Spectrograph*

Step number one in starting a night is to focus the spectrograph. Do this before it gets dark. Observing time is precious! One simple way of focusing is to take several exposures of emission lines from a lamp and change the focus until they are most sharp. This is not as good as using a Hartmann mask, however. All Hartmann masks obscure some portion

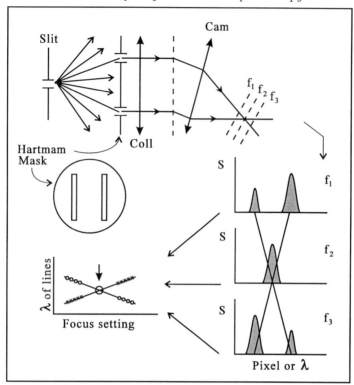

FIGURE 8. A two-slit Hartmann mask placed in from of the collimator is a simple device to aid in locating the correct camera focus. The diagram illustrates three exposures taken at the three focus positions f_1, f_2, and f_3. Several more positions are measured to give the plot in the lower left, from which interpolation yields the correct focus position.

of the beam, but the simplest version consists of two slits in a screen placed in front of the collimator. For a monochromatic source, this mask generates two 'beams' that naturally are supposed to come together at the camera focus. So take a few exposures starting inside true focus and finishing outside of true focus. Then make a plot like the one in Fig. 8. Interpolate to get the point where the lines cross, i.e., the focus setting for optimum focus. Remember to reset the focus to that point.

11.3. *Is the Alignment O.K.?*

Undoubtedly you will want to practice focussing well before the first night you wish to observe. While doing so, you could also check the alignment of the detector in the camera. For example, is the spectrum (assuming a single order and no cross dispersion) square with the detector, or does it make some crazy angle? If it is at an angle, that complicates the reduction because you cannot sum along each column. There should be some way to rotate the detector. Maybe this is something you should do, or maybe it is something under the jurisdiction of the resident staff. Likewise for misalignment of the detector with the focal plane. If the focus is not at the same position all across the field, the detector is probably tilted with respect to the focal plane of the camera. Getting this part right might be tricky, and it might be appropriate to ask the staff for help.

In some cases you might have to rotate the grating between exposures to get other wavelengths on the detector. If the spectrum changes its position perpendicular to the dispersion, it probably means that the grating is not aligned with the axis of rotation

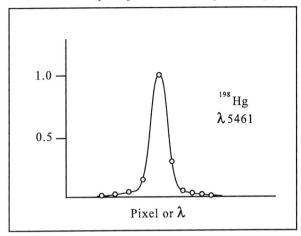

FIGURE 9. A typical measured instrumental profile.

of the grating, likely the grating is tipped in its cell causing the spectrum to be at an angle across the camera area of the spectrograph, but of course you see only a small portion of this slanted spectrum on the detector. The proper solution for this problem is to move the grating in its cell. This is very definitely not something you should attempt to do! Moving the grating in its cell is a delicate operation, and too much pressure on mounting pads could easily distort the grating causing multiple images or introducing focusing properties. Get help for this one.

Assuming there is no cross dispersion, check carefully the orientation of the spectral lines relative to the direction of dispersion. They should be perpendicular of course. This check is relatively easy to do when the spectrograph is outfitted with a Richardson image slicer or when a long portion of the slit is illuminated by a lamp, since then the spectrum will have sufficient height to allow you to see any slant in the lines. If the spectral lines are slanted, then the reduction is more complicated and likely to be less accurate. And if you were to sum columns without allowing for the slant, the spectral resolution would be badly damaged. Slanted lines occur when the entrance slit is not parallel to the rulings on the grating. This should be fixed by the observatory technical staff. Of course, if the spectrograph has a cross disperser, the spectral lines are always slanted, and this is one of the lesser problems associated with such configurations.

12. The Instrumental Profile

Is there such a thing as the instrumental profile (also called the delta-function response) for a spectrograph? Focal surfaces are rarely a true plane, nor are array detectors for that matter. Chances are the instrumental profile does change across the field, and you will have to check this point when you do the measurements. One would hope that the instrumental profile would be constant at least across the dimensions of the individual spectral lines of interest. In the best case, one instrumental profile will hold across the whole field of view. But the importance of any variation across the field will have to be assessed individually to see how the non-constancy might affect the analysis.

How are we going to measure the instrumental profile? The name δ-function response tells us, doesn't it. We need a narrow spectral feature, one that looks like a δ function compared to the instrumental profile. There are at least three such sources:

(a) an isotopic mercury lamp excited by microwave radiation to minimize the heating and hence the thermal width. One isotope avoids the closely spaced lines from different isotopes. ^{198}Hg is the most common.

(*b*) diffuse laser light works because the photons coming out of lasers are stimulated, hence in phase, and the effective length of the photons is long, i.e., the spectral lines are narrow. But the coherence of the photons causes strong diffraction that does not occur with starlight. Fortunately this coherence can be sufficiently garbled by scattering the light twice from ground glass placed in the beam.

(*c*) telluric lines in the spectrum of the sky, moon, or bright star are sometimes useful. They are not as good as the other sources because they have more intrinsic width, and one should be careful to choose telluric lines that appear to be relatively weak otherwise they may in fact be saturated and only look weak because of the spectrograph instrumental profile blurring (Griffin 1969a, b, c). For resolving power of ≈ 40000 and less they will often prove satisfactory.

For resolving power well above 100000, a correction should be applied for the thermal width of the mercury lamp lines, or if using a laser, a stabilized version might well be necessary.

Figure 9 shows an instrumental profile, a measurement of the λ 5461 mercury green line, from an isotopic lamp. One is struck by the few points outlining the profile, but upon reflection, that's exactly what we should expect if the sampling is correct and amounts to 2-3 pixels per resolution element.

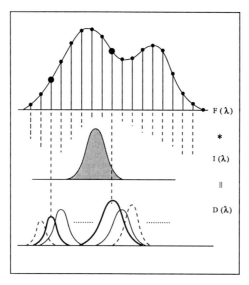

FIGURE 10. The convolution process is viewed here as the sum of many instrumental profiles each one coming from one of the δ functions used to represent the original spectrum, $F(\lambda)$.

Try and find some line in the δ-function source near the same grating angle as you plan to use with your observing. That way the diffraction pattern from the grating will be the same, and its contribution to the overall instrumental profile will not be altered. Since we generally expect the grating contribution to be a relatively small one, the grating angle doesn't have to be exactly the same. The pixel spacing across the measured instrumental profile can be translated into wavelength using the dispersion, $d\lambda/dx$, for the order used for the astronomical source. The dispersion itself can usually be found from the positions of spectral lines in an exposure of a cooler star (where there are plenty of lines who's positions can be measured). The non-constancy of the dispersion across the field varies from one spectrograph to the next, but often a cubic polynomial is adequate to relate λ to x.

The physical process we deal with between the true spectrum and the instrumental profile is the convolution,

$$D(\lambda) = F(\lambda) * I(\lambda), \tag{12.18}$$

where $D(\lambda)$ is the data we measure, $F(\lambda)$ is the true flux spectrum, and $I(\lambda)$ is the instrumental profile. Although this is no doubt a familiar operation to many of you, let me review the thinking that goes on here because it is useful in many contexts. Fundamentally, we think of $F(\lambda)$ being approximated by a series of δ functions like those shown in Fig. 10. Of course each δ function produces $I(\lambda)$ in the output spectrum since $I(\lambda)$ is the δ function response. In other words, the observed spectrum is the sum of many $I(\lambda)$'s, each one shifted to the position of one of the δ functions denoting $F(\lambda)$ and scaled to its strength. We can do this same δ-function breakdown in other situations too. It works because we are dealing with linear systems...what we can do for one δ function, we can do for all of them, and then simply add the individual results to get the complete answer. You might also care to look at *Photospheres*, Chapter 12 and Valenti et al. (1995).

13. The Problem of Scattered Light

First question: what's scattered light look light...how can I recognize it? To first order, it will appear as a zero-level off-set. In particular, the deep absorption lines will appear less deep than they really are.

Second question: how bad a problem is this? Well, differences of 30% in equivalent widths have been attributed to scattered light...that's pretty alright!

Third question: where does it come from? One component comes from the grating, most commonly from the far wings beyond where we normally measure the instrumental profile. A second component arises from more general scattering from dust in the air and on the optical surfaces, from 'random' aberrations and flaws in the grating, mirrors, and transmission optics. Consider Fig. 11, for example. Instrumental profiles are shown for the 'old' ESO spectrograph used with two different gratings. The first thing we see is the central spike in each. Both are sharp and narrow, as you would expect. These are the parts we generally measure at the telescope on the nights we observe. But there are other things here of critical interest. The lower panel shows the many ghosts and satellite lines that occur with the older ruled gratings. These all contribute to scattered light. But perhaps the more striking thing is the huge flat extended wings, almost like pedestals upon which the 'main' instrumental profile sits. We certainly would not normally measure this when routinely getting $I(\lambda)$. Look at the level of these wings (careful, logarithmic ordinate). They are down $\approx 10^{-7}$ and 10^{-6} from the peak. That means that if one lets into the spectrograph light from the whole visible window, a span of a few thousand Ångstroms, the scattered light level from these wings alone will be $\approx 10^{-3}$ to 10^{-2}. Another example is given in Fig. 12. Qualitatively the same story is seen here, but the 'pedestal' is now $10^{-4.5}$ or so... Take a look at some of the source material, Dravins (1994), Tull et al. (1995), Griffin (1968, 1969a, 1969b).

Fourth question: how do we measure it? Look at deep absorption features in a spectrum. These might be artificially created using a rejection-band filter or a Fabry-Perot, or they might be natural features such as the Ca II H & K line cores, or the O_2 bands in the near IR, as shown in Fig. 13. If the darkest O_2 lines are not at zero light, most likely

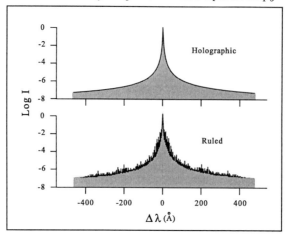

FIGURE 11. These remarkable instrumental profiles have been measured to extreme distances from the central peak. The flat broad 'wings' under each peak gives the light scattered in the direction of dispersion by each grating. Based on a figure in Dravins (1994).

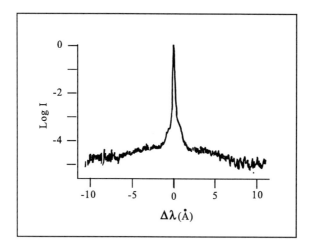

FIGURE 12. Compare the level of the broad wings here with those in the previous figure. Based on a figure in Tull et al. (1995).

the light you're measuring is scattered light. And don't forget to compare your deep core measurements with other people's results. Many of these strong absorptions are wide enough to make variations with resolving power second-order events in this game. Note also, that if you are measuring emission lines, that scattered light is not likely to be a problem.

Fifth question: how do we get rid of scattered light? Use high quality optical components, and keep their surfaces reasonable clean. But from the observer's perspective, the most effective action is to *reduce the bandpass*. Light you keep out of the spectrograph can't contribute to scattered light...it's as simple as that. For example, if you are measuring 75Å, use a bandpass filter with a bandwidth of ≈ 200 or 300Å, and you will eliminate the scattered-light problem. (Still narrower bandpasses start to suffer from reduced transmission...you shouldn't overdo a good thing.) Of course, if you are using a cross-dispersed echelle spectrograph, you will be stuck with the scattered-light problem. Cast a glance at some of the extensive efforts that have gone into handling scattered

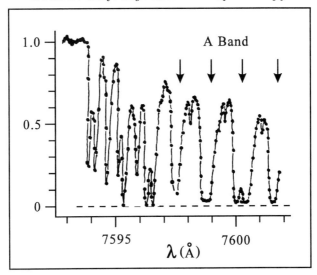

FIGURE 13. The atmospheric oxygen bands have very black sections useful for detecting scattered light. Based on a diagram in Tull et al. (1995).

light in echelle systems: Churchill and Allen (1995), Hall et al. (1994), Gehren and Ponz (1986), for example. I also suggest you read the summary discussion by Dravins (1994).

Sixth question: how do we correct for scattered light? If we can't eliminate it, we have to correct for it, or else we're producing inferior results. Some of the information you may need can be gotten from the references just mentioned. The simplest treatment assumes it is possible to ignore any spectral features that appear in the scattered light and simply define some mean value over the spectral region of interest, call it s. Then the observed flux level will be the true level less the amount scattered out of that wavelength plus the amount scattered in, or in symbols,

$$F_o = F_t - sF_t + s\langle F \rangle, \tag{13.19}$$

where $\langle F \rangle$ is the average value of light in the region being considered, typically reduced from the continuum by the presence of absorption lines. Write this equation down twice, once for a point in the spectral line and again for a point in the continuum, ratio them and solve for the true profile to get

$$\frac{F_t}{F_{t,c}} = \frac{F_o/F_{o,c} - s\langle F \rangle/F_{o,c}}{1 - s\langle F \rangle/F_{o,c}} \tag{13.20}$$

where the subscript c denotes continuum values. The last term in the numerator is the basic correction, while the denominator is simply a re-normalization to the new continuum.

14. Other System Light Problems

Beyond scattered light, one should look for reflections. Light can bounce off surfaces of lenses, windows, the detector, and so on. Internal reflections in the detector can cause Fabry-Perot-type interference fringing. Light reflected from the window of the dewar could go back through the camera, reflect from the grating facets, and come back again to the detector. Some reflections might be totally out of focus and appear as extra

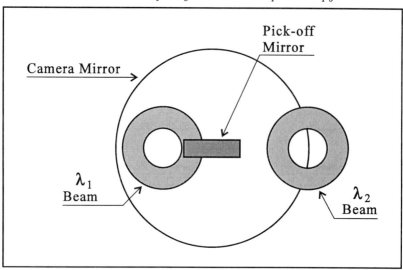

FIGURE 14. Central obstructions such as pick-off mirrors cause field errors because different beams (wavelengths) from the grating pass through different portions of the camera on their way to the detector. A Richardson image slicer can avoid this problem.

scattered light in some directions, but in other combinations, they might be focussed or nearly focused, and could be quite dramatic.

A simple fix for many reflections in a single-order configuration is to tip the detector slightly around an axis parallel to the dispersion. This tip is too small to sensibly alter the focus across the height of the spectrum, but of course, if you are using a cross-dispersed echelle, this option won't work.

Other field errors can also exist. Consider, for example, the common situation of a central obstruction such as a pick-off mirror located in the camera. Without an image slicer the beams of different color light will come into the camera as shown in Fig. 14. (The actual beams will more than likely be elliptical in cross-section because of the cosine terms in the equations we derived early on.) Obviously different fractions of each beam get through to the detector, meaning that a flat continuous spectrum going into the camera is not seen as flat by the detector. This kind of problem is greatly reduced or eliminated when an image slicer is used to produce a dark gap across the center of the beams.

15. Stability Checks

Not all spectrographs are nicely behaved. Keep an eye on the one you are using. Check the focus from time to time on the days before your observing run. You can't expect to get high resolution data if the spectrum is out of focus. If the focus is not stable, why has it changed? Is there a noise pickup on the hardware controlling the position of the detector? Is it thermal drift? Maybe too may people have spent too much time in the coude room and it has warmed up with the body heat. Maybe there is too little insulation and the outside temperature changes are penetrating the spectrograph...

Is the wavelength position stable? Many spectrographs used for measuring radial velocities are known to show substantial systematic velocity errors during the night. Could it be that the grating rotation is creeping slowly? Maybe some other component is expanding or contracting with thermal drifts.

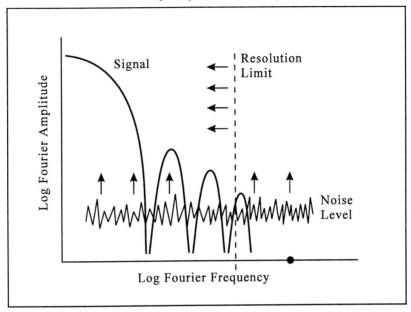

FIGURE 15. High spectral resolution and high signal-to-noise go together because both are needed in order to measure the spectral-line information that always lies in the lower right portion of the diagram.

How often do you have to take flat-field lamp exposures? Can you take a flat field at the beginning of the night and divide it by one taken at the end of the night and have no increase in error above the photon noise and have an otherwise flat response across the array? If not, then you have to do some homework and find out how often a new lamp has to be recorded.

Instabilities might also be introduced by collimation errors. The kind of thing I'm taking about is misalignment of telescope and relay mirrors such that the beam into the spectrograph is not exactly along the same axis when the telescope is pointing in different parts of the sky. Errors of this sort are easy to see near the edges of the field because the edges will have some vignetting (part of the beam from the grating for that wavelength doesn't make it into the camera aperture). The vignetting will be different for the astronomical source and for the flat-field lamp. You must ask yourself, does this affect my results?

16. Comments on Instrumentally-Induced Distorsions of Your Data

As most of you likely already know, the spectrograph affects the data rather drastically. Instead of a continuum of points, we get one point for each pixel in the detector, the integrated result of the small wavelength band corresponding to the width of the pixel, and we have the blurring of the instrumental profile, along with possible scattered light, and so on.

Most array detectors have a sensibly constant spacing between pixels, and this usually determines the wavelength spacing between data points. This sample spacing limits the highest Fourier frequency we can work to, given by $\sigma = 1/(2\Delta\lambda)$, and called the Nyquist frequency. Of course if we move the detector over by half a pixel width, take a second exposure, and interlace it into the first one, we have doubled the Nyquist frequency.

Again, we want to have the scale in the camera arranged to give about two pixel

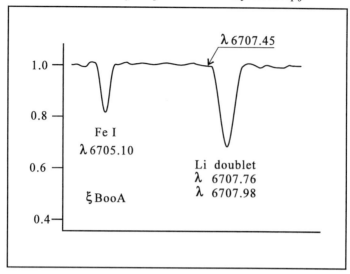

FIGURE 16. The lithium line in 2 Boo A is very strong. The blend on the left wing can cause serious errors when the lithium line is weak.

spacing over the width of the instrumental profile. If we spread the spectrum out more and sample more frequently, we will be increasing our exposure times and reducing our wavelength coverage. If we reduce the dispersion and sample less often, we might loose fine structure in the spectrum, and if so, then we might as well use a spectrograph of lower resolving power to begin with!

The effect of the instrumental profile can be seen as a blurring process, that is, the systematic removal of higher frequency information in the wavelength domain, or we can think in terms of a multiplicative filter being applied in the Fourier domain. In spectrograph cases, the filtering becomes monotonically more severe with higher Fourier frequencies.

Consider Fig. 15. I've drawn this to emphasize why high spectral resolving power and high signal-to-noise ratios generally go together. The Fourier transform of a typical spectral line is shown, and you see that there is a large main lobe at low frequencies and smaller sidelobes toward higher frequencies. At some level there will be measuring noise. Hopefully it will be white, meaning a statistically constant level with frequency, as indicated in the diagram. If the noise is high the sidelobe structure become buried, and the details of the line profile are lost.

But consider what happens with the resolution. High frequencies are preferentially filtered away, so that we can effectively work only on the low-frequency side of some resolution limit. Again, if the resolution is too low, we cannot measure the sidelobes and the details of the spectral line are lost.

So, in order to get at the sidelobes, we have to push toward the lower right portion of the Fourier domain. That takes both high spectral resolution and high signal-to-noise. These two go hand-in-hand together.

At this point, it is now time for us to expand our capabilities and turn from man-made machinery to the tools supplied by the stars.

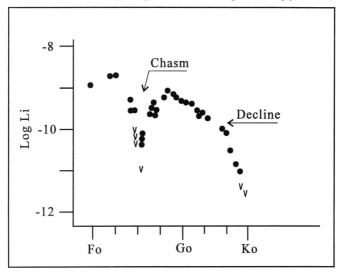

FIGURE 17. The abundance of lithium declines toward the cooler side of this diagram because deeper convection zones in cooler stars result in more lithium being destroyed. The remarkable lithium chasm remains to be explained, although there are several hypotheses.

17. Spectral Lines: an Extension of our Spectroscopic Tool Box

17.1. *Line Strengths - Low Resolution?*

What kinds of spectral lines are there, and what kinds of measurements can we do on spectral lines? I will restrict my discussion and examples to stellar absorption lines mainly, but the same principles apply rather generally. The simplest parameter we can think of for a line is its total strength or equivalent width. One often hears the partly correct statement that equivalent widths can be found without high spectral resolution. In a simpler world this might be true, but nature is usually more devious. The problem is simply that in many cases, especially for cooler stars, there are just too many lines and they overlap.

Blended lines must be at least partially resolved if we are to recognize the existence of the blend and if we are to accommodate it in the analysis. In this context, low resolution actually does two main things. First it blends the lines together, and the wing portion of the line often spreads into the neighboring line and is lost to the measurement. Second, as the numbers and widths of lines increase, less and less often can we identify the continuum. Low resolution lowers these small windows of continuum, causing us to place the observed continuum too low. Consider the simple situation in Fig. 16 where we see an unusually strong lithium line. Located on the left wing of the line is a small blend. At lower resolving power, this blend will be mistaken for part of the lithium line. You might not think the error would be terribly serious, and in this case it's not, but only because the lithium line is very strong. In cases more like the solar one, the lithium line is weaker than the blending line. Then including the blend can throw the apparent line strength off by a factor of two or three or more, and that is a serious error.

Still, in favorable cases, we can hope to get the equivalent widths of lines with modest resolving power of say 20000 or less. Since line strengths depends quite generally on temperature, pressure or surface gravity, and chemical abundance, each of these parameters can be found, although this usually requires using many different spectral having different sensitivities.

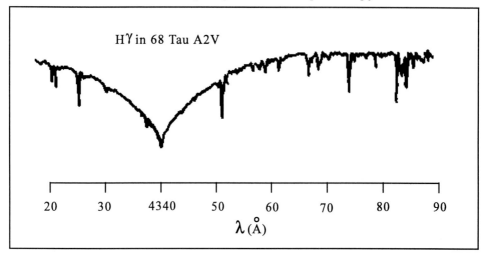

FIGURE 18. The widest lines in stars arise from the linear Stark effect in hydrogen.

There are numerous reasons to measure chemical abundances. For example, the processing of isotopes in stellar interiors gives us information on the nuclear processes going on there. The convection zone grows deeper as a star becomes a red giant, and some of this processed material is moved to the surface, and then we can bring our spectrographs into action. Numerous studies of isotope ratios illustrate the point. Here is a small sample: VandenBerg and Smith (1988), Langer et al. (1986), Sneden and Pilachowski (1986), Little-Marenin and Little (1984), Smith and Norris (1983).

Assuming generations of element building, one can seek out the past history of our galaxy and why it has its current structure. The study of population types has developed a venerable history. You can get a taste from: Luck (1982), Bessell and Norris (1984), Molaro and Castelli (1990), Ann and Kang (1986), Matteucci and Greggio (1986).

Or we might learn about the physics of the convective envelopes, or possibly chemical diffusion, when something like the lithium behavior, as shown in Fig. 17, is discovered. Theory tells us that we should expect a drop in apparent lithium abundance toward cooler stars because their convection zones are deeper, causing lithium in the envelope to be cycled through hotter temperatures in the lower reaches of the envelope where it is destroyed (see Spite and Spite 1982 for background). The cause of the dip or chasm at spectral types around F3 V is not so clear, and this chasm does not appear to exist in younger clusters (Boesgaard et al. 1988a, 1988b; Vauclair 1988). Are we looking at an aging effect, or is there some other fundamental difference from one star cluster to another?

There is one other basic thing we might get out of spectral lines with modest resolution spectroscopy, namely the projected rotation rate, $v \sin i$. As long as the resolving power, expressed in km/s, is smaller than the rotational broadening, we have a handle on the rotation rate. Most of the rotation rates measured in the first two-thirds of this century were done with modest resolution, and one is constantly running into rather high upper limits, typically 10-20 km/s. Incidentally, one should be careful on this point. Some authors and some compilers of catalogues use zero to denote values below the resolution limit. And even recently, erroneous conclusion have been drawn by averaging upper limits as if they were actual measurements of rotation (Rutten and Pylyser 1988, Schrijver 1994).

17.2. *Intrinsic Line Broadening*

One of the things higher resolution allows us to do is look at the detailed shapes, broadening, and asymmetries of spectral lines. There is a great deal of physical information here, but only a few basic mechanisms: 1) implicit widths of the atomic energy levels and distortions of the energy spacing between atomic levels caused by perturbations from outside the atom, 2) dramatic variations in the source function for strong lines, 3) Doppler broadening from thermal motions, velocity fields in the atmospheres of stars, and from rotation, and 4) magnetic splitting, with Zeeman patterns sometimes resolved and sometimes not. I'll take us briefly through the major aspects of this list, starting with the broadening from atomic sources.

The giant of all line broadening is seen in the hydrogen lines of A-type stars where the linear Stark effect stretches these lines over several tens of Ångstroms. Figure 18 shows what I'm talking about. Now you might think that since these lines are so broad, low resolution is all that is needed for their measurement, but look at all the weaker lines in the wings of the hydrogen lines. If you are careful to allow for these, then indeed low resolution might do it. Certainly this is not a job to be tackled with a cross-dispersed echelle...the hydrogen lines are wider than a couple of orders! Can you imagine trying to handle the blaze function with sufficient precision? No, this is a job for a low-order spectrograph. The cores of hydrogen lines can be quite sharp, at least for the small fraction of A stars having low rotation rates, and if they are to be measured, resolving power of 30000 or so is needed. The hydrogen lines depend on temperature and pressure, so both of these parameters can be measured using these lines.

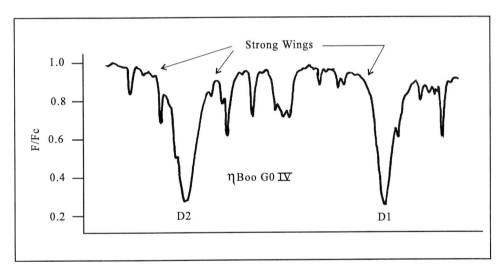

FIGURE 19. The wings of lines like these for the sodium D lines arise from pressure broadening, i.e., perturbations of the atomic levels by collisions with other particles in the star's photosphere. Most of the other lines arise from telluric water vapor.

Broadening by the quadratic Stark Effect and the van der Waals forces is commonly seen in the stronger non-hydrogen lines. One sees lines like the sodium D lines shown in Fig. 19. Here the wings are what we're after; they vary with temperature, pressure, and the chemical composition (for an example of an analysis, see Smith and Drake 1987). Notice also the numerous weaker lines in Fig. 19. Most of these are telluric water-vapor lines. I chose this exposure purposely to show you what kinds of problems can occur in

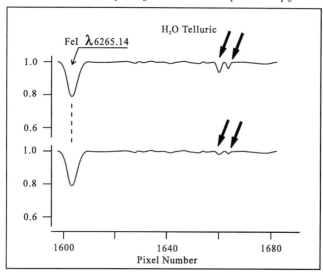

FIGURE 20. Here other telluric lines are illustrated. They can be used to measure the radial
velocity of the star with good precision.

wet climates. Resolving power of 40000 or so is probably adequate for studies of such
strong wings.

Incidentally, while I'm thinking about telluric lines, look at Fig. 20. Two exposures of
the same star are shown, taken on two night separated by a few weeks. Those weak lines
on the right have changed strength and position. They are from telluric water vapor.
Although such lines can be an annoyance, they can also serve as a reference frame for
measuring radial velocities. Some of the most precise velocity variations are measured
with high-resolution spectrographs in conjunction with such lines, either formed in the
atmosphere or by an absorption cell placed in front of the entrance slit of the spectrograph. Precision of 100 m/s can be achieved with a single telluric line; 10 m/s with
several lines from an absorption cell; and 1 m/s is claimed for specially designed radial-
velocity instruments. More information can be seen in Griffin (1969a, b, c), Campbell et
al. (1988), Marcy and Butler (1992), Larsen et al. (1993), Brown et al. (1994), Libbrecht
and Peri (1995), and also McMillan et al. (1992) who use a Fabry-Perot interferometer
rather than a spectrograph.

17.3. *Probing a Stellar Atmosphere*

Figure 21 illustrates what happens with the very strong H & K lines of Ca II. As the
top portion of the figure shows, the H & K lines are a lot stronger than the surrounding
'normal' lines. To understand what is happening here, it is a good idea to stop and
reflect on what spectral lines actually are. Why do stellar spectra show absorption lines?
When one looks in the continuum or in the wings of a line, the line absorption is small
and we see to deep photospheric layers. When we look at wavelengths closer to the core
of the line, there is more absorption, and our line of sight cannot penetrate as deeply,
we see mainly the radiation from higher layers. But the higher layers are cooler than
the deeper layers, so there is less light from them, and we see an 'absorption' line. Now
with the really strong lines like the H & K lines, the absorption is so strong toward the
core that our line of sight doesn't even penetrate as deeply as the photosphere. Instead
we can only see into the overlying chromospheric regions, but they are hotter than the
deeper layers, and so we see emission in the cores of these lines. There is a (non-linear)

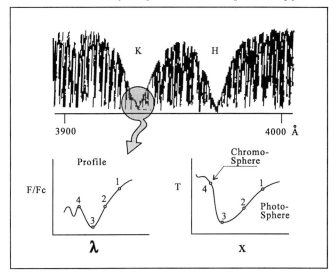

FIGURE 21. Super strong lines like these of Ca II allow us to probe the run of temperature with depth into the atmosphere. These lines are strong in cool stars, and virtually all cool stars have temperature inversions, chromospheres and corona, resulting in emission peaks in the cores of the lines. The profile is a sort of map of the temperature distribution. Depths of formation for four points in the profile are shown.

mapping between the temperature (source function, to be more precise) as a function of depth and the line profile, as shown qualitatively in Fig. 21. So with typical resolving power $\gtrsim$ 50000, we can do this kind of exploration of the temperature structure of stellar atmospheres. Higher resolving power is needed not for the calcium lines, but to handle all those weaker lines on top of them.

The emission bumps in the H & K line cores are of particular interest since they come from the region of the temperature inversion. Two source of power are thought to produce the chromospheric temperature rise, 1) dissipation of acoustic waves as shock waves and 2) magnetic fields. How the magnetic component works is still debated. We might have Alfven waves carrying the energy of granulation upward along the field lines, or maybe the magnetic fields themselves are being converted into thermal energy. Temperature inversions are common on the cool half of the H-R diagram, i.e., wherever there is rotation and a convective envelope. And there are variations in the chromospheric emission on several time scales from days to decades at least. One example is shown in Fig. 22. Here the K-giant Arcturus shows significant variations over a six-day interval. We still don't understand such variations.

Another chromospheric-type line is He I λ 10830, like the one shown in Fig. 23, taken from a study of cool binary stars. Many of these binaries are tidally coupled, resulting in the component stars rotating much more rapidly than comparable single stars. Invariably such stars show super-strong magnetic activity and strong chromospheres.

17.4. *Spectral Line Broadening by Rotation*

The velocity fields in normal stellar photospheres amount to a few km/s, and so when rotation is greater than about 10-15 km/s, it dominates the shaping of the line profiles. Compare the two exposures shown in Fig. 24. The rotational broadening in the *o* Dra exposure is immediately obvious, and we can even recognize the distinctive shape of rotation. In a sense, this is the 'easy' case because the one mechanism (rotation) is obvious

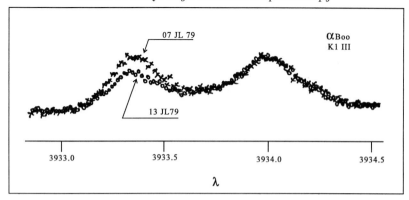

FIGURE 22. Variations of the chromospheric emission in the core of the K line of Arcturus occur on short times scales of a few days. Based on a figure from Gray (1980).

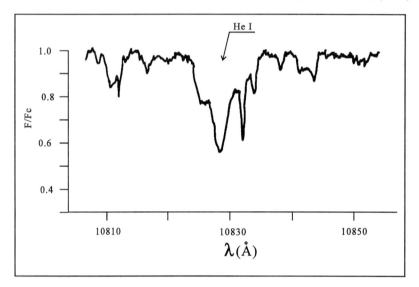

FIGURE 23. Another chromospheric line of interest is the He I line, λ 10830. It is fairly difficult to observe because the quantum efficiency of detectors is low in this spectral region. Based on a figure in Shcherbakov et al. (1996).

and essentially the only one we have to be concerned with. When we are confronted with lines like those in κ Cyg, the slower-rotating giant in Fig. 24, we have to be able to disentangle the Doppler broadening of rotation from the Doppler broadening of the atmospheric motions. Fourier analysis has proved the most successful way to do this, and these methods are reviewed and explained in *Lectures*. I recommend you read this material if your research leads in this direction.

A number of interesting results come out of studying rotation. Perhaps the most striking is shown in Fig. 25. High rotation is seen for the hotter stars, but only slow rotation for the cooler ones. There is essentially a Rotation Boundary in the middle of the H-R diagram. Furthermore, there is a Maxwell-Boltzmann distribution of rotation rates for the hotter stars, but rotation is a single-valued function of temperature for the cooler stars. The only confirmed exceptions are tidally coupled binaries. These stars have evolved off the main sequence, and since their masses are $\lesssim$ 3 solar masses, they are expected to pass once across the diagram from left to right. So in affect we have

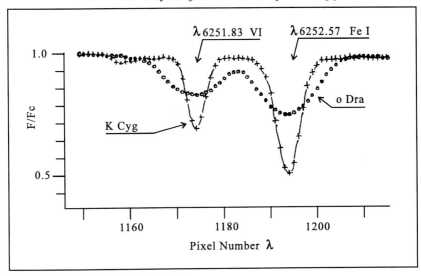

FIGURE 24. The rotational broadening of the lines of *o* Dra are obvious, while the rotation component in the broadening for *κ* Cyg's lines are much less so.

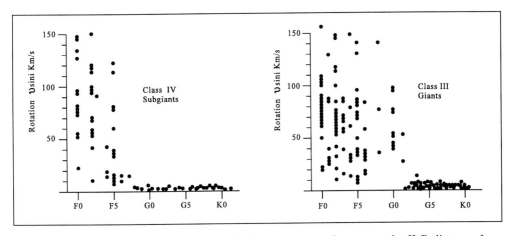

FIGURE 25. Rotation shows a very steep decline as stars evolve across the H-R diagram from left to right. Based on a diagram in Gray (1991).

a time progression displayed here. Shortly before the rotation drops so dramatically, these stars develop convective envelopes, so it seem inescapable that the discontinuous behavior in the rate of rotation is a result of magnetic braking. Calculated moment-of-inertia increases change more slowly and are an order of magnitude smaller than needed to account for Fig. 25. I think the rotation of the cooler stars is actually controlled by the convection zone and its parameters through the turning on and off of the magnetic brake. I call this the 'rotostat hypothesis.'

At higher luminosities, we see slow rotation that is completely accounted for by moment-of-inertia increases. It seems likely that the growing moment of inertia slows the rotation of these huge stars below a threshold for turn-on of a magnetic dynamo. Please refer to *Lectures* and to Gray (1991) for more details.

In any case, the H-R diagram is split down the middle with fast rotators on the left

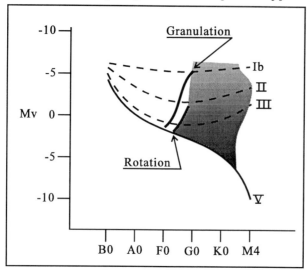

FIGURE 26. The H-R diagram is cut through the middle by two boundaries. Stars on the hot side are usually rapid rotators with rates of 100-300 km/s. Stars evolving from the main sequence first cross the granulation boundary and develop convective envelopes. Shortly thereafter, magnetic braking results in the steep drop of rotation as seen in the previous figure. More luminous stars of classes I and II experience a slower more continuous change arising from their increase in moment of inertia. Different shading is used to show these two populations in the diagram.

and slow rotators on the right, as shown in Fig. 26. Not far from the Rotation Boundary we see the Granulation Boundary, so let us turn our attention to stellar granulation.

17.5. Spectral-Line Asymmetries, Granulation's Mark

If you cast a casual glance at most spectral lines, even those taken with the highest spectral resolution, they look reasonably symmetric. For instance, Fig. 27 shows a typical case, the profile for λ 6252.57 Fe I in a normal G5 giant. As a measure of the asymmetry, we have come to use the line bisector. Just take several horizontal cuts across the profile and bisect them. The line connecting these midpoints is the line's bisector. Now you would expect a symmetrical line to have a straight vertical bisector, and the one shown in Fig. 27 looks pretty straight and vertical, doesn't it? The trick is to expand the wavelength scale a factor of 20 or 40 or so, as shown in the other part of the figure. Now we can see the shape of the bisector and it is definitely not straight nor vertical. This kind of bisector shape is normal for stars on the cool side of the Granulation Boundary. If the resolving power of the spectrograph is only as high as 50000 or less however, such asymmetries are usually wiped out by the spectrograph. The study of stellar granulation is the domain of truly high-resolution spectroscopy.

A word about what we think is happening. Consider the center portion of a stellar disk. Granulation is a convective motion, so hot material will be rising, and because it is hot, it will give a strong spectrum and it will be Doppler shifted toward the blue. Similarly, cool material will be falling, will give a weaker spectrum, and it will be Doppler shifted toward the red. In the disk-integrated light, these profiles are all combined together, but the net effect is a blueward shift of most of the line with the red wing depressed compared to the blue wing, hence the swing of the bisector to the red near the top of the profile.

The velocity span of bisectors, a measure of the strength of granulation, varies systematically across the H-R diagram, being larger toward hotter stars (up to the Granulation

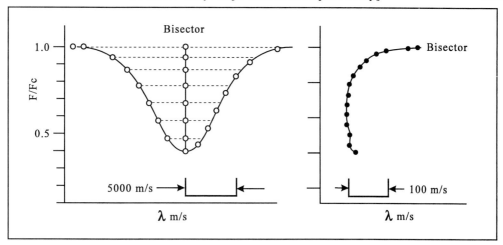

FIGURE 27. Bisectors of spectral lines show up the granulation in stellar photospheres of cool stars. The wavelength axis has to be greatly expanded to see the shapes of bisectors (note 40 times different scale on left versus right graphs). Bisectors of stars on the hot side of the granulation boundary are reversed in shape.

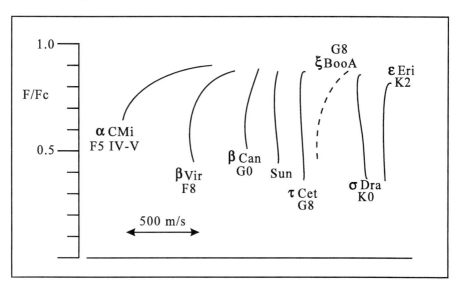

FIGURE 28. Line bisectors vary systematically across the H-R diagram. Here is a series for dwarfs running from near the granulation boundary well into the cooler portion of the diagram. The granulation in 2 Boo A seems to be anomalously strong.

Boundary) and larger toward more luminous stars. An example is shown in Fig. 28. The bisectors are arranged in order of effective temperature or spectral type, but remember, it is only the shapes of bisectors we can compare because we do not know the absolute zero velocity since every star has its own motion in space. If you look carefully at Fig. 28, you will see the bisector for ξ Boo A does not fit with the others, but shows the velocity span of a much hotter star. I don't know why this star has such strong granulation, but it is unusually active magnetically and it has a very strong lithium abundance. It also rotates several times faster than other G8 dwarfs. Maybe it is simply young, and all these things come with the hormones of youth.

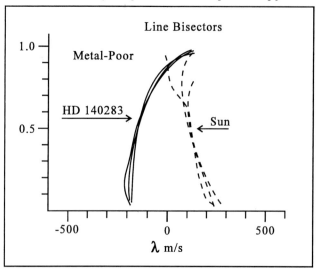

FIGURE 29. Shapes of line bisectors appear to depend on metallicity according to these observations. Based on data given in Prieto et al. (1996).

Figure 29 shows an investigation of a metal poor star. Since the continuous opacity in cool stars is dominated by the negative hydrogen ion, and since the negative hydrogen ion needs an extra electron to exist, and since these extra electrons come from the 'metals,' a metal-poor star has a very transparent atmosphere. We see deeper into the granulation layers, and so the spectral lines show stronger asymmetries. But these kinds of stars are faint, and this is where large telescopes with good (meaning high resolution, naturally!) spectrographs are necessary.

One big surprise: stars on the hot side of the granulation boundary show line asymmetries that go the opposite way, their bisectors sweep blueward toward the continuum! (See *Photospheres* and *Lectures* for details and further references.) These stars aren't even supposed to have convection zones, but they show stronger asymmetries than the cool stars. My but we still have a lot to learn.

17.6. *Disks and Pulsations of Hot Stars*

With resolving power of about 50000 or more one can see non-radial pulsations in early-type stars, and something of the disk material surrounding Be stars. All kinds of variations are going on. Figure 30 shows the kind of thing waiting to be watched by someone with patience and some good equipment.

17.7. *Rotational Modulation: Surface Features*

Want to see what the surface of a star (not the sun!) looks like? Well, you don't have to wait for those giant interferometers to get your view. You can use rotational mapping, alias Doppler imaging, right now. There was a whole conference in October '95 in Vienna dealing with stellar surface features, and a good portion of it dealt with the spectroscopic imaging.

First we should review briefly how it works. When we derive the profile for a rotating star, we first find that all points on a line across the stellar disk parallel to the rotation axis produce the same Doppler shift from the rotation (see *Photospheres* or *Lectures*). We therefore get the distribution of Doppler shifts induced by rotation by thinking of the disk divided into narrow strips parallel to the rotation axis. The weight for each

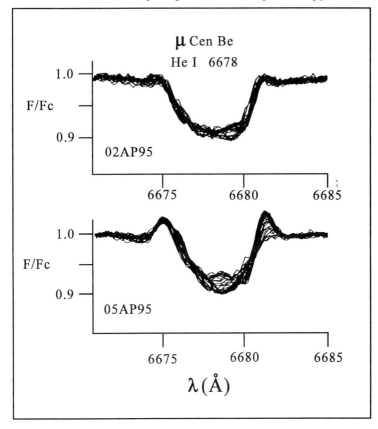

FIGURE 30. Variations in profiles of Be stars need to be studied in more detail. Based on a figure in Peters (1996).

Doppler shift is the amount of light contributed by the strip, and in the absence of limb darkening, that is just the length of the strip. This process is illustrated pictorially in the top part of Fig. 31.

Absorption lines in the rotating star's spectrum all have this shape impressed upon them. We are dealing with a convolution process; the Doppler-shift distribution is shifted to the position of the line, scaled to the equivalent width of the line, and turned up-side-down because it's an absorption line. The approaching limb has the largest blueward shift and corresponds to the left side of the profile. The center of the disk has no radial component of velocity from the rotation, and its light lies in the middle of the profile, and so on. Now consider a dark spot on the star's disk, as in the bottom part of Fig. 31.

The strips containing the spot have less light than before, and therefore the Doppler-shift distribution has a notch or depression at those Doppler shifts. When seen in an absorption line, with the Doppler-shift distribution turned over, the notch looks like an emission hump. Further, the hump migrates through the profile as rotation moves it across the stellar disk through the whole range of Doppler shifts. We therefore have resolution in longitude across the face of the star.

If the spot is located at the equator, it will cover the whole range of Doppler shifts, appearing at the blue edge of the profile and disappearing at the red edge. Then it is gone for half a rotation cycle. On the other hand, if the spot is at higher latitudes, its maximum Doppler shift will be reduced by the cosine of the latitude. That means that

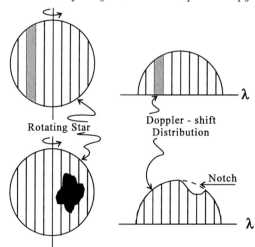

FIGURE 31. This shows the thought pattern in generating the shape of the rotation profile without surface features (upper) and with a surface dark spot (lower).

it will first appear somewhere between the left-most edge of the profile and the center, move across the center, and drop from view on the red side part way along, in a way symmetrical with its appearance. We therefore have latitude information on the spot.

One can expand the discussion to include an axis inclined at an arbitrary angle to the line of sight, limb darkening, differential rotation with latitude, and non-spherical disks even. But the point is, we can resolve the surface of the star spectroscopically. The higher the signal-to-noise we can reach, the smaller the light deficit we will detect. The higher the spectral resolving power compared to $v \sin i$, the more resolution we have on the surface of the star.

Many investigations have been done with rotational mapping, but the surface maps loose a great deal in reproduction and so are best viewed in the original publications. See the papers in the proceedings of IAU Symposium 176 (Strassmeier 1996a, 1996b), especially those by Ryabchikova et al. (1996), Hatzes et al. (1996), Kuschnig et al. (1996), Hempelmann et al. (1996), and Petrov et al. (1996).

One of the hardest parts in this work is getting good phase coverage around the whole rotation period.

17.8. *Stellar Activity Cycles*

As a last example of what high-resolution spectroscopy can do, let's turn our attention to magnetic-activity cycle-type variations in stars. These have characteristic time scales of decades, so you have to be prepared to have some dedication and equipment that has a stable configuration over this time base. We all know that the sun goes through a more-or-less periodic variation in magnetic activity, and many of us teach our elementary students about the variation of sunspots during the cycle, including the butterfly diagram and so on. But I think it came as a bit of a surprise when the satellite-bourne radiometers showed the power output of the sun to vary with the magnetic cycle. It takes only a moment of reflection on this remarkable discovery before our minds think of possible connections between terrestrial climate change and the sun. Now it turns out that the 0.1% variation during the last cycle (the only one observed by the radiometers), according to climate experts, is not likely to be large enough to have much influence on

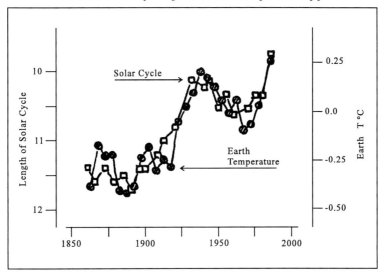

FIGURE 32. The temperature of the earth and the length of the solar activity cycle seem to track each other over the last century or so. Based on a diagram in Friss-Christensen and Lassen (1991).

earth's climate. But the ringer is that we know nothing of what the power variation was in former cycles, and certainly not in future cycles.

Then there is the equally startling comparison, first made by Friss-Christensen and Lassen (1991), between the variation of mean earth temperature and variation of the length of the solar cycle. Figure 32 is based on their work. There is some smoothing involved, and the temperatures in the southern hemisphere are not available, but even so, the agreement is unsettling.

Is the sun the major driver of earth climate? Hunting down the truth on this one could place us in a delicate position. You can just imagine what the chiefs of industry might say about pollution if the answer turns out to be yes. We have to tread carefully in such situations, but truth is nevertheless our target, so onward we must go.

I see two possible lines of attack: 1) watch the sun through many cycles and see what we can see, or 2) look at other solar-like stars and see what they do, and by inference deduce what the sun in likely to do. I'm not going to live long enough to implement plan one, nor is the pollution-global-warming problem likely to give us a long time to ponder. Consequently, off we go on plan two, leaving plan one to the solar division.

What can we measure relevant to activity cycles? Fortunately Olin Wilson (1968, 1978; see also Vaughan 1980) started monitoring the H & K line emission 30 years ago, and more recently the program has been going even more vigorously under S. Baliunas. At the Lowell Observatory about a third of these stars have been monitored photometrically. And last but not least, I have been taking spectroscopic data on a few of them since the early 1980's. So in the HK index we have a tag on the magnetic activity. From the photometry we have a measure of power output. From the spectroscopy we get two things, temperature and granulation measurements. High resolving power is necessary to measure the line bisectors and their variation, but high resolving power is also a powerful tool for measuring small changes in line depths arising from temperature changes. Lower resolution washes out such signals.

The procedure for temperature measurement is to calibrate ratios of spectral line depths against stars having a wide range of effective temperature (see Gray 1994a, 1994b,

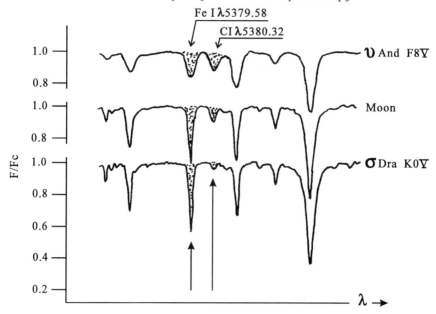

FIGURE 33. Pairs of spectral lines showing opposite sensitivity to temperature make good stellar thermometers.

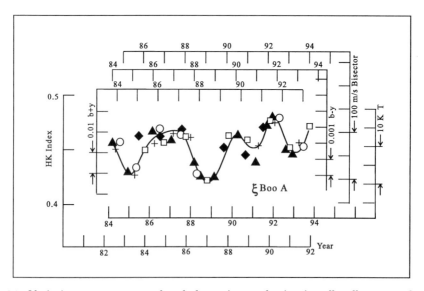

FIGURE 34. Variations are seen over decade-long time scales in virtually all measured parameters: HK emission (triangles), temperature (circles), magnitudes (squares), color index (diamonds), and granulation (crosses). Each shows the same pattern of variation, but they occur at different times!

Gray and Johanson 1991). The type of line variation one exploits is illustrated in Fig. 33. This particular calibration is from a project to measure the temperature changes of the sun spectroscopically, and it shows how the carbon line increases with temperature while the adjacent iron line decreases with temperature. The ratio of their depths is therefore a good thermometer. It turns out that for favorable cases, one can measure temperature variations with a precision of about 1 K.

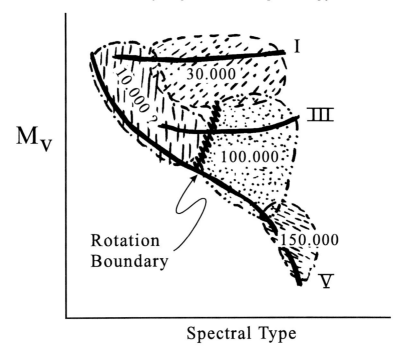

FIGURE 35. This schematic H-R diagram indicates the approximate resolving power needed to resolve line profiles of stars. When no other information is available, it can serve as a guide for initial investigations.

Activity variations of all sorts are observed. A relatively small fraction of cool stars shows small variations like the sun's or less, but most stars show variations considerably stronger than the sun's. To illustrate the interesting kind of thing that one stumbles across, consider Fig. 34 (based on a study by Gray et al. 1996). All the observed parameters: HK index, brightness, color index, spectroscopic temperature, and granulation (bisector velocity span) all show the same variation. But they do not occur in phase. Magnetic variations occur first, temperature and granulation variations are together coming in last, and brightness in between. In some stars, the time lags can amount to years, and they apparently depend on spectral type.

Well, it is obvious that high-resolution spectrographs are powerful tools, especially when they're coupled with the right spectral lines.

18. Choose the Right Equipment

The eternal struggle between exposure time on one side versus spectral resolution and signal-to-noise ratio on the other never goes away. These factors we must balance if we hope to succeed. If the investigation is an initial foray into some area, then you will have to take a guess at the resolution and signal-to-noise ratio you need. Some guidance might come from Fig. 35, where I've estimated approximate resolving power needed to work with the stellar lines in various parts of the H-R diagram. These are rough numbers, based on the behavior of rotation and macroturbulence across the diagram. Interstellar lines are usually much more narrow than stellar lines, whereas extragalactic objects often have wider lines.

Here are some points to ponder in this regard.

• Exposure times will increase as the inverse square of the resolution. We have to close the entrance slit to get higher resolution, but that also results in a smaller image of the entrance slit. Since the number of pixels is limited, going to higher resolution generally reduces the size of the field.

• Exposure times decrease only linearly with increasing telescope diameter unless an effective image slicer is used.

• Exposure times decrease with better seeing. Changes by a factor of 5 or more can occur in some locations, but always be careful to keep the entrance slit significantly smaller than the seeing disk or 'seeing noise' will plague your data.

• Time variable phenomena may introduce serious constraints on length of exposures. In most cases, the exposure should not exceed one tenth of a cycle, and preferably less.

• Efficiency of the grating across one blaze and from one blaze to the next can vary markedly. The maximum is usually specified, and one rarely works at that favored wavelength.

• When selecting spectral resolution, we need to ask if we are resolving the 'main' profile of structure within it.

• If we need the ultimate in accuracy, then we should use the highest resolving-power spectrograph we can lay our hands on, and use it in a single order with a bandpass filter to kill off scattered light.

• If we need wide wavelength coverage, use a cross-dispersed echelle spectrograph (but not for really wide lines).

• Is extra stability needed? Use a coude spectrograph, or second best, a bench mounted Cassegrain spectrograph.

In addition, practical reality causes us to ask what equipment is actually available to us? Can it meet the scientific requirements? How far can we go in lengthening the exposure times or in combining exposures?

Our extragalactic colleagues often forget that high spectral resolution can make even bright objects faint. Color filter photometry uses bandpasses $\approx$ 1000 Å wide; we use $\approx$ 0.03 Å. That's 11.3 magnitudes! Yes indeed high-resolution spectroscopists do have need for large telescopes.

REFERENCES

ANN, H.B. AND KANG, Y.H. 1986, *Ap. Space Sci.* 118, 325.

BESSELL, M.S. AND NORRIS, J. 1984, *Ap.J.* 285, 622.

BOESGAARD, A.M. ET AL. 1988a, *Ap.J.* 325, 749.

BOESGAARD, A.M. ET AL. 1988b, *Ap.J.* 327, 389.

BROWN, T.M. ET AL. 1994, *PASP* 106, 1285.

CAMPBELL, B., WALKER, G.A.H., AND YANG, S. 1988, *Ap.J.* 331, 902.

CHURCHILL, C.W. AND ALLEN, S.L. 1995, *PASP* 107, 193.

DRAVINS, D. 1994, *Impact of Long-Term Monitoring on Variable Star Research*, (Kluwer: Dordrecht), p. 269.

EPPS, H. AND VOGT, S.S. 1993, *Appl. Optics* 32, 6270.

FRISS-CHRISTENSEN, E. AND LASSEN, K. 1991, *Science* 254, 698.

GEHREN, T. AND PONZ, D. 1986, *A&A* 168, 386.

GRAY, D.F. 1968, *IAU Symposium* 118, (Reidel: Dordrecht), J.B Hearnshaw and P.L. Cotrell, eds., p. 401.

GRAY, D.F. 1980, *Ap.J.* 240, 125.

GRAY, D.F. 1988, *Lectures on Spectral-Line Analysis: F, G, and K Stars*, (The Publisher: Arva,

Ont.).

GRAY, D.F. 1991, *Angular Momentum Evolution of Young Stars*, (Kluwer: Dordrecht), S. Catalano and J.R. Stauffer, eds., p. 183.

GRAY, D.F. 1992, *The Observation and Analysis of Stellar Photospheres*, (Cambridge: Cambridge).

GRAY, D.F. 1994a, *PASP* 106, 1248.

GRAY, D.F. 1994b, *PASP* 107, 120.

GRAY, D.F., BALIUNAS, S.L., LOCKWOOD, G.W., AND SKIFF, B. A. 1996, *Ap. J.* 465, 945.

GRAY, D.F. AND JOHANSON, H.L. 1991, *PASP* 103, 439.

GRIFFIN, R.F. 1968, *A Photometric Atlas of the Spectrum of Arcturus* , (Cambridge Philosophical Society: Cambridge).

GRIFFIN, R.F. 1969a, *MNRAS* 143, 319.

GRIFFIN, R.F. 1969b, *MNRAS* 143, 349.

GRIFFIN, R.F. 1969c, *MNRAS* 143, 361.

HALL, J.C. ET AL. 1994, *PASP* 106, 315.

HATZES, A.P. ET AL. 1996, *IAU Symposium 176, Stellar Surface Structure, Poster Proc.*, (Inst. Astron.: Wien), K.G. Strassmeier, ed., p. 9.

HEMPELMANN ET AL. 1996, *IAU Symposium 176, Stellar Surface Structure, Poster Proc.*, (Inst. Astron.: Wien), K.G. Strassmeier, ed., p. 194.

KUSCHNIG ET AL. 1996, IAU SYMPOSIUM 176, STELLAR SURFACE STRUCTURE, POSTER PROC., (Inst. Astron.: Wien), K.G. Strassmeier, ed., p. 135.

LANGER, G.E. ET AL. 1986, PASP 98, 473.

LARSEN, A.M. ET AL. 1993, PASP 105, 825.

LIBBRECHT, K.G. AND PERI, M.L. 1995, *PASP* 107, 62.

LITTLE-MARENIN, I.R. AND LITTLE, S.J. 1984, *Ap.J.* 283, 188.

LUCK, R.E. 1982, *Ap.J.* 256, 177.

MARCY, G.W. AND BUTLER, R.P. 1992, *PASP* 104, 270.

MATTEUCCI, G. AND GREGGIO, L. 1986, *A&A* 154, 279.

McMILLAN, R.S. ET AL. 1992, *PASP* 104, 1173.

MOLARO, P. AND CASTELLI, F. 1990, *A&A* 228, 426.

PETERS, G.J. 1996, *IAU Symposium 176, Stellar Surface Structure, Poster Proc.*, (Inst. Astron.: Wien), K.G. Strassmeier, ed., p. 212.

PETROV, P.P. ET AL. 1996, *IAU Symposium 176, Stellar Surface Structure,Poster Proc.*, (Inst. Astron.: Wien), K.G. Strassmeier, ed., p. 217.

PILACHOWSKI, C. ET AL. 1995, *PASP* 107, 983.

PRIETO, C.A. ET AL. 1996, *IAU Symposium 176, Stellar Surface Structure, Poster Proc.*, (Inst. Astron.: Wien), K.G. Strassmeier, ed., p. 107.

RICHARDSON, E.H. 1966, *PASP* 78, 436.

VOGT, S.S. 1992, *ESO Workshop No. 40, High Resolution Spectroscopy with the VLT*, (ESO: Garching), M.-H. Ulrich, ed., p. 223.

RUTTEN, R.G.M. AND PYLYSER, E. 1988, *A&A* 191, 227.

RYABCHIKOVA, T. ET AL. 1996, *IAU Symposium 176, Stellar Surface Structure, Poster Proc.*, (Inst. Astron.: Wien), K.G. Strassmeier, ed., p. 133.

SCHRIJVER, C.J. 1994, *Cool Stars, Stellar Systems, and the Sun, 8th Cambridge Workshop on Cool Stars, Astron. Soc. Pacific Conf. Ser.* 64, 328.

SHCHERBAKOV, A.G. ET AL. 1996, *IAU Symposium 176, Stellar Surface Structure, Poster Proc.*, (Inst. Astron.: Wien), K.G. Strassmeier, ed., p. 181.

SMITH, G. AND DRAKE, J.J. 1987, *A&A* 181, 103.

SMITH, G.H. AND NORRIS, J. 1983, *Ap.J.* 264, 215.

SNEDEN, C. AND PILOCHOWSKI, C.A. 1986, *Ap.J.* 301, 860.

SPITE, F. AND SPITE, M. 1982, *A&A* 115, 357.

STRASSMEIER, K.G., (ed.) 1996a, *IAU Symposium 176, Stellar Surface Structure, Poster Proc.*, (Inst. Astron.: Wien).

STRASSMEIER, K.G., (ed.) 1996b, *IAU Symposium 176, Stellar Surface Structure, Invited Presentations*, (Kluwer: Dordrecht)

TULL, R.G. ET AL. 1995, *PASP* 107, 251.

VALENTI, J.A. ET AL. 1995, *PASP* 107, 966.

VANDENBERG, D.A. AND SMITH, G.H. 1988, *PASP* 100, 314.

VAUCLAIR, S. 1988, *Ap.J.* 335, 971.

VAUGHAN, A.H. 1980, *PASP* 92, 392.

VOGT, S.S. 1992, *ESO Workshop 40, High Resolution Spectroscopy with the VLT*, M.-H. Ulrich, ed., p.223.

VOGT, S.S. ET AL. 1994, *SPIE Conf. 2198, Instrumentation in Astronomy VII*, D.L. Crawford and E.R. Craine, eds., p. 362.

WILSON, O.C. 1968, *Ap.J.* 153, 221.

WILSON, O.C. 1978, *Ap.J.* 226, 379.

Near Infrared Instrumentation for Large Telescopes

By IAN S. McLEAN

Department of Physics and Astronomy, University of California Los Angeles (UCLA), Los Angeles, CA 90095, USA

This paper reviews near infrared instrumentation for large telescopes. Modern instrumentation for near infrared astronomy is dominated by systems which employ state-of-the-art infrared array detectors. Following a general introduction to the near infrared wavebands and transmission features of the atmosphere, a description of the latest detector technology is given. Matching of these detectors to large telescopes is then discussed in the context of imaging and spectroscopic instruments. Both the seeing-limited and diffraction-limited cases are considered. Practical considerations (e.g. the impact of operation in a vacuum cryogenic environment) that enter into the design of infrared cameras and spectrographs are explored in more detail and specific examples are described. One of these is a 2-channel IR camera and the other is a NIR echelle spectrograph, both of which are designed for the f/15 focus of the 10-m W. M. Keck Telescope.

1. The Near Infrared Waveband

In the last ten years there has been tremendous growth in the field of Infrared Astronomy. This growth has been stimulated in large part by the development of very sensitive imaging devices called *infrared arrays*. These detectors are similar, but not identical, to the better-known silicon charge-coupled device or CCD, which is limited to wavelengths shorter than 1.1 μm. In particular, near infrared array detectors are now sufficiently sensitive that images of comparable depth to those obtained with visible-light CCDs can be achieved from 1.0 μm to 2.4 μm and high resolution IR spectrographs are now feasible. Coincidentally, this breakthrough in detector technology has come at the same time as the development of very large telescopes. Indeed, the prospect of near diffraction-limited imaging in the near infrared is a major justification for the construction of very large telescopes. Before describing these detectors and their uses in instrumentation for very large telescopes, it is useful to begin with a brief review and an explanation of the terminology of the subject.

1.1. *Ground-based near infrared astronomy – the problems*

Modern infrared astronomy really began in the 1950s — the era of the transistor — when simple, photoelectric detectors made from semiconductor crystals first became possible. The lead sulphide cell was used by Johnson to extend the young field of photoelectric photometry of stars, and by Neugebauer and Leighton at Caltech for a huge survey of the sky at a wavelength of 2 μm with an angular resolution of about 4 arcminutes. When Frank Low introduced the gallium-doped germanium bolometer to astronomy in the early 1960's, this opened up the study of much longer infrared wavelengths, especially with the aid of rocket, balloon and airplane surveys. In the seventies, lead sulphide cells were replaced by photodiodes of indium antimonide, while improved bolometers and detectors using suitably doped silicon were introduced for longer wavelengths. By 1979, a new generation of large telescopes dedicated to infrared astronomy were coming into operation, including the United Kingdom 3.8-meter Infrared Telescope (UKIRT) and the NASA 3-m Infrared Telescope Facility (IRTF), both in Hawaii. With the launch of the

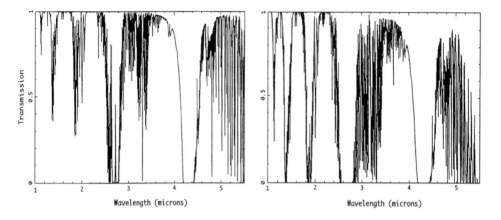

FIGURE 1. Atmospheric transmission spectra for Mauna Kea, Hawaii for two extreme cases:
(a) 0.3 mm water vapour; zenith and (b) 4.8 mm water vapour; altitude $= 30°$.

Infrared Astronomical Satellite (IRAS), by the USA, Britain and the Netherlands, in
1983, the science of infrared astronomy took another leap forward. This very successful
mission mapped the entire sky at wavelengths of 12, 25, 60 and 100 μm until its on-board
supply of liquid helium was exhausted and the telescope and the detectors warmed-up
and lost their sensitivity. Then, as we will see, the final boost came a few years later with
the introduction of two-dimensional detector arrays tailored for astronomy applications.

The near-infrared (NIR) is generally taken to be the interval from 1 to 5 μm, with some
astronomers (and industry) using the term short wavelength infrared (SWIR) to indicate
the interval from 1 to 2.5 μm. Mid-infrared (MIR) is a region which extends from 5
to 25 μm or so, and the far-infrared (FIR) stretches from about 30 to 200 μm. Wavelengths longer than about 350 μm are now referred to as "sub-millimeter", and although
some of the observational techniques are similar, (e.g. continuum bolometers), much of
the instrumentation is really more closely related to radio astronomy (e.g. heterodyne
receivers).

There are three fundamental problems facing ground-based astronomy at infrared
wavelengths. One is the detector technology already mentioned. Apart from the lack of
panoramic detectors until recently, existing devices were often "noisy" and not limited
by photon noise statistics. The second problem is that everything is warm, including the
telescope, and thus there is an emitted "background" radiation against which faint astronomical measurements must be made. This effect dominates at the longer wavelengths.
At shorter infrared wavelengths we must contend with a different background, this one
due to line emission by molecules in the upper atmosphere. Finally, the atmosphere is
not uniformly transparent at infrared wavelengths leading to the notion of atmospheric
"windows". All of these factors lead to complications for designers of infrared instruments.

1.2. Atmospheric transmission

Water vapour and carbon dioxide do an efficient job of blocking out a lot of infrared
radiation. Figure 1, adapted from McCaughrean (1988), shows a simplified transmission
spectrum of the atmosphere from 1 μm to 5.5 μm.

The water vapour content is particularly destructive for the longer wavelengths, but it

TABLE 1. Standard infrared windows and passbands.

Wavelength (μm)	Symbol	Width† (μm)
1.25	J	0.3
1.62	H	0.3
2.2	K	0.4
3.5	L	0.6
3.8	L′	0.6
4.8	M	0.6

† Approximate.

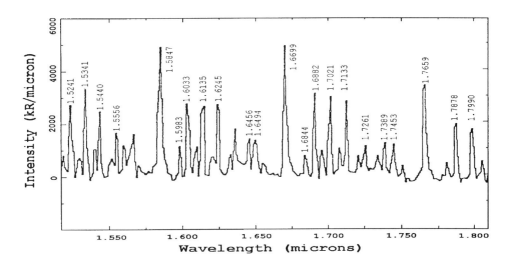

FIGURE 2. OH emission lines at R = 1,000 obtained with IRSPEC at ESO, Chile.

is sensitive to the altitude of the observatory, and that is why most infrared observatories are located at high, very dry sites. The absorption by these molecules occurs in certain wavelength intervals or bands, between which the atmosphere is remarkably transparent. These wavelength intervals are called atmospheric "windows" of transparency.

The standard near infrared windows are listed in Table 1. Interference filters can be manufactured to match these atmospheric windows (e.g. by OCLI or Barr Associates, USA) and, of course, narrower passbands can be defined within each window. Filter sets and detector response functions are not identical and therefore care must be taken to determine the exact effective wavelength and passband for each set.

1.3. *Atmospheric emissions*

There are two major sources of non-thermal radiation that dominate the near-infrared night sky from 1–2.5 μm. The first is the polar aurora, due mainly to emission from

N_2 molecules, but which is negligible at mid-latitude sites such as Hawaii, the Canary Islands and Chile. The dominant problem is "airglow" which has three components:
- OH vibration-rotation bands
- O_2 IR atmospheric bands
- the near-infrared nightglow continuum

Of these, the strongest emission comes from the hydroxyl (OH) molecule which produces a dense "forest" of emission lines, especially in the 1–2.5 μm region. Figure 2, adapted from Moorwood (1987), shows a typical spectrum of OH emission lines in the near infrared. These features are barely resolved at a spectral resolving power of $R = \lambda/\Delta\lambda = 500$ and so become a major impediment to near infrared spectroscopy.

1.4. *Thermal emission*

Since the atmosphere is at a finite temperature, it also emits thermal (blackbody) radiation with an *emissivity* ϵ which depends on the opacity in the atmosphere at that wavelength; the emissivity will be 1.0 (or 100%) if the atmosphere is totally opaque, but ϵ may be less than 0.1 in good windows where absorption is low. Again, water vapour is the main problem, as it is responsible for much of the absorption from 3–5 μm, where the thermal emission of the atmosphere rises steeply. At wavelengths less than 13 μm however, it is thermal blackbody emission by the telescope and warm optics which dominate the background since both are at least 20 K warmer than the effective water vapour temperature.

To predict the thermal emission from the telescope optics and any other warm optics in the beam we need to know two parameters: the temperature T (K) which determines the spectrum of blackbody radiation from the Planck function $B(\lambda)$, and the emissivity ϵ of each component which determines the fraction of blackbody radiation added to the beam. The emissivity of the telescope mirrors (due to absorption by the coating and from dust on the surface) may be found from Kirchoff's law, by taking one *minus* the measured spectral reflectivity.

When the thermal and non-thermal components of background are added together we get the total background flux entering the instrument. The resulting spectrum is very steep (see Figure 3) and has two
easily recognized portions, one due to OH emission for $\lambda < 2.5$ μm, and one towards longer wavelengths due to blackbody emission from the warm telescope and optics. Note that the background changes by many orders of magnitude between 1 and 5 μm.

1.5. *Detector technology*

Infrared detectors are classified as *photon* detectors or *thermal* detectors. In photon detectors, individual incident photons interact with electrons within the detector material. For example, if a photon frees an electron from the material the process is called *photoemission*. This, of course, is how a standard photomultiplier tube (PMT) works and it is also the underlying principle of the Schottky-Barrier diode (e.g. platinum on silicon), except that in that case the free electron escapes from the semiconductor into a metal rather than into a vacuum. When absorption of photons increases the number of charge carriers in a material, or changes their mobility, the process is called *photoconductivity*. Such devices are usually operated with an external voltage across them to separate the photogenerated electron-hole pairs before they can recombine. If absorption of a photon leads to the production of a voltage difference across a junction between differently "doped" semiconductors, the process is known as the *photovoltaic effect*. The interface is normally a simple **pn** junction which provides an internal electric field to separate

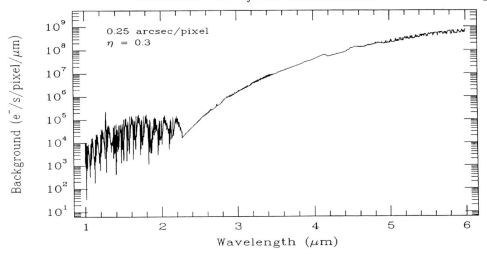

FIGURE 3. Total background as a function of wavelength for a typical infrared instrument on a large telescopes at a good (dry) site.

electron-hole pairs. In general, the pn junction (also known as a photodiode) is operated with an externally applied voltage to cause a condition known as a "reverse bias".

There are two classes of semiconductors used in photon detection; intrinsic and extrinsic. In an intrinsic semiconductor crystal, a photon with sufficient energy creates an electron-hole pair when the electron is excited to the conduction band, leaving a hole in the valence band. Extrinsic semiconductors are doped with impurities such that a photon with insufficient energy to excite an electron-hole pair intrinsically, can still cause an excitation from an energy level associated with the impurity atom.

In thermal detectors, the absorption of photons leads to a temperature change of the detector material, which may be observed as a change in the electrical resistance of the material as in the *bolometer*; a voltage difference across a junction between two different conductors as in the *thermopile*; or a change of internal dipole moment in a temperature-sensitive ferro-electric crystal, as in the *pyroelectric* detector. Of these, only the bolometer has been widely used in infrared astronomy. The germanium bolometer is commonly used for wide band energy detection at wavelengths longer than 10 μm, but silicon bolometers have also been used successfully in recent years. In some instruments, several individual bolometers have been arranged in a grid pattern to produce an "array" of individual detectors.

Obviously, single element detectors can only measure the infrared signal from one small patch of sky at a time. Usually, that patch will contain the source plus some unwanted background. However, the infrared background is bright and variable compared to astronomical sources and therefore some method of checking the background level very frequently is needed to ensure that the correct amount of background can be subtracted. The method used is called "chopping". The infrared beam is rapidly switched between the source position on the sky and a nearby reference position, by the use of a small plane mirror near the focal plane or a "wobbling" secondary mirror in the telescope. Chopping typically takes place at frequencies of about 10-20 Hz and the output signal is fed to a phase-locked amplifier with a reference signal from the chopper so that only the AC signal component representing the source flux is amplified.

The infrared detectors described above do not function well unless they, and their local environment, are cooled to very low temperatures. Traditionally, liquid cryogens such as liquid nitrogen (LN$_2$) or liquid helium (LHe) are used to obtain temperatures of 77 K and 4 K respectively. In more modern instruments, closed-cycle refrigerators (CCRs) have been used successfully. Cooling achieves two important results. Firstly, the heat or infrared emission from all the filters, lenses and supporting metal structures surrounding the infrared detector is eliminated, which greatly reduces the local background contribution. Secondly, at low temperatures the detector's own internal, thermally-generated background or *dark current* is greatly reduced which implies a large gain in sensitivity.

Of course, in order to cool the detector and all surrounding metal and glass (e.g. filters, optics) the entire unit must be placed inside a vacuum chamber with an infrared transmitting window, and the cryogenic components must be carefully isolated from the warm (radiating) walls of the enclosure and the warm (conducting) drive shafts needed to operate mechanisms such as filter wheels inside the vacuum chamber. All infrared instruments are therefore contained within a vacuum chamber which also includes a cooling system; these units are called *cryostats* or *dewars* and are commercially available (e.g. IRLabs, USA and Oxford Instruments, UK).

The kinds of infrared detector systems available to infrared astronomers up until about 1983, although fairly sensitive, were not really ideal for making extensive infrared pictures of the sky with the fine seeing-limited detail of optical images. There were no infrared "photographic" plates or infrared "TV tubes"! All published infrared "images" of celestial objects at that time were really "maps" made by scanning the infrared scene point-by-point, back and forth in a sweeping pattern, called a *raster*, using a single detector having a small angular field of view on the sky which was determined by an aperture within the instrument; typically, the field was a circle of 2–3 arcseconds in diameter. Imagine using just *one* pixel of a CCD (charge-coupled device) which had 500×500 pixels and then moving the entire telescope (or the secondary mirror) to scan the image of the scene over that pixel; the term *pixel* is short for picture element, and refers to the basic photo-sensitive "unit cell" in a 2-dimensional array detector. It is easy to appreciate the excitement that was felt when infrared devices similar to CCDs were introduced into astronomy. Having worked with some of the earliest CCDs in the late seventies while a post-doc at Steward Observatory and again in the early eighties at the Royal Observatory Edinburgh (ROE), I became very interested in the possibilities of their infrared analogues. After two years of research and extensive discussions with vendors, I initiated the IRCAM project in 1984, on behalf of the ROE and UKIRT, to develop with Dr. Alan Hoffman of SBRC the first 62×58 InSb array specifically designed for NIR astronomy applications.

By late 1986 the first astronomical results from the new generation of arrays had begun to appear. An historic moment, a "turning point", occurred in March 1987 at the first conference to focus on the applications of these new infrared array detectors for ground-based astronomy. This workshop, which attracted over 200 people from all over the world, was held in Hawaii and hosted by the University of Hawaii, the UK Infrared Telescope and the Canada-France-Hawaii telescope. Although now out of date, the results presented at that meeting demonstrated clearly and dramatically to all that were present, that a New Era in infrared astronomy had begun. In July 1993 a "next generation" conference was held at UCLA (McLean 1994) and it was immediately clear from the science presented that IR astronomy had changed forever. (For popular accounts with colour pictures see McLean (1995)).

Today, there are several kinds of new generation arrays in use with formats of 256×256 pixels, and even larger detectors — the *Next Generation* — are planned for the next 3 years. Those new arrays will have formats of 1024×1024 pixels. From one to one million pixels in such a short period of time is amazing.

1.6. *Large telescope technology*

Coincidentally, the last decade has also seen the evolution of telescope technology, with the development of "spin-cast" mirrors, thin "meniscus" mirrors and segmented mirrors. Together with the technology of Adaptive Optics, these innovations mean that we are now witnessing the construction of huge telescopes in the 8 – 10-m class. Already fully operational since January 1994, the vanguard of this new era is the Keck I 10-m telescope on Mauna Kea, Hawaii. The brainchild of Dr. Jerry Nelson, the Keck Observatory is operated by the California Association for Research in Astronomy on behalf of the University of California, Caltech, the University of Hawaii and (recently) NASA. These large telescopes pose constraints on instrument designs, such as matching image scales to available detectors, or dealing with the field rotation that comes with alt-az telescope mountings, but they also provide great opportunities such as diffraction-limited imaging in the mid- and near-infrared.

Having briefly introduced the wavebands and the basic techniques of IR astronomy in Section 1, I will describe the array detectors and IR instrumentation in more detail in the following parts.

2. Infrared Array Detectors

2.1. *Basic Principles*

Infrared array detectors are extremely important in the effort to develop instrumentation which can capitalize on the light-gathering power of very large telescopes for high resolution near infrared spectroscopy, as well as near diffraction-limited imaging. The advantages of infrared array detectors over single element devices are:

• very time-efficient because of large numbers of elements, which also implies they can reach fainter detection limits
• higher spatial resolution easily obtained; usually seeing-limited, but can be diffraction-limited in some cases
• very high sensitivity, partly due to small detector size
• high resolution spectroscopy is now viable
• "sky on the frame" introduces possibility of eliminating chopping, at least in the near infrared.

The latter point is extremely significant; infrared astronomy at wavelengths less than 4 μm is now generally performed in a "staring" mode without the use of a chopping secondary mirror. It is still necessary to "nod" the entire telescope fairly frequently (e.g. every few minutes or so) to obtain good sky differences, since the backgrounds are so large and variable.

Photon detectors are characterized by a cut-off wavelength λ_c for the absorption of infrared photons which is given in terms of the band-gap energy E_g in electron volts (eV) by:

$$\lambda_c \ (\mu m) = \frac{1.24}{E_g} \tag{2.1}$$

Cut-off wavelengths, detector materials, photo-absorption method and array formats are tabulated for currently available NIR astronomy detectors in Table 2.

Notice that, in the case of HgCdTe, the ratio of mercury to cadmium determines the effective energy band gap; with 55% Hg the longest wavelength of sensitivity is 2.5 μm whereas with 80% Hg the long wavelength limit extends to 12.4 μm. Currently available arrays in HgCdTe for near IR astronomy include the 256×256 array from Rockwell International known as the NICMOS 3 array (because it was developed for the University

TABLE 2. Current near infrared array detectors.

Material	Symbol	Band Gap[†] (eV)	λ_c (μm)	Type	Formats
Indium antimonide	InSb	0.23	5.4	PV	256×256 1024×1024
Mercury cadmium telluride	$Hg_xCd_{1-x}Te$	0.50 ($x = 0.554$)	2.5	PV	256×256 1024×1024
Platinum silicide	PtSi	-	$\sim 4\mu$m	Schottky	256×256 640×480

† At 77 K approximately.

of Arizona's Hubble Space Telescope NICMOS project) and the Rockwell "HAWAII" 1024×1024 array. For InSb arrays, there is the 256×256 astronomy FPA from SBRC together with their 1024×1024 ALADDIN array which is still under development. Platinum silicide arrays can be obtained from Hughes, Kodak, Mitsubishi and other companies, but simply do not have the quantum efficiency required for many applications.

2.1.1. *The "hybrid" structure; comparison with CCDs*

Infrared array detectors are not based on the charge-coupling principle of the silicon CCD. This is an important distinction which has some practical implications. For example, IR arrays do not "bleed" along columns when a pixel saturates and bad pixels do not block off others in the same column. Also, since the pixel charge does not move, non-destructive readout schemes are possible and very effective. On the other hand, on-chip charge-binning and "drift-scanning" are not possible.

In an infrared array, the role of detecting infrared photons is separated from the role of multiplexing the resultant electronic signal from a pixel to the outside system. To achieve this, each device is made in two parts, an upper slab and a lower slab. The upper slab is made of the IR sensitive material and effectively subdivided into a grid of pixels by construction of a pattern of tiny photodiodes or photoconductors. In the lower slab, made of silicon, there is a matching grid but each "unit cell" contains a silicon field-effect transistor (FET) which is used as a source-follower amplifier to act as a "buffer" for the accumulated charge in the infrared pixel. Interconnecting the two slabs are tiny columns of indium, called " bumps", and the slabs are literally pressed together in a process called "bump-bonding". Finally, in the case of InSb the upper slab is thinned to about 10 μm to enable photons to penetrate to the pixel locations on the underside or, in the case of HgCdTe arrays, the IR detector is deposited as a thin or "epitaxial" layer on an infrared transparent substrate. The entire structure is called a "hybrid", and it is shown schematically in Figure 4. Other names for this structure include Sensor Chip Assembly (SCA) and Focal Plane Array (FPA).

In an infrared array detector, the absorption of a photon with a wavelength shorter than the cut-off wavelength (λ_c) generates an electron-hole pair in the semiconductor at the location of the pixel. The electron-hole pair is immediately separated by an electric field which can be applied externally or internally or both. For example, in the large

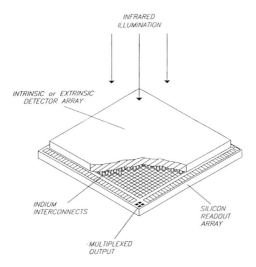

FIGURE 4. Schematic representation of a hybrid infrared detector.

near infrared arrays of HgCdTe and InSb available today, the electric field is produced by a reverse-biased pn junction. The depletion region produced by the reverse-biased junction acts like a capacitor which is *discharged* from the initial reverse bias voltage by the migration of the electrons and holes; in effect, photogenerated charges are being "stored" locally at each pixel. The voltage change across the detector capacitance is applied directly to the input of the silicon FET which in turn relays the voltage change to one or more output lines when it is switched (or addressed) to do so. There is *no charge-coupling* between pixels since each pixel has its own FET, and there is no overflow or "charge bleeding" since the worst that can happen is that the diode becomes completely de-biased and no further integration occurs, in other words, the pixel is saturated.

Infrared arrays are backside illuminated and, with the exception of platinum silicide (PtSi) Schottky Barrier devices, all achieve excellent quantum efficiencies, typically $\geq$ 60%. The PtSi devices partially compensate for a very low quantum efficiency by having good uniformity and large formats (e.g. 640×480 and larger already demonstrated).

2.1.2. *Charge detection and storage*

An equivalent circuit for a typical IR array with a photodiode detector is shown in Figure 5 and the details of the process are summarized below:
- internal photoelectric effect $\rightarrow$ electron-hole pairs
- electric field separates electrons and holes
- migration of electrons across junction decreases the reverse bias, like discharging a capacitor
- amount of charge is $Q = CV/e$ electrons where $e = 1.6 \times 10^{-19}$ Coulombs per electron, V is the voltage across the detector and C is the effective capacitance
- each detector is connected to a source follower amplifier whose output voltage follows the input voltage with a small loss of gain; $V_{OUT} = A_{SF}V_{IN} \simeq 0.7V_{IN}$
- the output voltage of the source follower can be sampled or "read" without affecting the input
- after sampling, the voltage across the diode can be RESET to the full, original reverse bias in readiness for the next integration.

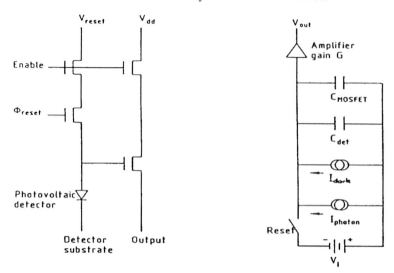

FIGURE 5. Typical unit cell structure for a near IR array detector.

The reset action is accomplished with another FET acting as a simple on/off switch in response to a pulse applied to its gate. It is customary to refer to the combination of the detector node, the source follower FET and the reset FET as the "unit cell". Actually, the equivalent circuit of the unit cell reveals that there are two sources of capacitance — the pn junction and the gate of the FET. Also, there are two sources of current to drive the discharge of the reverse bias, namely photoelectrons and "dark current" electrons. There is an obvious potential for non-linearity between photon flux and output voltage in these detectors because the detector capacitance is *not* fixed, but does in fact depend on the width of the depletion region which in turn depends on the value of the reverse bias voltage. Unfortunately, this voltage changes continuously as the unit cell integrates either photogenerated charges or dark current charges. Using $Q = CV$ and taking both C and V as functions of V then

$$dQ = (C + \frac{\partial C}{\partial V}V)dV \equiv I_{det}dt \tag{2.2}$$

The rate of change of voltage with time ($\frac{dV}{dt}$) is not linear with detector current I_{det} because of the term $\frac{\partial C}{\partial V}$. The effect is small however, and slowly varying ($< 10\%$ worst case) and it is easily calibrated to high precision.

2.2. *Performance of NIR arrays*

2.2.1. *Dark Current*

In the total absence of photons, an accumulation of electrons still occurs in each pixel and this effect is therefore known as *dark current*. There are at least three sources of dark current; diffusion, *thermal* generation-recombination (G-R) of charges within the semiconductor and leakage currents. The latter are determined mainly by manufacturing processes and applied voltages, but diffusion currents and G-R currents are both very strong (exponential) functions of temperature and can be dramatically reduced by cooling the detector to a low temperature. Dark currents below 1 electron/minute/pixel have been achieved in the most recent HgCdTe arrays at 77 K and $\sim$ 0.1 electron/s/pixel in InSb arrays with low bias and temperatures $\leq$ 35 K. Note that lower band-gaps imply

Schematic of Single-Sampled Readout

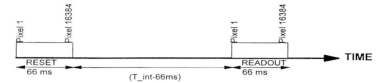

Schematic of Double-Sampled Readout

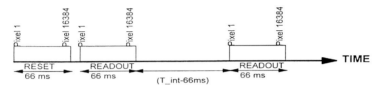

FIGURE 6. Typical readout and clocking strategies for 256^2 IR arrays.

lower temperatures to obtain the same dark current, and that the observed residual dark current is probably due to leakage currents in most cases.

2.2.2. *Readout noise*

Resetting the detector capacitance and sampling the output voltage cannot be done without a penalty. The term "readout noise" describes the random fluctuations in voltage (V_{noise}) which are added to the true signal and photon noise. Readout noise is always converted to an equivalent number of electrons (R) at the detector by using the effective capacitance (C), therefore

$$R = \frac{CV_{noise}}{e} \quad electrons \qquad (2.3)$$

One potentially serious component of the readout noise can be eliminated easily. This is the reset or "kTC noise" associated with resetting the detector capacitance. When the reset transistor is on, the voltage across the detector increases exponentially to the reset value V_{RD} with a time constant of $R_{on}C$, where R_{on} is the "on" resistance of the transistor; this time constant is very short. Random noise fluctuations at $t \gg R_{on}C$ have a root mean square (rms) voltage noise of $\sqrt{kT/C}$ or an equivalent charge noise of $\sqrt{kTC}$, where k = Boltzmann's constant (1.38×10^{-23} joules per kelvin) and T is the absolute temperature (K). After the reset switch is closed, the time constant for the decay of the (uncertain) reset voltage on the capacitor is $R_{off}C$, but $R_{off} \gg R_{on}$ and therefore the noise on the reset level will still be present long after the end of the pixel time. The unknown offset in voltage can be eliminated by taking the difference between the output voltages before and after reset. This method is called Correlated Double Sampling (CDS).

There are several possible strategies for reading out infrared array detectors. Figure 6 shows some of these. In the first scheme, called Single Sampling, each pixel is reset to a fixed level (within the reset noise) to begin the integration, and each pixel is read and digitized (without resets) at the end of the preset time denoted T_int. When the reset pulse is released however, there is usually always a small shift to another level called the "pedestal" level before the detector begins to discharge with photocurrent. Hence this mode does not remove kTC-noise. Nevertheless, single sampling is probably the best mode to use for 3–5 μm imaging since it is fast and because photon noise will be the dominant noise source. The second scheme provides digitization of the pedestal level by immediately following the reset with a non-destructive read cycle from which the preset integration period begins. By forming the difference of the beginning and ending readouts (which is done digitally in computer memory) we get a correlated double sample and kTC-noise is eliminated. This method of resetting the entire array first and then performing a non-destructive read is sometimes called "Fowler" sampling and is less noisy than the alternative of performing both a read and reset as each pixel is addressed. Finally, multiple readout schemes can be implemented simply by applying several non-destructive frame readouts at the beginning, during or the end of each integration.

Using a few multiple reads, typical noise values for the 256×256 pixel arrays are $\sim$30–50 electrons, but $\sim$5–10 electrons is expected from the new large format arrays.

2.2.3. *Quantum efficiency*

The InSb and HgCdTe devices used in NIR astronomy have very good intrinsic quantum efficiency (QE) due to the nature of the charge collection process. One limitation is reflection losses due to the high refractive indices. For the InSb arrays from SBRC, this effect is overcome using an anti-reflection (AR) coating. HgCdTe arrays are not AR-coated and, in common with some InSb devices from other manufacturers, they also suffer from some sort of surface passivation problem which sharply reduces the QE towards shorter wavelengths. For example, NICMOS 3 arrays have 40% QE at J and 60% QE at K, whereas the QE of the SBRC arrays is about 80% across the entire 1–5 μm region.

Platinum silicide arrays are intrinsically low QE due in large part to the poor electron collecting efficiency and threshold effects of the Schottky junction. Values less than 2% are typical, although some research is under way in both Japan and the USA to achieve values above 10%.

2.2.4. *Unusual effects*

Infrared array technology is still developing and even so-called "science-grade" detectors can exhibit flaws. It is not surprising that some detectors show a variety of unusual and unwanted effects. For example, the NICMOS 3 arrays suffer from "residual" images on subsequent readouts after saturation by a bright source due to charge persistence, whereas the InSb devices exhibit many "hot" pixels with high dark current or "hot" edges, especially after exposure to air; the latter effect can be eliminated by baking the detector in a vacuum oven at 80 °C for about two weeks. Both devices can show "glow" due to the readout amplifiers, but it is particularly noticeable in the NICMOS devices when using multiple reads. The SBRC 256×256 InSb arrays suffer from a stray capacitance problem which results in the loss of effective bias across the detector during integration. Finally, thermal cycling and bump bonding is more problematical with the new generation of mega-pixel arrays.

2.3. *Practicalities of Use*

Of course, as already indicated, these devices must be used in vacuum dewars at low operating temperatures. Stability may be an issue for certain kinds of work such as narrow band imaging or grism spectroscopy where the dark current signal may be significant. Since the quantum efficiency (QE) of the detector is also dependent on temperature, it is important to maintain thermal stability to better than ±0.1 K for good photometry.

The removal of pixel-to-pixel variations in quantum efficiency to a tiny fraction of one percent of the mean sky background is a non-trivial, but necessary, exercise. In principle, the methods used at infrared wavelengths are the same as for optical CCDs. Each pixel on the array must be illuminated with the same flux level; such a uniform illumination is called a "flat field". Basically, a flat field source is obtained either by pointing the telescope at the inside of the dome, which is so far out of focus that it gives a fairly uniform illumination across the detector (in principle), or by pointing the telescope at blank sky, including twilight sky. Dome flats are achievable at J and H and sometimes at K, but the thermal emission and high emissivity of the dome rapidly produce too much flux for longer wavelengths. In general, the InSb arrays are more uniform than the HgCdTe devices. The latter have no particular pattern to the variations whereas the InSb arrays exhibit curved striations, like tree-rings.

The "full well" charge capacity of each pixel is given approximately by CV_{rb}/e where V_{rb} is the actual reverse bias across the junction. Setting $V = 1\mu V$, we can also derive the number of electrons per microvolt or its inverse, the number of microvolts per electron (typically $\sim 2.5\mu V/e$). A sequence of flats from zero exposure time to saturation is an extremely valuable tool since it can be used to calibrate linearity and one can extract a variance (noise-squared) versus signal plot whose slope yields the conversion between counts and electrons.

The technical feasibility of deep infrared imaging to limiting magnitudes fainter than $K = 24$ requires extremely precise flat-fielding at a level of 1 in 10^5 in each pixel. It was not clear until the work of Cowie (1990) using IRCAM on UKIRT, that such a precision could be achieved. The solution is a technique based on *median-filtered* sky flats.

The median sky flattening technique has been used widely in optical CCD work where high precision flat-fielding is required (e.g. Tyson 1988). It consists of obtaining multiple exposures, each moved by a few arcseconds in a non-replicating pattern. The individual exposure time is optimized by making each exposure sky-background limited but otherwise as short as possible to maximize the rate of acquisition of frames while not making the overhead from telescope slewing too large a fraction of the total time. Median values of a large number of exposures on either side of an individual exposure are then used to generate the sky flat for the exposure and the exposures are then accurately registered and added to form the final frame during which cosmic rays and other defects are simply filtered out. The individual frames also provide sub-pixel sampling of the final image, which allows us to obtain higher spatial resolution than would be expected from an application of the Nyquist criterion to single frames.

A second quantity which can be optimized to obtain maximum depth at 2.2 μm is the background in the K band filter. Part of this background is often produced by thermal background within the telescope and instrument. Maps of the background versus airmass obtained with the UKIRT IRCAM showed that of the measured 12.6 mag/arcsec^2 of background in that instrument, most is local thermal emission while only 14.2 mag/arcsec^2 on average arises from the sky. (This sky background is very time variable.) Mapping the K band through superimposed narrow-band filters also showed that almost all of this background was concentrated at wavelengths > 2.4 μm and could be removed by using a sharp cut-off K band filter. This led to the introduction of the Mauna Kea K' filter (1.95 – 2.30 μm). Since water vapour affects the transmission at 1.95 μm at most sites other than Mauna Kea, another option is the K_{short} filter from 2.00 – 2.30 μm.

2.4. Signal-to-Noise Calculations

Only a very simplified analysis is given here. First we determine the signal and then identify all the noise contributions. An estimate of the total number of photoelectrons per second detected in the instrument from a star of magnitude **m** is:

$$S = (hc)^{-1} \tau \eta \lambda \Delta\lambda A F_\lambda(0) \times 10^{-0.4m} \quad e/s \tag{2.4}$$

TABLE 3. Flux levels for a zeroth magnitude star in the near infrared.

Wavelength (μm)	Symbol	$F_\lambda(0)$ ($W\ cm^{-2}\mu m^{-1}$)	$F_\nu(0)^\dagger$ (Jy)
1.22	J	3.31×10^{-13}	1630
1.65	H	1.15×10^{-13}	1050
2.2	K	4.14×10^{-14}	655
3.5	L	6.59×10^{-15}	276
4.7	M	2.11×10^{-15}	160

$\dagger$ One jansky (Jy) $= 10^{-26}$ W m^{-2}Hz^{-1} $= 3\times 10^{-16}\lambda^{-2}$ W cm$^{-2}\mu m^{-1}$ where λ is in μm.

where

h = Planck's constant and c = speed of light; $(hc)^{-1} = 5.03 \times 10^{18}$ J$^{-1}\mu m^{-1}$

τ = total transmission of optics

η = quantum efficiency of detector

λ = wavelength (μm)

$\Delta\lambda$ = passband of filter (μm)

A = collecting area of the telescope (cm^2)

$F_\lambda(0)$ = flux from a magnitude zero star (Vega) in W cm$^{-2}\mu m^{-1}$. (See Table 3 which is based on Cohen (1992).)

m = magnitude in the given passband (corrected for atmospheric extinction)

In practice we should integrate over the filter profile and take account of the spectral energy curve of the source.

The signal (S) is really spread over n_{pix} pixels, each of size $\theta_{pix} \times \theta_{pix}$ arcseconds. There is a "background" signal (B e/s/pixel) on each pixel which can be predicted from first principles as the sum of OH emission and thermal components from the sky and telescope. Alternatively, we can estimated the background using a similar expression to Eq. 2.4, if m is replaced with the *observed* sky magnitude per square arcsecond (m_{sky}) and the whole expression is multiplied by $(\theta_{pix})^2$ to account for the area on the sky subtended by each pixel; m_{sky} is ~16 at J and ~13 at K. Dark current can be predicted from theory in some cases, but it is usually a measured quantity (D e/s/pixel). Each of these sources of "signal" (S, B and D) also produce a noise term which follows Poisson statistics and varies as the square root of the total number of electrons, whether generated by dark or photon processes. Of course, the readout noise (R) is an additional noise source and we assume that all the noise sources add in quadrature.

2.4.1. The "array equation"

A full derivation of the signal-to-noise (S/N) ratio of an array measurement is straightforward but tedious (see, for example, McLean 1989). The calculation is based on the assumption that the data reduction procedure is of the form:

$$FinalFrame = \frac{Source\ Frame - Dark\ Frame}{Flat\ Frame - Dark\ Frame} \qquad (2.5)$$

Strictly speaking, we should take into account the number of source frames (N_S), dark frames (N_D) and flat field frames (N_F) actually used in the reduction and follow the propagation of errors resulting from the measurement of each kind of frame. However, if we assume that flat-fielding is nearly perfect and the dark current is small and/or well-determined then the expression for S/N can be simplified. If each individual on-chip "exposure" of the source requires t seconds and N_S exposures are "co-added" to form the final frame with total integration time of $N_S t$ then,

$$\frac{S}{N} = \frac{S(N_S t)^{\frac{1}{2}}}{[S + n_{pix}\{B + D + \frac{R^2}{t}\}]^{\frac{1}{2}}} \tag{2.6}$$

where the total signal S (e/s) is spread over n_{pix} pixels each of which contributes a background of B (e/s), a dark current of D (e/s) and a readout noise of R electrons. This result is known as the "array equation" (also called the CCD equation) and is sufficient for most purposes provided that the observer takes enough calibration data (flats and darks) to ensure that the assumptions are valid.

2.4.2. *Background-limited (BLIP) case*

An important simplification of the array equation occurs when

$$B \gg \frac{R^2}{t} \tag{2.7}$$

Notice that R^2 is the equivalent signal strength that would have a Poisson shot noise equal to the observed readout noise. This condition is known as the Background Limited Performance or BLIP condition and the time $t_{BLIP} = R^2/B$ is a useful figure of merit. When B$\gg$ S, D, and R^2/t then the array equation (Eq. 2.6) can be simplified and re-arranged to give

$$S = \frac{S}{N}\sqrt{\frac{n_{pix}B}{N_S t}} \quad e/s \tag{2.8}$$

and we can convert to measured counts (data numbers, DN) by dividing the right hand side by g electrons/DN. Finally, this simplified expression can be used to predict the limiting magnitude of the measurement from

$$m = m_{ZP} - 2.5\log\{\frac{1}{g}\frac{S}{N}\sqrt{\frac{n_{pix}B}{N_S t}}\} \tag{2.9}$$

where m_{ZP} is the "zeropoint" magnitude for the system. The zeropoint is the magnitude corresponding to 1 DN/s and can be obtained from measurements of a star of known magnitude using $m_{ZP} = m_{true} + 2.5\log\{S(DN/s)\}$.

3. Matching to Large Telescopes

Obviously, the starting point of any design is a set of specifications related to the science goals. For example, which parameters of the infrared radiation are to be measured? What signal-to-noise ratios are required? What spectral resolution, spatial resolution and time resolution are needed? In reality of course, the design will be constrained by many practical considerations such as the size, weight, cost, and time-to-completion.

3.1. *Problems and Trade-offs*

The key factor in every optical design is to "match" the spatial or spectral resolution element to the size of the detector pixel with the following constraints:
- maximum observing efficiency

• best photometry

For normal imaging, or imaging of a relatively wide field of view, the angular resolution even at the best sites is usually "seeing-limited" by atmospheric turbulence to ~ 0.5–1 arcsecond, unless extremely short exposure times are used or some other image sharpening method is employed. To ensure the best photometry and astrometry, the image of a star must be well-sampled. Good "centroids" imply about 5 pixels across image. The minimum limit (often called the Nyquist limit) is 2 pixels across the image, and this may be adequate if the goal is to obtain a wide field of view.

In a spectrometer, the *width* of the entrance slit is the determining factor. A narrow slit implies higher spectral resolution, but the highest efficiency is achieved when the slit is large enough to accept all of the seeing disk.

3.2. *Imaging*

To understand the problem of optical matching to a telescope, consider first the "plate scale" of a telescope given in arcseconds per mm ($''$/mm) by:

$$(ps)_{tel} = 206265/F_{tel} \quad (''/mm) \tag{3.10}$$

where F_{tel} is the focal length of the telescope and the focal ratio (or f/number) is simply F_{tel}/D_{tel}. NOTE: f/no. (or $f/\#$) is a single symbol and not a ratio! (The numerical factor is just the number of arcseconds in 1 radian.)

The angle on the sky subtended by the detector pixel is,

$$\theta_{pix} = (ps)_{tel} d_{pix} \tag{3.11}$$

Typically, $d_{pix} \sim 0.030$ mm

To determine the optical magnification factor we can follow these steps:

• choose required pixel size in arcseconds θ_{pix}
• given size of detector (d_{pix}), derive the plate scale on the detector

$$(ps)_{det} = \frac{\theta_{pix}}{d_{pix}} \tag{3.12}$$

• the required magnification is

$$m = \frac{(ps)_{tel}}{(ps)_{det}} \tag{3.13}$$

NOTE: m also defines an Effective Focal Length ($EFL = mF_{tel}$) for the new optical system. For $m < 1$, the optics are called a FOCAL REDUCER.

From optical principles, such as the Étendue (Area-Solid angle product) of the system or from the Lagrange Invariant, we can relate the pixel size in arcseconds to the f-number of the focal reducer (also known simply as the "camera").

$$\theta_{pix} = 206265 \frac{d_{pix}}{D_{tel}(f/\#)_{cam}} \tag{3.14}$$

where $(f/\#)_{cam} = F_{cam}/D_{cam}$.

EXAMPLE: For $d_{pix} = 27\mu$m and $D_{tel} = 10$-m: $\theta_{pix} = 0.56''/(f/\#)_{cam}$. Assuming seeing of 0.5 arcseconds on Mauna Kea, we need pixels of about 0.25$''$ (2 pixel sampling) which implies that $(f/\#)_{cam} = 2.2$, which is very fast. Remember that the optical system must be well-corrected. For even smaller pixels, e.g. 18.5 μm, we would need an f/1.5 camera! Over-sampling to three or four pixels instead of two, or having smaller image sizes due to adaptive optics, makes things much easier.

3.3. *Spectroscopy*

A brief summary of some useful relations are given here for reference. The *angular dispersion* A of a grating follows from the grating equation:

$$m\lambda = d(\sin\theta + \sin i) \tag{3.15}$$

$$A = \frac{d\theta}{d\lambda} = \frac{m}{d\cos\theta} \tag{3.16}$$

therefore,

$$A = \frac{\sin\theta + \sin i}{\lambda\cos\theta} \tag{3.17}$$

where λ is the wavelength, θ is the angle of diffraction, i is the angle of incidence, m is the order of diffraction and d is the spacing between the grooves of the grating. Many combinations of m and d yield the same A provided the grating angles remain unchanged. Coarsely ruled reflection gratings (large d) can achieve high angular dispersion by making i and θ large, typically $\sim 60°$. Such gratings, called *echelles*, have groove densities from 30–300 lines/mm with values of m in the range 10–100. Typical first order gratings have 300–1200 lines/mm. The grooves of a diffraction grating can be thought of as a series of small wedges with a tilt angle of θ_B called the *blaze angle*. In the special case when $i = \theta = \theta_B$ the configuration is called Littrow and

$$\frac{d\theta}{d\lambda} = \frac{2\tan\theta_B}{\lambda} \tag{3.18}$$

How does the resolution of a spectrometer depend on its size, and on the size of the telescope? From the definition of resolving power we have:

$$R = \frac{\lambda}{\Delta\lambda} = (\lambda\frac{d\theta}{d\lambda})\frac{f_{cam}}{\Delta l} \tag{3.19}$$

where Δl is the projected slit width on the detector and f_{cam} is the focal length of the camera system. But $\Delta l = \theta_{res}/(ps)_{det}$ where θ_{res} is the number of arcseconds on the sky corresponding to the slit width. This angle will be matched to two or more pixels on the detector. We already know how to relate $(ps)_{det}$ to the EFL and, by definition

$$\frac{f_{cam}}{f_{coll}} = \frac{EFL}{F_{tel}} \tag{3.20}$$

and

$$\frac{D_{coll}}{f_{coll}} = \frac{D_{tel}}{F_{tel}} \tag{3.21}$$

therefore,

$$R = (\lambda\frac{d\theta}{d\lambda})\frac{206265}{D_{tel}}\frac{D_{coll}}{\theta_{res}} \tag{3.22}$$

For a given angular slit size, to retain a given resolving power (R) as the diameter of the telescope (D_{tel}) increases, the diameter of the collimator (D_{coll}) and hence the entire spectrometer must increase in proportion. NOTE: if W = width of grating, then $D_{coll} = W\cos i$. This equation explains why spectrometers on large telescopes are so big, and it also implies problems for infrared instrument builders who must put everything inside a vacuum chamber and cool down the entire spectrometer.

3.4. *Diffraction-limited performance*

The image of a very distant point-source object produced by a perfect telescope with a circular entrance aperture (D) should have a bright core surrounded by fainter rings.

This pattern is called the Airy diffraction disk and the first dark minimum corresponds to an angular radius (in radians) of

$$\theta_{diff} = 1.22 \frac{\lambda}{D} \qquad (3.23)$$

which can be related to a physical size by multiplying by the focal length (F) of the telescope and using $F/\# = F/D$,

$$d_{diff} = 1.22\lambda(F/\#). \qquad (3.24)$$

From these equations we see that the diameter of a diffraction-limited image from a f/3 camera at $\lambda = 1$ μm is 7.3 μm, and at $\lambda = 4$ μm this grows to 29.2 μm, or about 1 pixel size. For a 10-m telescope at a wavelength of 1 μm the diffraction disk corresponds to an angular diameter of only 0.05″, which is much smaller than the typical seeing disk whose size can be predicted as λ/r_0, where the Fried parameter r_0 is the length over which the incoming wavefront is not significantly disturbed by motions in the atmosphere. The Fried parameter r_0 is larger at a better site, but it is generally much smaller (< 1 m) than the 8-10 metre apertures of very large telescopes and hence the image resolution is limited to about 0.5″. Adaptive optics systems are designed to compensate for effects of the atmosphere and all large modern telescopes will employ some kind of AO system. Therefore, it is essential to design cameras and spectrometers to be diffraction-limited. For large telescopes, this implies very small physical dimensions for slits and very tight constraints on alignment. For example, for an f/15 telescope at a wavelength of 1 μm, the diameter corresponding to the first dark ring in the Airy pattern is less than four hundredths of 1 mm. It is difficult to make such small slits, and even small temperature variations might change the optical alignment.

3.5. *Field rotation*

One of the fundamental issues to be faced when designing instruments for very large telescopes is the fact that all of these new telescopes are constructed with *alt-az* mounts and hence suffer from field rotation.

The basic equation of field rotation is

$$\frac{dp}{dt} = \Omega \cos A \frac{\cos \phi}{\sin z} \qquad (3.25)$$

where
$\frac{dp}{dt}$ = field rotation rate in radians per second
Ω = sidereal rate = 7.2925×10^{-5} rad/s
A = azimuth and z = zenith distance respectively, and
ϕ = latitude of observatory.

Field rotation can be compensated by counter-rotating the entire instrument at a variable rate, or one can compensate optically by counter-rotating a smaller optical sub-system such as a K-mirror. Both methods are used in practice and both have advantages and disadvantages. It is not very attractive to rotate a large dewar with liquid cryogens, but rotating mirrors add to the light loss and emissivity of IR instruments.

So these are some of the basic issues. Next we will investigate actual designs and some real instruments.

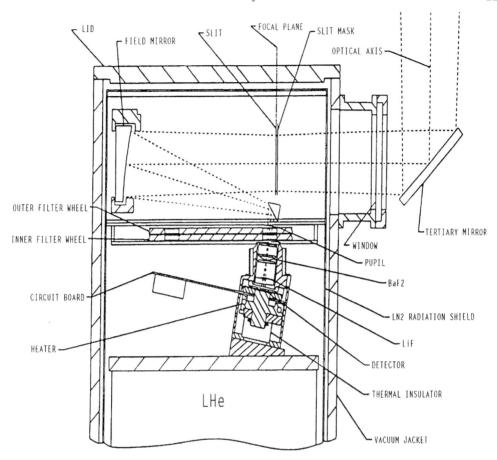

FIGURE 7. Schematic layout of the Caltech Near Infrared Camera (NIRC) developed by Keith Matthews for the Keck telescope.

4. Imaging Systems

4.1. *Typical Requirements*

The optical design must match the detector to the telescope to achieve the required field-of-view, sampling, encircled energy, scattering and stray-light rejection. As already noted, infrared arrays must be operated at temperatures well below room temperature to minimize intrinsic dark current. Moreover, the optical components in the beam, as well as *all* the metal parts around the optics, must also be cooled to low temperatures to eliminate thermal radiation which would otherwise saturate the infrared detector. To achieve these goals, the detector is mounted against a small heat sink of copper (or aluminium) which is connected to a low temperature source. Similarly, all optical and mechanical parts are heat sunk to a low temperature reservoir, which may be different from that of the detector. The complete assembly is placed inside a vacuum chamber with an entrance window made of an IR transmitting material. Electronics and extensive software is required to operate the array, digitize and co-add the data frames, archive, display and manipulate the final images, as well as control the motorized mechanisms associated with the instrument.

In general, modern instruments for large telescopes require "project teams" for efficient

TABLE 4. Near IR facility instruments proposed or operational for large telescopes.

Telescope	Primary	Instrument	Detector	Modes/Features
Keck I[†]	10 m	NIRC	256^2 InSb	imaging $38 \times 38 \times 0.15''$
				$R \sim 120$ grisms
Keck II	10 m	NIRC 2	1024^2 InSb	diff. lim. imaging (AO)
				$R \sim 2,000$ grisms
		NIRSPEC	1024^2 InSb	$R = 25,000$; $0.4''$; echelles
				$R \simeq 2,000$
VLT	8 m	ISAAC	1024^2 InSb	$0.15 - 0.5''$/pixel
				$R \simeq 500 - 3,000$
		CONICA	1024^2 InSb	diff. lim. (A0)
				$R \sim 1,000$ grisms
		CRIRES	1024^2 InSb	$R \sim 10^5$ echelle
		NIRMOS	1024^2 HgCdTe	120 slits; $R \geq 2,000$
				$\lambda < 1.8\mu$m; vis. also
GEMINI (N)	8 m	NIR IMAGER	1024^2 InSb	$0.02 - 0.12$ asec/pixel
		NIR SPECT.	1024^2 InSb	$R = 2,000 - 8,000$
SUBARU	8 m	MIC	1024^2 PtSi	10×10 mosaic; prime focus
			1024^2 InSb	3×3 mosaic; cass focus
		IRCS	1024^2 InSb	$0.05 - 0.125''$/pix; R660–1600
		OHS	1024^2 HgCdTe	R=100; OH suppressed

† fully operational.

and timely construction and it is therefore advisable for students to become familiar with basic project management techniques, such as planning charts, budget preparation and progress reports (see McLean 1989).

4.2. *Designs*

Since the wavelength range of near IR instruments is large ($1 - 5 \mu$m), it is advisable to use all-reflective optical designs if possible. For instruments that work only over the SWIR range ($1 - 2.5 \mu$m), or twin-channel instruments like the UCLA camera, it is possible to develop successful refractive designs which are sufficiently achromatic and for which suitable anti-reflection coatings can be obtained. Some instruments use a combination, e.g. the Keck Near Infrared Camera (NIRC) developed by Keith Matthews at Caltech.

Several designs for cameras and spectrographs for very large telescopes were presented at the SPIE meeting on Infrared Detectors and Instrumentation in Orlando, Florida 1995 (Fowler 1995). Table 4 summarizes current plans.

4.2.1. *NIR instruments for the Keck telescopes*

Already in full operation on Keck I is the Caltech Near Infrared Camera (NIRC). This instrument employs the SBRC 256×256 InSb array in a very efficient configuration with collimation by an off-axis parabolic mirror and re-imaging using a state-of-the-art lens doublet of LiF and BaF with four aspheric diamond-machined surfaces (see Figure 7).

The pixel scale is fixed at 0.15 ''/pixel giving a field of view of about 38''×38''. In

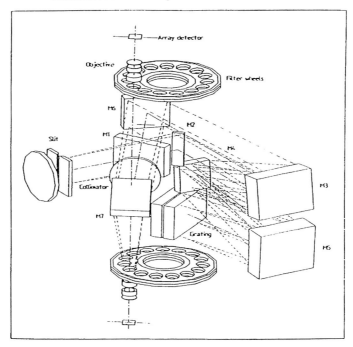

FIGURE 8. Optical layout of the ISAAC infrared imaging and spectrometer instrument for the
ESO VLT.

addition to a large selection of filters, this instrument also provides grism spectroscopy
with R~150. NIRC is located in the "forward cassegrain" position at the focus of the
gold-coated, f/25 IR chopping secondary. The measured throughput of this instrument
is excellent at about 42% and the observed K background is only about 14 magnitudes
per square arcsecond.

Planned for Keck II are two new instruments, NIRC2 and NIRSPEC. NIRC2, which
will also be built at Caltech, is a diffraction-limited imager based on the SBRC Aladdin
1024×1024 array and will work in conjunction with the proposed adaptive optics (AO)
bench at the f/15 Nasmyth focus. NIRSPEC is a cryogenic, cross-dispersed echelle
spectrograph for resolving powers ~25,000 and is intended for either AO or non-AO
applications. This instrument is being built at UCLA and I will give more details later
by using it as a case study.

4.2.2. *NIR instruments for the ESO VLT*

Five infrared instruments are proposed or already under construction for ESO's Very
Large Telescope project. ISAAC (Infrared Spectrometer and Array Camera) is a 1–5 μm
instrument designed for imaging and long slit low-to-medium resolution spectroscopy.
It is expected to be the "first light" instrument on the Unit 1 telescope. A schematic
optical layout is shown in Figure 8.

CONICA is a high resolution near IR camera intended primarily for 1–5 μm diffraction-
limited imaging using an adaptive optics system at one of the f/15 Nasmyth foci of UT1.

CRIRES (Cryogenic IR Echelle Spectrometer) is a proposed instrument designed for
very high resolution (R ~ 100,000) using an echelle or an immersion grating.

NIRMOS (NIR Multi-Object Spectrometer) is another proposed instrument which will perform multi-slit spectroscopy from visible to about 1.8 μm using a combination of CCDs and SWIR arrays.

Note the optical complexity of these large, general-purpose facility-class instruments.

4.2.3. *NIR instruments for the GEMINI North telescope*

Both an IR imager and an IR spectrograph are planned for the infrared optimized GEMINI North telescope. The imager, which will be build by the University of Hawaii, will have multiple image scales to allow diffraction-limited imaging as well as seeing-limited imaging over a wider field. NOAO will build the spectrograph which will be optimized for resolving powers in the range R = 2,000 – 6,000. Both instruments expect to use the 1024×1024 InSb (Aladdin) chips from SBRC.

4.2.4. *NIR instruments for Subaru*

Plans include a medium-to-high resolution camera/ spectrograph combination (IRCS) which will have modes to benefit from adaptive optics. There are ambitious plans for a Mosaic Infrared Camera (MIC) which would use multiple detectors — a 10×10 mosaic of 1K×1K PtSi chips at prime focus and a 3×3 mosaic of Aladdin detectors at cass focus. Finally, there will also be an OH suppression spectrograph (see Section 5) with a final resolution of R = 100 for the 1–2 μm region.

4.2.5. *Other novel instruments*

Although not commissioned as facility instruments for any of the 8–10 metre telescopes, several unique IR instruments have been developed recently which have been used very successfully on 3 – 4 m telescopes. Included in this category are a suite of instruments developed at the Max Planck Institutes in Germany. One of these (3-D), uses a technique called integral field mode imaging to provide imaging spectroscopy, and another (SHARP) uses rapid shift-and-add techniques and/or speckle imaging to improve image quality. The 3-D instrument employs an image slicer, also known as an integral field device, in the focal plane to dissect the image into 16 strips and to direct these strips to another mirror with multiple facets which stacks them end-to-end along the entrance slit of a spectrometer. Although the field of view of such an instrument is usually fairly small (8″×8″ in this case), this technique provides both spatial and spectral information simultaneously without any sequential scanning, as might be done using a conventional long slit spectrometer or a Fabry-Perot etalon. For more details see Fowler 1995.

Our own unique instrument developed at UCLA, although designed as a visitor instrument for the Keck telescope, has re-vitalized the Lick Observatory 3-m telescope — a classic, long focal length optical telescope — for infrared work because it has *twin-channels* and multiple modes (imaging, spectroscopy and polarimetry). Constructing IR instruments with multiple detectors provides a way of increasing the efficiency and competitiveness of smaller telescopes.

Now let's consider in general some of the technical issues facing designers and builders of near infrared instruments.

4.2.6. *Practical issues*

Whether mirrors or lenses are used, it is essential to consider how the optical components will be affected by operation at cryogenic temperatures. Two things need to be calculated: the variation of refractive index with temperature must be known and the differential thermal contraction between the optical component and the mounting structure. Some method of spring-loading in the metal mounts is required to protect the optics. A way to avoid such complications is to make the instrument "athermal" by constructing everything from the same material, such as aluminium (Al). The optics must be reflective of course, and can be manufactured to high precision using diamond machining methods, with or without additional post-polishing.

There are two methods commonly used to obtain the low temperature conditions required in IR instruments: liquid cryogens (liquid nitrogen (LN$_2$) which gives 77 K and liquid helium (LHe) which gives 4 K), or closed cycle refrigerators (CCRs) which are electrically-operated heat pumps — similar to a refrigerator in the home but using the expansion of 99.999% pure helium gas instead of freon — with two temperature stages, typically 77 K and 10 K. One of the smaller, least expensive units (e.g. the CTI 350) can provide 20 watts of power at 77 K and 5 W at 10 K. Larger units can provide about three times this power. Extracting this level of performance from these devices is not as easy as it might seem however. Whereas a liquid cryogen always makes good contact with the surface to be cooled, the CCR head must be connected by copper braids to the required location and this implies a thermal impedance due to losses down the length of the braid and across the contact face. Many braids are required and very high clamping pressures are needed.

For liquid cryogens the cooling ability is expressed in terms of the product of the density (ρ) and the latent heat of vaporization (L$_V$):

$$(\rho L_V)_{LHe} = 0.74 \ W \ hr \ l^{-1} \tag{4.26}$$

$$(\rho L_V)_{LN_2} = 44.7 \ W \ hr \ l^{-1} \tag{4.27}$$

For example, a 10 W heat load boils away 1 litre (l) of LHe in 0.07 hours whereas 1l of LN$_2$ lasts 4.5 hr; the boil-off rate can be measured directly in l/hr with a flow meter. Since liquid helium boils very rapidly and is harder to handle than liquid nitrogen, infrared instruments developed over the last five years or so tend to incorporate CCRs in preference to LHe.

The primary heat loads in a near IR vacuum-cryogenic instrument are radiation from the walls and conduction through wires and fibreglass support frames. For shiny walls ($\epsilon \sim 3\%$), a typical 300 K - 77 K radiation loading is 10 Wm^{-2}. For conduction, everything can be simplified to the following expression for the heat flow (Q_H) in watts

$$Q_H = \frac{A}{L}TCI \tag{4.28}$$

where TCI is the difference of the thermal conductivity integrals I_{Th} and I_{Tc} for the hot (Tc) and cold (Tc) ends of the flow, A is the cross-sectional area of the conductor and L is its length. Useful TCI values in watts/cm for 300 K and 77 K respectively are: copper (1520, 586), aluminium (613, 158), stainless steel (30.6, 3.17), fibreglass (0.995, 0.112).

Short wavelength near infrared arrays such as the NICMOS III HgCdTe arrays have a relatively large bandgap and will yield dark currents of less than 1 e/s/pixel if cooled to 77 K. This temperature is also adequate to eliminate thermal emission from the optics and mechanical structure too. For these detectors a simple, single-cryogen vessel is fine. Indium antimonide arrays are sensitive out to 5 μm and the detector must be cooled to around 30 K to achieve comparable dark currents. For most applications the remainder of the system can be cooled to LN$_2$ temperature.

Infrared arrays, like CCDs, are digitally controlled devices. To operate each device the electronics system must supply 3 things: (i) a small number of very stable, low noise, direct current (DC) voltages to provide bias and power; (ii) several input lines carrying voltages which switch between two fixed levels in a precise and repeatable pattern known as *clocks*; (iii) a low noise amplifier and analogue-to-digital conversion system to handle the stream of voltage signals occurring at the output in response to the clocking of each pixel. There are many ways in which such systems can be implemented and it is beyond the scope of this text to describe and analyze them all (see McLean 1989). The reader is referred to the publications of the SPIE which carry technical descriptions of astronomical instrumentation.

We have already mentioned that the analysis of infrared array data is similar to that for CCDs and involves the numerical manipulation of image frames which are merely two-dimensional arrays of numbers representing the total number of photoelectrons detected in each pixel. As with CCDs, a combined (Dark + Bias) frame must be subtracted to eliminate those additive terms — a Bias frame is one with no illumination and almost zero exposure time and a Dark frame is one with no illumination but an exposure time equal to that used in the source frame — and then the new frame must be divided by a normalized flat-field frame to correct for pixel to pixel variations. Finally, the observed signal levels in counts must be related to magnitudes or absolute flux levels.

Photometry with an array camera is of course performed in software by creating a circular aperture around the source and summing the enclosed signal and then subtracting an equivalent area of background, either by using an annulus around the source or a blank area offset from the source. To calibrate the process the same sized aperture must be used on the standard star of known magnitude (Elias 1982, Cas 1992) or, alternatively, a "curve of growth" giving magnitude as a function of aperture size can be obtained using the standard star and then an "aperture correction" in magnitudes can be applied if a different aperture size is used for the unknown source. There are many internationally recognized software packages which provide facilities to perform numerical processes such as this on digitized image data. For example, IRAF (USA), STARLINK (UK) and MIDAS (Europe). Commercial packages such as IDL can also be used and are becoming very popular.

4.3. *Real Instruments*

Figure 9 shows the UCLA twin-channel camera on the Lick Observatory 3-m telescope at Mt. Hamilton. This instrument is the only general-purpose IR camera in the world with two channels, each containing high QE detectors. NOAO (Tucson, Arizona) has a 4-channel, fixed-filter camera system called SQIID which employs platinum silicide detectors with very low quantum efficiency ($\leq 2\%$). What are the implications of a two-channel camera?

Assuming that observations are always made in a background-limited mode, we can derive the integration time which gives the same S/N ratio in any two different wavebands, such as J and K, and then form the ratio of these integration times as a function of some characteristic or effective temperature of the source. Results of such a calculation show that simultaneous observations at J, H and K are only strictly compatible for effective temperatures around 2,000 K, but the ratio remains within a factor of 10 of unity in either direction over quite a wide range. On the other hand, there is very poor compatibility with the L and M bands. The dependency on other factors was also examined including reddening and redshift. In almost all circumstances, the longest wavelength observed will require the most integration time, except for extremely cool or very heavily obscured objects, suggesting that either the J, H and K measurements are "nested" within the L measurement time, or that the J and H integrations are nested within the K integration.

Our twin-channel system is divided both physically and conceptually into two parts: the cryostat and electronics at the telescope, and the instrument computer and display system in a remote control room. The telescope end includes all necessary electronics to operate the arrays, convert the analog outputs to digital data and co-add images. A motor control unit and a temperature control system are all packaged alongside the

FIGURE 9. The UCLA twin-channel camera on the Lick Observatory 3-m telescope on Mt. Hamilton, California.

camera itself, with the entire assembly held in a frame which supports the instrument on the telescope and during handling. Observers interact with the camera system at the keyboard of a fast PC located in a remote control room. Images from one or both channels can be manipulated and displayed on a second screen.

The main constraint for the optical design is to match the detector pixel sizes to the f/15 image scale of the W. M. Keck telescope (1.375 ″/mm) to yield 0.25″ per pixel and a 64″×64″ field size. Our design, shown in Figure 10, employs transmission optics (mainly fluorides) to provide the following features.

(*a*) An image of the telescope primary mirror is formed inside the cryostat to provide a location for a cold stop with a diameter of 16 mm. This size gives a large number of illuminated grooves for grism work and makes the Lyot stop easier to construct.

(*b*) The focus of the telescope falls about 160 mm inside the cryostat to ensure that focal plane apertures, including slits and occulting spots, are cold.

(*c*) A collimated beam is formed to feed the 45° dichroic beam-splitter and allow both narrow and broad band filters of different optical thicknesses to be placed in the beam using a filter wheel; two dichroics provide the option to select a split at either 1.9 or 2.9 μm.

(*d*) Telecentric re-imaging optics give a uniform plate scale in each beam and an achromatic focus on the detector with spot sizes well within the constraints of current pixel sizes. NOTE: a telecentric system means that the chief ray from each field point is parallel to the optical axis when it reaches the detector and therefore refocusing does not change the scale.

(*e*) The multiplets used in the collimator and imagers are low index fluorides with

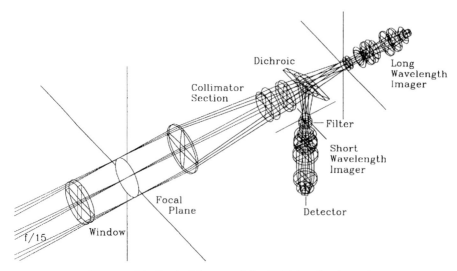

FIGURE 10. Optical layout of the UCLA twin-channel camera.

each optical channel anti-reflection coated for the waveband of interest to give good throughput.

(*f*) A polarimeter module with a quartz-MgF$_2$ achromatic, rotatable half-wave plate is located immediately in front of the large entrance window, and a second aperture wheel carries polarizing components.

The axial length of the collimator and camera sections define the length and width of the cryostat, and the beam size at the dichroic (near the Lyot stop) defines the height needed for the two-position dichroic slide.

We followed the approach of building the instrument from the "inside-out" by customizing the vacuum chamber around the optical design, rather than folding the design to fit a given cryostat. Essentially, the basic concept is that of an optical bench or Optical Base Plate, large enough to accommodate the optical layout, with each identifiable unit, such as Detector Assembly, Filter Wheel, Aperture Wheel, Dichroic Slide, Collimator lens and Camera lens, being treated as a "module" with a discrete location on the plate. To minimize mechanical flexure, the Optical Base Plate is an exceptionally flat, 1 inch thick slab of 6061-T6 aluminum which is 28 in long by 15.5 in wide (see Figure 11). The unloaded weight of the Base Plate is 15 kg and with all modules installed the total weight becomes 27 kg. The optical plate is mounted directly to one wall (top plate) of the vacuum enclosure and is surrounded by a light-tight radiation shield. To thermally isolate the optical bench from heat conduction from the outer wall of the chamber, a mounting arrangement is used which consists of four G-10 fibreglass rods attached to four corresponding "A" frame trusses of aluminium located at the center of the four sides of the plate. This structure is rigid yet allows motion in the required direction due to thermal contraction of the base plate.

All vacuum fittings, electrical connections (vacuum feedthroughs for detectors, motors and status lines) and cooling systems (CCR head, LN$_2$ ports) are located on the same top surface plate that supports the internal optical bench. The entire optical assembly, sealed with a single o-ring by the top plate, fits within a large chamber with a single opening at one end for the entrance window. The vacuum enclosure is constructed of 1/2 inch thick 6061-T6 aluminum plate welded by the standard heliarc process into a

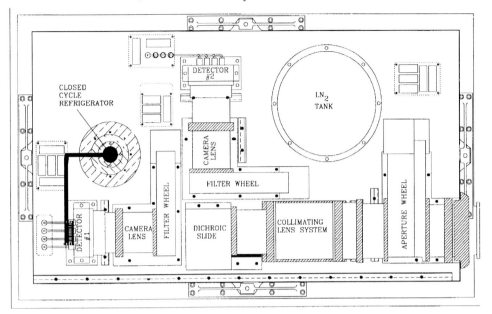

FIGURE 11. Mechanical layout of the UCLA twin-channel IR camera system showing the modular design.

box-shape with dimensions of $33.0 \times 20.5 \times 13.25$ inches. Except for the window, the five remaining sides of the chamber are a single unit, with no openings, which can be lowered away when the top plate is supported by the handling rig therefore providing easy access to all components of the instrument, both interior and exterior.

Motion control mechanisms inside the dewar are operated by internal stepper motors, with no mechanical feedthroughs in the dewar walls. This approach makes dismantling and assembly easier, eliminates a source of backlash and aids in achieving a compact, modular design. In order to minimize heat input to the cryogenic section of the instrument, all internal mechanisms were designed to maintain position without "holding" current being applied to the motors. When a mechanism has been moved to a desired position, the motor driver is powered down by software. Model M-162-03 motors from American Precision Industries, Buffalo, NY, were selected. (Other suitable motors are made by Berger-Lahr and Escap.) These are standard size-17, full or half-step motors with a static torque rating of 25 oz-in. To operate under vacuum cryogenic conditions, the stepper motors must be modified to use dry lubrication. The original sealed bearings were removed one at a time using a specially constructed tool. Each bearing was replaced by an open cage bearing of identical size which had been degreased and burnished repeatedly with molybdenum disulfide (MoS_2) powder. There are 5 cryo-motors within the UCLA camera, but only two basic mechanisms, four wheels and one slide.

The dichroic selector is a slide which is driven by a ball screw mechanism. An aluminium carrier holding the two dichroics is mounted on two parallel polished steel rails by split rings made of Vespel. This plastic material, made by DuPont, is available in a form impregnated with molybdenum disulfide for use in bearing applications (Type SP3). The Vespel split rings allow the carrier to slide smoothly on the steel rails. A steel lead screw is driven via a 1:1 gear from the stepper motor, and a ball nut drives the slide along its rails.

Each wheel is driven at its circumference by a worm gear. The gear is cut directly in the aluminium wheel, while the worm is made of SP3 Vespel. The worm is attached via a flexible coupling directly to the motor shaft, and rests in bushings of SP3 Vespel. Position sensing is via two cryogenically-tested microswitches mounted diametrically opposite to each other near the edge of the wheel. In the edge of the wheel are a notch and a raised ball. The switches are positioned radially so that when the wheel is at its zero position both switches are actuated, but neither operates at the 180° position. As in the dichroic slide, positioning is achieved by counting steps from the zero point; using two switches provides redundancy in the datum information.

As already mentioned, cooling is achieved by using a closed-cycle cooler (CCR), Model 350, CTI-Cryogenics, Waltham, MA, which operates in any orientation, and by using a liquid nitrogen can with a fill-tube whose length ensures that the can will always be half full irrespective of orientation.

The greatest heat load comes from radiation emitted by the large surface area of warm (300 K) walls of the vacuum chamber. Radiation loading is minimized by using polished aluminium sheeting to line all the interior surfaces of the chamber and to make the light-tight shield around the optical base plate. Assuming the emissivity of the sheeting to be 0.03, then the radiative heat load is about 12 W.

Although the total heat load is within the capability of the 1st stage of the CCR once the system has reached equilibrium at 77 K, the thermal mass of 27 kg is too great for the initial cool down, hence the need for the LN_2 supply.

Vibration from the mechanical pump in the CTI 350 head is efficiently damped out using four supports of neoprene rubber and stainless steel bellows as recommended by the manufacturer, and the compressor which is needed to supply the high pressure lines carrying the working fluid (high purity helium) should not be attached directly to the telescope structure.

Both infrared arrays used in the camera have four separate outputs; on the NICMOS 3 array, each output corresponds to a different quadrant of the device. On the SBRC array, outputs are connected to alternating columns in groups of four. Output number one is connected to columns 1, 5, 9, 13 etc., output number 2 to columns 2, 6, 10, 14, and so on. Each output corresponds to 16,384 pixels. If the 14-bit range of the ADC is mapped by our electronics to the well depth of the detector, probably about 200,000 electrons for the SBRC array, and about 200,000 for the NICMOS 3 array, this results in an overall system gain of 24 electrons per data number (DN) for the SBRC array, and 19 electrons per DN for the NICMOS 3 array.

In order to control the camera and collect data from the arrays, we have designed and constructed a front end system based on the Inmos transputer. The transputer is a microprocessor with on-chip RAM and four high-speed serial links which allow it to be easily networked for use in parallel applications. In addition, the transputer is capable of running multiple parallel processes, and can be programmed in its own high-level language, Occam, which has many constructs to facilitate development of parallel applications. Our system consists of sixteen transputers.

One transputer (the "root" processor) is situated on a PC-bus card in our host computer where it receives commands from the PC, passes them over the serial links to the three front end subsystems: the long-wavelength (fast) channel, the short wavelength (slow) channel, and the stepper motor controller.

Each of the two channels consists of a clock generator transputer and a number of data acquisition transputers. The clock generator is responsible for producing the waveforms required to drive the array and triggering the A/D converters at the appropriate points in the clock sequence. The acquisition transputers collect the data from the A/D converters and perform operations such as co-adding of successive frames and correlated sampling.

NOTE: An observation on one channel can be started or completed at any time regardless of what the other channel is doing, but some noise will be picked up if a stepper motor is running during readout of a detector.

The SBRC array requires six negative voltage bias lines while the NICMOS 3 array requires two positive voltage bias lines. This assumes that all lines requiring 5 V on the NICMOS 3 array are tied together at the array and are sourced by one 5 V bias line. The bias board can generate six voltage levels, all positive or all negative, which are set by adjusting low noise wirewound potentiometers; these potentiometers can later be replaced by low noise metal film resistors. Each line is heavily filtered and buffered with an OP-27 op-amp. The board is populated with an AD587, a high precision 10 V reference, configured to provide levels between 0 V and -10 V for the SBRC array; or it is populated with an AD586, a high precision 5 V reference, to provide levels between 0 V and 5 V for the NICMOS 3 array.

Seven negative voltage clock lines are needed by the SBRC array, while the NICMOS 3 array requires six positive voltage clock lines. Our level shifter board can generate 14 voltage levels, all positive or all negative, which are set by adjusting low noise wirewound potentiometers which can later be replaced by low noise metal film resistors. Each line is heavily filtered and buffered with an HA5002 op-amp. Each of the 14 lines is connected to the pole of a SPST analog switch. Four HI-201HS's, also made by Harris Semiconductor, are used to give a total of 16 independent SPST switches, and two of the switches are not used.

The array output stages consist of amplifier FETs tied in a source-follower configuration, and they require current from a load resistor or current source. The preamp contains both a load resistor and a current source, based on the low noise 2N4393 FET, for each array output; either of these can be engaged via a jumper. All four array output voltages are simultaneously transmitted through four identical preamp channels. Each channel has a gain op-amp, the CLC400, for the SBRC channel and the OPA620, for the NICMOS 3 channel. An active filter of the 2-pole Bessel type, based upon the OPA620, can be engaged at the preamp outputs for bandwidth limiting, and thus, noise reduction.

The Gemini control software consists of a large number of discrete processes running on separate processors, but can be divided first into two broad categories: the low-level programs running on the various transputers of the front end system, and the high-level control program running on the host PC. The most important design goal was complete modularity of the two channels of the front end. There is no interdependence (but a lot of similarity) between the software running on the transputers for each channel. The high-level software on the PC must satisfy the usual requirements of being easy and efficient to use, while coordinating control of the separate channels. An overriding design goal for the transputer systems was sufficient processing speed to be able to generate clock signals and acquire digitized data at the required pixel rates.

All but one of the transputers in our system run single processes, the exception being the motor control transputer which has multiple interleaved processes controlling separate motors.

The current host computer is a 486-based PC, programmed in C using the Watcom compiler. The Watcom compiler is a 32-bit compiler and includes a DOS extender, so we can use the full memory space of the PC. The PC program is responsible for the user interface, image display, image analysis, and communicating with the root transputer. An important feature of the program is that the observer may use any of the data display or analysis functions while integrations take place in the front-end system. When an integration is complete, operation pauses for a second or two, while the FITS data file is written to disk, then resumes. This is achieved by subverting the normal PC timer interrupt to poll the PC-transputer link, for incoming messages or data, several times a second.

The program has an excellent graphical user interface designed for easy operation by inexperienced observers using the HiScreen Pro software package. A script interpreter allows standard, repetitive observing tasks to be stored as script files and executed repeatedly. This script language allows access to all camera functions, telescope control functions at sites that permit our software to move the telescope, and simple loops and iterations. In addition, the program supports quick-look image analysis, including automated background subtraction and flat-fielding and simple aperture photometry.

From this rather detailed example, you should be able to recognize all of the features in any IR camera system.

5. Spectrometers

With the advent of sensitive, large-format near IR arrays, infrared spectroscopy can now be extended with almost the same sophistication and precision as optical spectroscopy throughout the $1 - 5$ μm region. The near infrared provides a unique window on the universe at high redshifts, it opens up the study of many species of molecules, and it enables atomic transitions to be observed with little or no extinction due to dust. Many new studies requiring line strengths, velocities or line profiles will now be possible, ranging from stellar physics and the chemistry of the interstellar medium to the most distant extragalactic sources.

5.1. *Typical Requirements*

To obtain an infrared spectrum with substantially higher spectral resolution than provided by a grism, requires an optimized design. Just like optical wavelengths, infrared spectrometers rely on diffraction gratings to provide the spectral dispersion. Unlike optical spectrographs however, the grating — which can be quite large — and all surrounding optics and metal must be cooled to cryogenic temperatures!

One of the most significant advances in infrared instrumentation in the last 5 years or so has been the development of very efficient infrared spectrometers. To obtain the required image quality and field-of-view (long slit) often requires the use of complex, aspheric, optical surfaces. Such surfaces can now be machined directly into aluminium substrates by diamond-turning techniques, and the finished surface is then gold-coated to give a very high reflectance in the infrared. The use of properly treated and thermally-cycled aluminium substrates eliminates concern about differences in the thermal contraction of glass optics and metal mountings.

5.2. *Designs*

The basic elements of any spectrometer are well-known to every student of physics: an entrance slit, a collimating lens, a prism or diffraction grating, a camera lens and a screen or detector on which the dispersed spectrum appears.

5.2.1. *Grisms*

The simplest way to produce an IR spectrograph is to "convert" an existing IR camera by installing a grism. A grism is a right-angled prism with a diffraction grating either ruled directly into the prism or replicated in a resin deposited on the hypotenuse face. The latter approach is cheaper, but the available resins absorb beyond 3 μm. The basic equations to be solved are

$$m\lambda_c T = (n - 1)sin\phi \qquad (5.29)$$

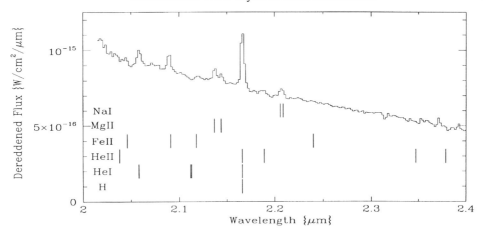

FIGURE 12. R = 550 grism spectrum of the sky emission in the K band with the UCLA camera.

and

$$R = \frac{EFL}{2p}(n-1)tan\phi \qquad (5.30)$$

where λ_c is the central wavelength, T is the number of lines per mm of the grating, n is the refractive index of the prism material and ϕ is the prism angle. EFL is the effective focal length of the camera system and p is the pixel size. The factor of 2 assumes that two pixels are matched to the slit width. In practice, the number of free parameters is constrained by available materials and grating rulings and given conditions in the camera system. Typical resolving powers (2 pixels) are $R \sim 500$, which is much better than circular variable filters (CVFs), but is not really ideal since the OH lines are not well-resolved. Figure 12 shows an example of a grism spectrum of the sky emission in the K band obtained with the UCLA instrument.

5.2.2. *OH Suppresion Instruments*

Since OH emission lines are a major problem for NIR spectroscopy, it is important that this issue be addressed. An interesting new technique for near infrared work is a class of instruments known as OH SUPPRESSION spectrometers. The goal of these instruments is to eliminate the OH lines from the final spectrum. Pioneering work in this field was done by Maihara and colleagues at the University of Hawaii (Mai 1993). In this instrument a high dispersion spectrograph is used to generate a 250 mm long spectrum from which the OH lines are physically removed by a blackened mask or comb with a pattern corresponding to the OH lines. The "OH suppressed" light is then recombined back to white light and then dispersed once more at much lower resolution, typically R $\sim$ few hundred, to yield a faint object spectrograph suitable for redshift surveys. The reduction in background can be as much as a factor of 20. A similar instrument is planned for the Subaru telescope. Note that these particular OH suppression instruments are not implemented in cryogenic form and are mounted on the telescope.

Another similar but more recent development is the Cambridge OH Suppression Instrument (COHSI), led by Dr. Ian Parry, which provides a number of improvements on the original concept by using IR transmitting optical fibres.

COHSI will provide $R = 500$ suppressed spectra simultaneously for both J and H using an integral field fibre feed of about 100 fibres. A multi-object spectroscopy upgrade and an OH suppression imaging upgrade are also planned. COHSI works by forming the spectra at $R = 6,000$ at a mask where the 200 OH emission lines are masked out. The spectra are then undispersed to a polychromatic suppressed slit image and then dispersed to $R = 500$ by a cryogenically cooled spectrograph.

The OH suppression technique will give a sensitivity gain of 1 mag in J and 1.5 mag in H. So, for instance, COHSI on the 3.8-m UKIRT would be 0.5 mag more sensitive in H than an unsuppressed $R = 500$ spectrograph on the 10-m Keck telescope. This is therefore a good technology to enhance the power of 4-m class telescopes in the era of very large telescopes and is ideal if the only role of the spectrograph is to obtain redshifts. Of course, as mentioned below, the proposed spectrographs for the Keck telescopes will not have such low resolutions, but will take advantage of the light gathering power of the Keck to work at much higher spectral resolution ($R = 25,000$) or with diffraction-limited spatial resolution at $R = 2,000$.

COHSI is an improvement on the Maihara system for the University of Hawaii's 2.2-m telescope for several reasons:

(a) it will not have flexure problems, because it's fibre fed.

(b) it has higher final spectral resolution because it uses 2 NICMOS-3 array detectors. Maihara's system has a final R=100 which is not ideal for redshift work.

(c) it has a higher throughput because the gratings work in first order, not 3rd and 4th.

(d) the suppressor is a Schmidt which has a bigger field of view which can be used with the image reformatting properties of fibres to do OH suppression imaging.

The use of IR transmitting fibres is likely to increase in the years ahead. COHSI is expected to see first light on UKIRT in early 1997.

5.3. *Real and proposed instruments*

Most IR spectrographs proposed for large telescopes are cooled grating spectrometers and seek to achieve either very high R or excellent spatial resolution or both. Once again to provide a case study, I will summarize a design we are working on at UCLA.

Analysis of potential science programs for the Keck 10-m telescope strongly favors the highest possible spectral resolution, consistent with practical exposures times. The instrument must be a workhorse for all kinds of atomic and molecular spectroscopy and be capable of more than being a redshift detector. Typical values of resolving power are in the range $R \sim 25,000$ resolution at 2 μm. It is also apparent that observations are not restricted to just one of the near infrared atmospheric windows, and it is therefore important to achieve this resolution at least over the J, H, K and L bands to a wavelength of 4.05 μm (Brackett α).

As we saw in an earlier section, for any reasonable size of spectrometer, the spectral resolution will be slit-limited (i.e. $R < mN$). Typical image quality being achieved now at the Keck telescope without adaptive optics is 0.4 - 0.6″ and it is reasonable to extrapolate to routine images of 0.4″ FWHM in the near future. This slit width drives the size of the collimated beam. From the basic equations we get $R = 25,000$ as required, with $D_{coll} = 0.12$ m and $\theta_{res} = 0.4''$ for an echelle in Littrow mode (2tan $\theta_B = 4$) and a 10-m telescope. The task is therefore to match 0.2″at the chosen telescope focus to the detector pixel size of 27 microns. The pixel size on the sky is related to the true pixel size and the telescope diameter by the f/number of the camera system: $(f/\#)_{cam}\theta_{pix} = 206265d_{pix}/D_{tel} = 0.56''$ after substituting the numbers for the Keck telescope and the SBRC array. Hence, $(f/\#)_{cam} \sim 2.8$; the camera must be faster (about 25%) in practice to accept the beam-spread from the dispersing elements.

For an instrument of the size and complexity expected, the stationary f/15 Nasmyth focus of Keck II is ideal, provided there is a means of image de-rotation. Even at f/15, the spectrograph is a formidable cryogenic instrument and therefore we have elected to convert the input beam to f/10.

While it is not optimal to cross disperse an echelle if the free spectral range of an order (λ/m) exceeds the size of the detector array — the advent of 1024×1024 arrays is critically important in this context — it is also true that very few astrophysical measurements can rely on observations of just one spectral feature, which is all we are likely to get in a single order. For example, emission from molecular hydrogen shocks or from photo-dissociation regions result in about 10 H_2 lines in the K-band; the spectrum of a K5V star contains 15 atomic lines and five molecular band heads in the same spectral region; four coronal lines typical of novae and active galactic nuclei also fall in this wavelength range. There is therefore a distinct speed advantage, of at least an order of magnitude, in using cross-dispersion for these studies.

There are two inescapable penalties of cross dispersion, namely, reduced throughput due to the additional grating required, and greater cost. Taken together with the huge collecting area of the Keck telescope however, we concluded that these disadvantages are offset by the gain in efficiency mentioned above.

The proposed design for NIRSPEC is all reflecting and unobscured, with about 80% of the light from all positions along the slit, and at all wavelengths dispersed on the detector, falling within one 27 micron pixel. NIRSPEC is predicated on the availability of 1024×1024 InSb arrays from SBRC. The f/15 Nasmyth focus is selected for NIRSPEC to enable the large cryostat to remain stationary, but a track and rail system will allow convenient changes with the optical spectrograph (DEIMOS) and permit moves to the Adaptive Optics bench being designed for the opposite Nasmyth platform. We propose to perform image de-rotation with a cryogenic "K-mirror" rotating inside the vacuum enclosure. The incoming f/15 beam is also collimated by the de-rotator to produce a critical Lyot stop and focal plane is then re-imaged at f/10 onto the entrance slit of the spectrograph section. The spectrometer has a 120 mm beam and an f/2.8 camera. A HgCdTe 256x256 array camera with a 46″×46″ field and 0.18″ pixels is included for slit-viewing. (See Figure 13.)

Much of the cryo-mechanical design and infrared array control system is a derivative of experience gained in the development of the UCLA twin-channel IR camera. The transputer-based system will provide 32 channels of low noise data collection for the 1024^2 array and an additional 4 channels for the 256×256 HgCdTe slit-viewing camera. All mechanisms will be driven with cryogenic stepper motors and the large vacuum vessel will be cooled with LN_2 and a powerful closed cycle refrigerator.

The performance of the detector is critical. Low dark currents, low read out noise and high quantum efficiency are absolutely essential. We have set the following stringent goals. For the detector dark current (and/or scattered light background inside the dewar) we seek 0.1 electrons/s/pixel and for the readout noise we assume 10 electrons rms (with multiple read techniques if required). First results with the SBRC detector (A. Fowler, private communication) are encouraging.

The size of the input optics is driven by the slit viewing camera's field of view, rather than by the slit length. The telescope focal plane occurs just outside the entrance window for future accessibility. The diverging f/15 beam is collimated by an arrangement of three mirrors, two of which are flats, to produce a pupil image of diameter 27 mm which becomes the primary cold (Lyot) stop for the instrument. By rotating the three-mirror collimator about the optical axis it is possible to achieve two effects. First, the image of an extended object can be oriented to an arbitrary angle relative to the fixed slit without rotating the instrument. Secondly, by continuous movement of the rotator it is

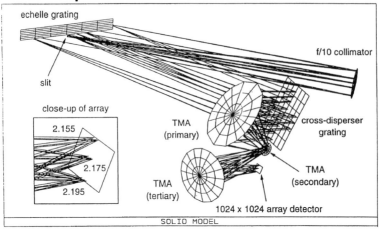

FIGURE 13. The basic optical design concept of NIRSPEC, a cryogenic cross-dispersed echelle grating spectrograph for the Keck II telescope.

possible to cancel the field rotation caused by the alt-az nature of the telescope mounting. Order sorting and other filters are located in a wheel just after the Lyot stop and then the beam is refocused at f/10 onto the entrance slit of the spectrometer section using another "powered K-mirror". Slit width will be variable, but only in discrete amounts, from 0.1″ up to about 1.0″. The image scale at the slit is approximately 2″/mm (500 μm per arcsec).

Of course, the post-slit dispersion system will induce anamorphic magnification, with the slit magnification being, in general, dependent on wavelength. The baseline design assumes an echelle grating with 31.6 l/mm and $\theta_{blaze} = 63.5°$ and a cross-disperser grating with 75 l/mm and $\theta_{blaze} = 10°$. The diffraction-limited resolution in the baseline design ranges from 111,405 at M (m=12) to 417,769 at J (m=45) and therefore it will be possible to upgrade to even higher resolution when AO is available. A useful low resolution mode (R = 2,000) can be achieved by using a flat mirror to bypass the echelle grating and employs only the cross-disperser grating.

The relative efficiency of a grating can be expressed as the detected energy of a given wavelength in the order of interest divided by the total diffracted energy of that wavelength in all orders. At the wavelength and order of interest, the energy in the output beam will be attenuated according to the phase difference of light across a single facet and the resulting intensity pattern is modelled by the blaze function. The greatest efficiency is calculated for the case where the grating is used in the Littrow configuration, i.e. where the input, output and blaze angles are equal; but such a mounting is impractical unless introducing an out-of-plane tilt (γ) so that the standard grating equation becomes

$$m\lambda = d(\sin\theta_i + \sin\theta_o)\cos\gamma \qquad (5.31)$$

where m = order of diffraction, λ is the wavelength, d is the groove spacing and θ_i and θ_o are the angles of incidence and diffraction respectively. This is known as the quasi-Littrow mode (QLM), and the slit image is slightly rotated as a function of λ at the detector. The effect is relatively small and can be handled in the data analysis.

Comparing designs for the near-Littrow mode (NLM) in which the input and output beams are allowed to move off the blaze angle by about 4°–6°and for the QLM mode

TABLE 5. Transmission factors for NIRSPEC.

Optical Element Description	Transmission	Transmission
Window (CaF2: coated for NIR)	98%	
3 mirror rotator/collimator		
– gold coated, 98% each	94%	
Filters (lower limit)	70%	
3 mirror f/convertor		
– gold coated, 98% each	94%	
f/10 collimator (gold)	98%	
Grating — echelle		
– gold-coated replica (QLM)	70%	
(Optional flat – gold coated)		(98%)
Grating — cross disperser		
– gold coated replica	60%	
TMA — three mirror anastigmat		
– gold coated, 98% each	94%	
TOTAL OPTICAL TRANSMISSION	24%	(38%)
Detector QE (2 μm)	80%	
TOTAL SYSTEM EFFICIENCY	19%	(30%)

with $\gamma = 4°$ we find that the predicted efficiency is 70% for QLM with a variable slit image rotation of $10°$–$15°$ across the free spectral range, whereas in the NLM mode with a tilt of $6°$— the minimum required for clearance issues — the expected efficiency is only 53%. In calculations of sensitivity we have assumed the QLM case.

5.3.1. *Predicted performance*

To predict the performance of NIRSPEC we have used a model. Apart from an underestimate of the OH emission flux in the H band of about 0.4 magnitudes, which we correct for empirically, the agreement with measurements is good. Table 5 lists the transmission estimates for the instrument.

Assuming that the primary, secondary and tertiary mirrors of Keck II are all silver-coated, which is the eventual goal although the mirrors are currently being aluminized for installation, and that each reflects 96% respectively then the telescope transmission is 88.5%. This reduces the overall throughput to about 16%. This is the figure we adopt in all our estimates of performance. Table 6 summarizes the parameters assumed in calculating NIRSPEC sensitivity.

Assuming that $S/N = 10$ and that the total integration time is $t = 3,600s$ (1hr), we derive the limiting magnitudes shown in Table 7. Note that since the resolving power in our preliminary designs ranges from about 24,000 to 27,000 we have used the higher value as a limiting case. Sensitivities for a 0.6″ slit are also given. For R = 2,000, we assume the OH background is contributed by either the averaged OH emission lines, or by the continuum flux between the OH lines.

Figure 14 shows the expected limiting magnitudes as a function of wavelength for different resolutions. Typical astronomical objects are shown for comparison.

In summary, NIRSPEC will provide a baseline resolving power of R = 25,000 (12 km/s)

TABLE 6. Factors in sensitivity model.

Parameter	Assumed Value
Background: OH $< 2.2\ \mu$m	Model + data
telescope thermal $> 2.2\ \mu$m	T $\sim$ 273 K
Emissivity:	
(3 mirrors (Ag): 4%+4%+4%)	12%
Throughput: tel. + NIRSPEC	16%
after slit loss (80%)	13%
Readout noise: (InSb)	10 e^-
Dark current + dewar background	$\leq$0.1 e^-/s/pixel

TABLE 7. 10σ, 1hr limiting magnitudes for different configurations.

$\lambda(\mu$m)	R = 27,000	R = 18,000	R = 2,000[†]	R = 2,000
1.25	19.1	19.5	20.2	22.2
1.65	18.6	19.0	19.4	21.5
1.8	18.5	18.9	19.2	21.3
2.0	18.2	18.6	19.4	21.1
2.2	17.9	18.2	19.1	20.1
2.4	17.1	17.3	18.6	19.9
3.0	14.9	15.2	16.7	16.7
3.5	13.4	13.6	15.2	15.2
4.0	12.4	12.6	14.2	14.2

† with contamination by OH lines.

in cross-dispersed mode with a 2.6-pixel slit width of 0.4″and a slit length of about 20″, depending on wavelength. Between 60–90% of an atmospheric window will be captured in a single observation. A low resolution survey mode with R = 2,000 is also provided. At a spectral resolving power large enough to isolate the atmospheric OH emission lines in the near IR, the predicted sensitivity for NIRSPEC is dark current limited at short wavelengths ($\lambda < 2.3\mu$m) and background limited at longer wavelengths due to thermal emission from the telescope. For instance, typical 2 μm continuum fluxes of high-z radio galaxies are about 20 μJy or $K = 18.8$ (corrected for line contamination). NIRSPEC, in its low resolution mode, will reach this limit with a signal-to-noise ratio (SNR) of about 16 in only one hour.

   NIRSPEC full scale construction work is currently ongoing.

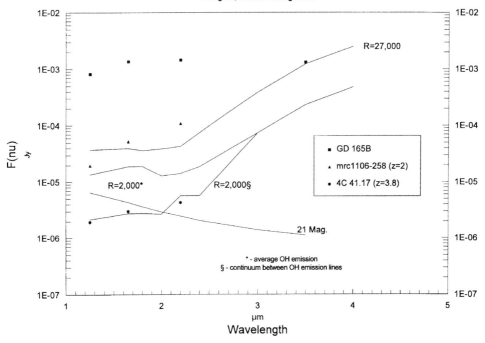

FIGURE 14. Predicted sensitivity as a function of wavelength and resolution for NIRSPEC, together with typical astronomical source spectra.

5.4. Future prospects

Finally, let us compare estimates of the performance of a truly IR optimized telescope with predictions for the Space Infrared Telescope Facility (SIRTF).

The limiting sensitivity of an optimized infrared telescope has been calculated for three bands where the atmospheric transmission is high and atmospheric emission is low. The telescope is assumed to have an average emissivity as low as 0.03 at a temperature of only 273 K; the Gemini North telescope has a goal of 2% emissivity. The atmosphere is assumed to have an average emissivity of 0.01 at the same temperature. Array detectors are assumed to be background limited at all wavelengths and the telescope is taken as an 8-m aperture working at the diffraction limit with the aid of low background adaptive optics. Table 8 gives the one sigma flux and magnitude limits for a one hour integration on the source in each of three wavebands given the assumed system efficiencies in column two; theta is the angular diameter of the diffraction disk in arcseconds.

If one compares these estimates with those for an 8-m telescope designed merely to be "infrared friendly" rather than "IR optimized" (IRO) one finds that the optimized telescope is 3 – 16 times more efficient in exposure time. For broadband observations in the atmospheric windows at 3.65 μm and 11.5 μm, SIRTF will be about ten times faster for point sources than the IRO telescope but will probably have array detectors no larger than 256×256 pixels. For spectroscopy with a resolving power of R = 1,000 in the atmospheric windows the infrared optimized 8-m telescope will be from 10 to 1,000 times faster than SIRTF.

In conclusion, it seems clear that with the advent of IR arrays, large infrared optimized telescopes using adaptive optics techniques to achieve diffraction-limited performance will

TABLE 8. The 1σ 1 hr flux and magnitude limits for an IR optimized telescope.

Bandpass (μm)	Efficiency	Theta ($''$)	Flux (microJy)	Mag
2.25 – 2.40	0.40	0.07	0.01	27
2.50 – 3.80	0.40	0.11	0.4	22.1
11.0 – 12.0	0.30	0.35	28	15.3

be very competitive with space telescopes at the shortest IR wavelengths for broad band imaging and offer tremendous advantages for moderate to high resolution spectroscopy. Building cryogenic instruments for these large telescopes will be challenging and will require many inter-disciplinary skills. These are surely exciting times to be a young astronomer!

REFERENCES

COHEN ET AL. 1992 *Astron. J.*, **104**, 1650.

COWIE, L.L., LILLY, S.J., GARDNER, J.P. AND MCLEAN, I.S. 1988 *Ap. J. Letters* 332, L29.

ELIAS, J.H., FROGEL, J.A., MATTHEWS, K., AND NEUGEBAUER, G. 1982 *Astron. J.*, **87**, 1029-1034.

FOWLER, A. M. 1995 *Infrared Detectors and Instrumentation for Astronomy*, A. M. Fowler, ed., SPIE Vol.2475, Bellingham, Washington.

MAIHARA, T., ET AL. 1993 *PASP*, **105**, 940.

MCCAUGHREAN, M. J. 1988 The astronomical applications of infrared array detectors. PhD thesis, University of Edinburgh.

MCLEAN, I. S. 1989 *Electronic and Computer-aided Astronomy:from eyes to electronic sensors*, Ellis Horwood Limited. Chichester.(UK).

MCLEAN, I. S. 1994 Infrared astronomy: from one to one million pixels", *Infrared Astronomy with Arrays; The Next Generation*, I. McLean ed., Kluwer Academic Publisher, Dordrecht, Holland (1994).

MCLEAN, I. S. 1995 Infrared Arrays: The Next Generation, *Sky & Telescope*, vol. 89, No. 6, June 1995, pp. 18-24

MOORWOOD, A. F. M. 1987 IRSPEC:design, performance and first results. In *Infrared Astronomy with Arrays* (eds. C.G. wynn-Williams & E.E. Becklin), pp. 379-387. Institute for Astronomy, University of Hawaii.

TYSON, J.A. 1988 Techniques for Faint Object Imagin at 1 micron. In *Infrared Astronomy with Arrays* (eds. C.G. Wynn-Williams & E.E. Becklin), pp. 483-488. Institute for Astronomy, University of Hawaii.

Mid-IR Astronomy with Large Telescopes

By BARBARA JONES

Center for Astrophysics and Space Sciences, University of California, San Diego
La Jolla, CA, 92093-0111, USA

This lecture introduces the opportunities presented by ground-based telescopes for new discoveries in the thermal infrared, and discusses techniques used to make sensitive observations in an environment with high background flux levels from atmospheric emission and from the telescope structure and mirrors.

1. Mid-IR astronomy—opportunities and problems

The capability now exists to observe mid-IR astronomical objects with spatial resolution of a third of an arcsecond and sensitivities reaching well below a mJy. Both imaging and spectroscopy with new array instruments on optimized large telescopes are producing new data on sources from comets, to active galactic nuclei. With sensitivity to emission from cool dust, diagnostic lines from ionized gas and molecular species, and the capability to look through clouds opaque in the visible, many new results are appearing, and many more can be anticipated. In particular, our understanding of the star formation process should improve significantly in the next decade. Yet all of this is achieved operating through the earth's atmosphere which absorbs and distorts the signals, and which, together with the telescope structure itself, radiates into the beam up to a million times the power detected from the source. The problems encountered, and the techniques used to make ground based mid-IR observations will be discussed here.

IRAS (Infrared Astronomical Satellite) revealed how fascinating and complex the IR sky is at wavelengths of 12, 25, 60 and 100 μm. The *IRAS* mission lasted for 300 days in 1983 completing an all sky survey with a 57-cm diameter cooled telescope. *ISO (Infrared Space Observatory)* is now in orbit (launched 12/95). This is a 60-cm diameter cooled telescope with imagers and spectrometers operating from 2.5 to 240 μm. We hope that *SIRTF (Space Infrared Telescope Facility)* will be launched in the next few years.

Comparisons of the capabilities of large ground-based facilities with cooled telescopes above the atmosphere should be made to achieve optimum uses of each. Background radiation (which limits the sensitivity of almost all ground-based observations) is lower by a factor of about 10^7 from a cooled orbiting telescope. This more than offsets the ground-based gains from larger primary mirrors, allowing better point source sensitivities for broad-band imaging. Clearly, space instruments can work at all wavelengths, whereas from the ground observations are limited to the atmospheric windows. The capability to observe in "stare" mode gives orbiting instruments an advantage in observing extended emission in complex regions. Not all the advantages are in favor of space however. Ground-based instruments, with diffraction-limited images, offer much higher spatial resolution. They also gain a significant advantage from their large collecting areas in "photon starved regimes", e.g. at higher spectral resolution. A practical advantage in favor of ground-based instruments is the shorter timescale between the conception and operation of a new instrument. Ground-based instruments can use the latest available technology, an important advantage in this rapidly developing field. The following references are useful for general information. *The Infrared Handbook* (ed. Wolfe & Zizzis) is a fount of technical information. Optical design is discussed in *Astronomical Optics*

241

Window	λ cut-on (μm)	λ cut-off (μm)	Main absorber
K	1.96	2.48	H_2O, CO_2
L	3.06	4.03	H_2O, CH_4
M	4.5	5.1	H_2O
N	8.1	13.1	H_2O, CO_2
Q	15.5	26.5	H_2O
X	≈ 30	≈ 35	H_2O

TABLE 1. Mid-infrared windows and main absorbers.

by Schroeder, and observational techniques by Walker in *Astronomical Observations* and by Lena in *Observational Astrophysics*. Recent conference proceedings which give good surveys of current instrumentation and observations (e.g. McLean 1994; Elston 1991; Kwok 1993; Mampaso, Prieto & Sánchez 1993; Kaldeich 1988).

1.1. *The Atmosphere*

Thermal IR astronomy is somewhat loosely interpreted as the regime from about 2.5 μm to 35 μm where ground-based observations are dominated by thermal background from the atmosphere and telescope. At 273 K, typical night-time temperatures at a ground-based site, the Planck function peaks in the 8 to 13 μm atmospheric window. This thermal background emission provides the fundamental limit on the sensitivity of almost all mid-IR ground-based observations, due to the random fluctuations in the arriving photon stream. Understanding and minimizing this thermal background is essential. Building a telescope on a mountain top increases transmission and decreases emission. A high, cold dry site is a prerequisite for sensitive measurements. Where the atmosphere is opaque the emissivity is 100% of a black body. In the best "windows" the atmospheric transmission is remarkably good($t \geq 99\%$), usually implying an emissivity $\epsilon \leq 1\%$ of a blackbody at ambient temperature. Even in this condition random fluctuations in the background radiation dominate the system noise.

1.2. *Atmospheric transmission*

Figure 1 shows the principal molecular absorbers and a solar spectrum at low resolution (taken from Wolfe and Zizzis 1989). Water vapor is one of the strongest absorbers. The scale height for water vapor is small, so observing from mountain tops can significantly reduce the amount of water vapor in the line of sight. A good site should offer a high proportion of nights below 1 mm of "precipitable" atmospheric water vapor. Carbon dioxide is the other main absorber. CO_2 has a larger scale height, giving smaller variation from one site to another. Ozone has an absorption feature centered at 9.6 μm, awkwardly placed in the middle of the 8 to 13 μm window. Typical transmission values in the peak of the O_3 feature drop to about 30%. However since O_3 is concentrated high in the atmosphere it does not contribute as severely to the background emission. Table 1 shows the passbands generally adopted to standardize measurements for photometry (Low 1974; Campins et al. 1985; Rieke et al. 1985).

These windows are usually defined by cooled multi-layer interference filters in the optical path of the instrument. To achieve consistent measurements from site to site in varying observing conditions, the cut-on and cut-off wavelengths should be well into the region of good atmospheric transmission. However, to maximize signal, using a larger bandwidth may be an advantage. In background-limited conditions a more careful anal-

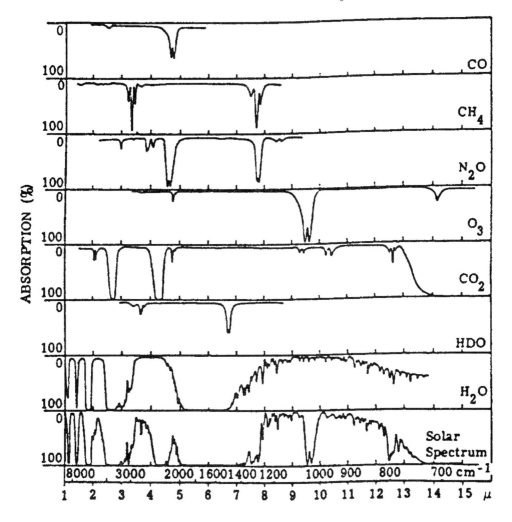

FIGURE 1. Low-resolution solar spectrum from 1 to 15 μm.

ysis is needed to optimize the signal-to-noise ratio, since the increase in signal resulting from an increase in bandwidth may be more than offset by the increase in noise from the extra background (Milone & Stagg 1993).

Detailed calculations can be made of the atmospheric transmission using the ATRAN program (Lord 1992). Figure 2 presents the calculated transmission from ATRAN for Mauna Kea with 1 mm precipitable water vapor.

1.3. *Atmospheric emission*

For the given area and solid angle ($A\Omega$) of an observation, defined by the telescope and instrument, there is an amount of background radiation, emitted by a variety of molecular species in the atmosphere, incident on the instrument. Assuming all emitters are at the same temperature, the atmospheric emissivity is usually defined as the measured atmospheric emission compared to the emission from a blackbody in the same $A\Omega$. When the atmospheric transmission is high, the emissivity can be approximated by $\epsilon = 1 - transmission$. The assumption that the emitters are all at the same effective

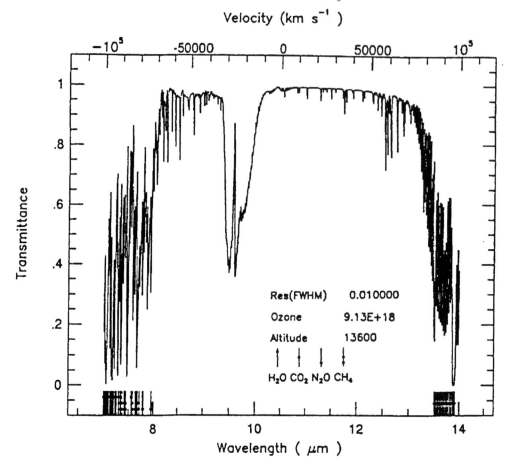

FIGURE 2. ATRAN calculations of the atmospheric transmission above Mauna Kea for 1 mm precipitable water vapor.

temperature, which will be close to the temperature of the telescope structure, is good for water vapor which has a small scale height. For CO_2 it is not so good, and for O_3 it clearly is inappropriate. To calculate the energy reaching the telescope focal plane from atmospheric emission let: telescope area $= A = \pi D^2/4$ (m^2); effective atmospheric emissivity in the appropriate bandwidth $= \epsilon$; effective atmospheric temperature $= T$ (K); wavelength of observation $= \lambda$ (μm); bandwidth $= \delta\lambda$ (μm); Planck function $= B(\lambda, T)$ (W m^{-2} ster^{-1} μm^{-1}); solid angle $= \Omega$ (1 arcsec^2 $= 2.34 \times 10^{-11}$ ster), and background power from atmospheric emission $= P_{atm} = \epsilon\, B(\lambda, T)\delta\lambda A\Omega$ (W).

Taking some typical values to assess the order of magnitude, let $\epsilon = 0.01$ (for the best part of the 10 μm window), $T = 273$ K, $\lambda = 10$ μm, and $\delta\lambda = 1$ μm. Then $B(\lambda, T) = 6.1$ W m^{-2} ster^{-1} μm^{-1}. The power per *sq arcsec* is: $P_{atm} = 1.1 \times 10^{-12}$ D^2 (W arcsec^{-2}) or $N = 5 \times 10^7 D^2$ (photon s^{-1} arcsec^{-2}), where N is the photon arrival rate. Calculating typical numbers for a 3-m diameter telescope this gives: $P_{atm} = 10^{-11}$ (W arcsec^{-2}) or $N = 5 \times 10^8$ (photon s^{-1} arcsec^{-2}), corresponding to a 10-μm sky brightness of 0 mag arcsec^{-2} in the best part of the atmospheric window. (For comparison, for a good dark site, $V = 22$ mag arcsec^{-2} on a moonless night.) For a pixel size of 0.35 arcsec (chosen to sample the diffraction limit of the 3-m telescope) this flux is spread over 8.5 pixels, giving 6×10^7 photon s^{-1} pixel^{-1} just from the atmosphere.

A typical number for the photon arrival rate must allow for telescope emission also, and should use a more typical atmospheric window average emissivity. The lowest reported (telescope + atmosphere) emissivities are about 10%, with 20–30% not unusual. The higher numbers can result from many problems including warm wet weather, poor IR telescope design, poor instrument baffling, dirty mirrors, etc. For an effective emissivity of 10% the photon arrival rate is $N = 6 \times 10^8$ photon s^{-1} pixel^{-1}. For a typical BIB detector with $\eta G = 0.6$ and $\beta G = 2.2$ (see *Detectors* chapter) the number of collected electrons $(\eta G N)$ is 3.6×10^8 e$^-$s^{-1} pixel^{-1}. This requires a frame rate of 360 Hz for a well depth of 1×10^6 e$^-$, and 12 Hz for the deep wells of 3×10^7 e$^-$. For the higher rate, assuming a 128×128 element array, this gives about 150 ns to read each pixel. Fortunately the chips have multiple read lines, allowing parallel processing of the data stream.

The minimum noise that can be achieved is determined by the statistics in the electron collection rate. We can calculate the noise equivalent power (NEP) for this system given that $S/N = \sqrt{(\eta/\beta)N}$. For $t = 1$ s, we get $S/N = 1.3 \times 10^4$. To determine the effective NEP we need $S/N = 1$. This corresponds to a photon flux per pixel of 4.6×10^4 photon s^{-1}, or a power of 9×10^{-16} W pixel^{-2}. The NEP is thus 9×10^{-16} W $/\sqrt{\text{Hz}}$, determined by the background photon arrival rate N and by η/β for the detector.

1.4. Telescope design

Many criteria important for optical performance are also essential for IR performance, such as good pointing, tracking, and image quality, but many others are unique to the IR and may conflict with optical requirements. Typical optical instruments at the secondary focus need a fast beam and a wide field. This requires a large secondary mirror, with its associated massive support, and a large central hole in the primary. Baffling to block unwanted skylight from reaching the optical detector is almost always done in the telescope, not in the instrument. For infrared use the thermal emission from the telescope should be made as low as possible. The most critical atmospheric window is from 8–13 μm, where the atmospheric emission can be as low as 1% and the Planck function for typical night-time temperatures peaks. Thus the level of emission from the telescope is the critical factor in determining the performance for background limited observations. This requires minimizing, or removing wherever possible, baffles and support structure which radiate as blackbodies. In addition, for thermal IR work, a chopping secondary is essential. This is usually much smaller than the optical secondary to be able to achieve reasonable performance. It should be supported by spiders of minimal cross section, and be undersized for the on-axis beam. The telescope should be capable of nodding with an amplitude and position angle matched to the chop throw.

In general any structure in or close to the beam may increase the effective emissivity and provide non-statistical fluctuations that may not be rejected by chopping and nodding. These non-statistical fluctuations must be made smaller than the statistical $(\sqrt{N(\text{photons})})$ fluctuations from the background radiation incident on a pixel, measured over both the chop and nod timescales.

1.5. The secondary mirror

The secondary should, at a minimum, be capable of chopping several tens of arcseconds on the sky at frequencies up to 20 Hz. This requires that the transition time from one beam to the other be shorter than 5 ms for reasonable efficiency. The telescope should be capable of precise nodding, with an amplitude matching the chop, at time intervals

Telescope/primary diam.	Chopping secondary diam.	Focal ratio
IRTF 3.0 m	0.24 m	$f/35$
IKIRT 3.8 m	0.31 m	$f/35$
Keck 10.0 m	0.5 m	$f/25$
Gemini 8.0 m	1 m	$f/16$

TABLE 2. A Sample of chopping secondary diameters and focal ratios.

of 10 to 30 seconds. Autoguiding should not be confused by the chopped, nodded beam. In order to achieve a "square wave" chop with a low rise time and a reasonable throw, without using large amounts of power, the moment of inertia of the mirror needs to be small. At a fixed amplitude and frequency, the power consumption goes as D^4, where D is the secondary diameter. This requires a small secondary mirror (see Table 2), preferably thin and lightweighted. The resulting IR beam is slow (i.e. high f ratio).

The Keck secondary at 0.5 m diameter is a very lightweight, rigid structure. It requires active cooling to remove heat generated by high-frequency, large-amplitude chopping. The IR beam at Keck has been made somewhat faster by putting the focal point 4 m forward of the vertex of the primary. The Gemini telescopes, with their $f/16$ focal ratio and 1m diameter secondaries, are being optimized for thermal IR observing.

The secondary mirror should be mounted using thin spiders to reduce the thermal background radiation into the beam. It should not have low-frequency oscillation modes from the support structure, and should not couple energy into the telescope at the chop frequency. Reactionless choppers, which use an equal mass, oscillating out of phase, mounted immediately behind the mirror are commonly used. The mirror should be mounted from behind, with no structure that would radiate into the beam at larger radii than the mirror (i.e. around the edge of the secondary the detector should see low-emissivity sky).

Frequently a central hole is used in the chopping secondary mirror, large enough to mask the detector's view of the central hole in the primary, or the secondary "cage" (after reflection in the primary). This ensures that, close to the telescope axis, the detector views the sky. (Alternatively a reflecting cone can be mounted at the center of the secondary to reflect sky from the primary). The secondary should also be marginally undersized. An oversized secondary (if not perfectly baffled at a pupil in the instrument) will reflect the structure immediately surrounding the primary mirror into the beam. This structure will radiate as a blackbody, increasing the thermal background. Even when a cold baffle is placed at an image of the secondary mirror within the instrument optics, the baffling is never perfect. A combination of optical aberrations and diffraction always produce some "wings" on the beam from the dewar. If the secondary is undersized these "wings" produce very little extra background as they view the sky beyond the mirror.

The secondary spiders need to provide good support for the mirror without obscuring the incoming beam. This is true for both optical and IR performance. They typically have a long thin rectangular cross section for rigidity. The control signals for a chopping secondary must be routed and hidden from view, over a spider strut. Oscillations in the spiders can modify the thermal emission of the telescope. A reactionless secondary helps reduce this, but wind forced vibrations must also be considered. A small "lip" on the edge closer to the primary can significantly reduce any thermal oscillations from torsion in the spiders. A full analysis of the frequencies and amplitudes of the oscillation modes is necessary to assess the modulation of the background in the IR beam.

1.6. *Level and stability of telescope emission*

Other sources of emission from the telescope include baffles, the mirror surfaces, scattered light, beam flare hitting the dome, and emission from the local atmosphere. The Keck telescopes, with segmented primary mirrors, also have emission from the gaps between the segments.

Ideally the shutter opening is significantly larger than the primary size. Automatic dome rotation should keep the shutter centered on the telescope, minimizing variations in background caused by the beam flare getting close to the edge of the shutter. Vignetting the beam with the dome causes huge changes in background, usually requiring an observation to be repeated.

For the highest reflectivity in the IR gold coatings are used ($R_{gold}(10\mu m) \approx 99\%$. Since the chopping secondary is usually exclusively used for IR, little controversy arises by using a gold coating, but for a telescope which is also used in the optical, a gold coated primary severely reduces efficiency at the blue end of the spectrum. Bare aluminum gives about 98% reflectivity at 10 μm, silver is a little better at the cost of a few UV photons. These differences in reflectivity are not important, but the emissivity of a 98% reflecting surface is twice that of a 99% reflecting surface. This also illustrates how important it is to keep the surfaces clean. Dirt radiates efficiently.

1.7. *The chop and nod Technique*

Since the sky emission is much larger than the signal expected from a typical source, it must be cancelled by chopping the telescope secondary mirror (typically at 3–10 Hz) and nodding the telescope (typically every 20 sec), see e.g. Piña (1993). Figure 3 shows both chop positions for each nod position with the astronomical source shown in *nod 1/chop 1*, and *nod 2/chop 2*.

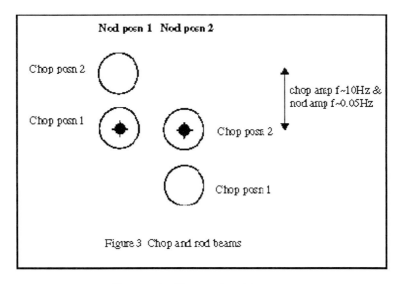

FIGURE 3. Chop and nod beams

The telescope background is slightly different in the two chop positions, $B_{tel,1}$ and $B_{tel,2}$. The sky background also has a small spatial gradient, so that B_{sky} is different in the three sky positions in this pattern. The flux per pixel in the different positions can be expressed as follows:

In nod position 1, chop position 1: $F_{1,1}(x) = S + B_{sky}(x) + B_{tel,1}$

In nod position 1, chop position 2: $F_{1,2}(x - \Delta x) = B_{\text{sky}}(x - \Delta x) + B_{\text{tel},2}$

The "chopped" beam for nod position 1 is the difference between these terms. This can be written as

$$C_1(x) = F_{1,1}(x) - F_{1,2}(x - \Delta x) = S + B_{\text{tel},1} - B_{\text{tel},2} + (\frac{dB_{\text{sky}}(x)}{dx})\Delta x \qquad (1.1)$$

The term $B_{\text{tel},1} - B_{\text{tel},2}$ represents the imbalance in the telescope beams, the last term is the linear part of the sky gradient. These terms usually dominate the source term. For the second nod position the fluxes can be written as Nod position 2, chop position 2: $F_{2,2}(x) = S + B_{\text{sky}}(x) + B_{\text{tel},2}$

Nod position 2, chop position 1: $F_{2,1}(x + \Delta x) = B_{\text{sky}}(x + \Delta x) + B_{tel,1}$

Chopped beam, nod position 2:

$$C_2(x) = F_{2,2}(x) - F_{2,1}(x - \Delta x) = S - B_{\text{tel},1} + B_{\text{tel},2} - (\frac{dB_{\text{sky}}(x)}{dx})\Delta x \qquad (1.2)$$

Adding the chopped beams for the 2 nod positions then gives the source term only

$$SIGNAL(x) = C_1(x) + C_2(x) = 2S \qquad (1.3)$$

A useful diagnostic can be made by subtracting the 2 "central" $F_{1,1}$ and $F_{2,2}$ beams to isolate the telescope inbalance signal

$$TELBAL = F_{1,1} - F_{2,2} = B_{\text{tel},1} - B_{\text{tel},2} \qquad (1.4)$$

The quality of the signal extraction depends on the temporal stability of the sky and telescope backgrounds. Chopping allows a fast subtraction of the emission in adjacent sky beams, nodding provides a slower subtraction of the imbalance of the two beams through the telescope. This process is remarkably successful. It allows for measurements of signal of about 1 photon in 10^6 over long integrations. Variations in sky emission at the chop or nod frequencies are not rejected by this technique and occasionally provide the limiting noise source.

Changes in emission from day to night are typically less than 50% and are caused primarily by daytime warming and higher daytime dew points. Day-time thermal IR observing can be done with reasonable sensitivity. The most usual problems are the poorer day time seeing (see below), the need to keep the telescope and dome cold to reach equilibrium for optimum night work, and the lack of optical acquisition and guiding capabilities.

1.8. *Special factors important for a segmented-mirror telescope*

This section refers to the Keck I telescope. Keck II will be identical except for the performance requirements for the IR secondary, and possible future use of alternative mirror coatings. The Keck primary has 36 hexagonal segments, each 1.8 m across, for a total area of 75 m^2 reflecting surface. The segments are phased to about 20 nm to form a coherent surface. The 5-mm gaps between the segments are about 0.5% of the reflecting area and radiate as a blackbody, adding 0.5% to the overall telescope emission. Approximate values for the emissivity are: spiders 0.5%, segment gaps 0.5%, secondary cage obscuration 4.5%, and mirror surfaces 3% (1 gold and one aluminum surface). The secondary cage obscuration is masked by a central hole in the secondary, giving a total design emissivity of about 4%. The telescope has an altitude-azimuth mount, and an overall hexagonal symmetry. The chopping secondary is also hexagonal, about 5% undersized, gold coated, with 2 axis control. To maintain a fixed position angle on the sky the instrument and chop/nod angle rotate continuously during an observation. The cold pupil in the instrument is ideally also a rotating hexagon. However all Keck IR instruments to date use an inscribed circular pupil. The primary mirror control system updates at 2 Hz, and provides a potential modulation of the background unique to a segmented design. Calculations of the typical motions of the segments at each update show that non-statistical fluctuations from a phased primary mirror do not present a problem. Advantages of a segmented design include the ease of removing and recoating segments to keep the primary surface clean, and of potentially applying exotic coatings over a somewhat smaller area than the full primary. Another advantage for image optimization is that of separating the 36 individual foci for optical diagnostic tests.

1.9. *"Seeing" (image quality) in the thermal IR*

As for optical observations, the fewer the number of pixels needed to collect the signal from a point source the larger the signal to noise in a given time, and the larger the information gained on the spatial structure of extended sources. For optical wavelengths image size is limited by atmospheric seeing, not by diffraction. For an optical point source many speckles are formed, each with the diffraction limited image size. These speckles are contained in the much larger "seeing disk". The size of the seeing disk is determined by the atmospheric turbulent coherence length, r_0, which scales as $(\lambda/\lambda_0)^{1.2}$ (Stock & Keller, 1960; Korff 1973). For a good site $r_0(0.5\mu m) \approx 30$ cm, implying $r_0(10\mu m) \approx 11$ m. Comparing r_0 with the telescope diameter, D, determines approximately how many speckles in the image, and $\alpha = \frac{\lambda}{r_0}$ gives the diameter of the circle containing the speckles. Thus α scales as $(\lambda/\lambda_0)^{0.2}$, giving about a factor of 2 improvement in seeing from V to mid-IR. The timescale for coherent variations in the image also increases with wavelength, observations suggest a dependance approximately as $\lambda^{0.5}$ (Selby et al. 1979).

Experience with our cameras at the 1.5-m Mt. Lemmon telescope near Tucson, AZ, and at Keck (10 m) suggests that the mid-IR images are similar to these predictions. At Mt. Lemmon the visible seeing is frequently poor, yet we usually achieve images very close to the diffraction limit at 10 μm. At Keck, under fairly good visible seeing, we are also very close to the diffraction limit. The Keck images at 12 μm remain stable over periods of minutes, with a FWHM of 0.35 arcsec.

Since the IR seeing is so good, it is likely that many telescopes will get their best image quality in the mid-IR. Thus infrared requirements are as critically important as those at shorter wavelengths. The optical quality of telescope and instrument must be excellent, and every effort must be made to preserve good seeing.

2. Mid-IR imaging

This lecture discusses some of the major design features of mid-IR cameras using 2-dimensional arrays. Limits on sensitivity are calculated for a background-limited situation where the pixel size is set by the diffraction-limited image. A list of instruments has been assembled from the published literature.

2.1. *Introduction*

Imaging through bandpass filters at the diffraction limit of the telescope has been the first successful application of mid-IR arrays to astronomy. Although the field of view of these instruments is limited compared to optical imagers, the sensitivity and spatial resolution are excellent. As new, large telescopes are completed on good sites and with emphasis on achieving the ultimate image quality, the mid-IR data will be spectacular. Tables 3 and 4 give a list of operational 2D mid-IR array cameras. The length of these lists shows the scientific interest in this technique.

Instrument Name	PI/Org	Array	Telescope & FOV	Sensitivity	Comments
Spectro-Cam-10	Houck @ Cornell	128x128 As:Si BIBIB Rockwell	5m f/70, 0.25"/pix 15x15"fov for im	@0mJy pt src t=100s 1μm @10μm	all refl 8-13μm spectrometer
MIRAC2	Hoffman @ U.Az & Fazio @ Harvard	128x128 As:Si BIBIB Rockwell $3x10 e^-$ wells	IRTF 3m & Steward 2.3m zoom .25-.50"/pix	IRTF .34"/pix 30mJy/ sq"t=1m 1μm @ 12μm	2-26μm 2 FW & CVF zoom camera
LWS	Jones & Puetter @ UCSD	96x96 Si:As MC^2(BIB)AESC	Keck 0.12"/pix 11.8"fov	28mJy pt src t=1s 1μm @12μm	all refl 3-26μm spectrometer
Golden Gopher	Jones & Puetter @ UCSD	20x64 Si:As MC^2(BIB)AESC	UM/UCSD Mt Lemmon 1.5m 0.83"/pix	25mJy/sq" t=1m	all refl 5-26μm
NASA	Gezari @ GSFC	58x62 Hughes IBC Si:As	IRPTF 3m 0.26"/pix		cryocooler
AIR Camera	Roellig @ NASA AMES	128x128 As:Si BIBIB Rockwell	1.5 NASA/UA @Mt Lemmon 0.73"/pix	0.04Jy/sq" t=1m 0.3μm@10μm	single lens 2 FW & CVF 8-22μm
Berkcam	Arens @ SSL(UCB)	10x64 Si:Ga Hughes BIB	UKIRT/IRTF 0.39"/pix	25mJy/sq" t=1m 1μm @ 10μm	lenses CVF
MIRLIN	Werner @ JPL	128x128 Si:As BIBIB Rockwell	Pal 5m 0.15"/pix IRPTF 3m 0.5"/pix		Asph cam mirror & ZnTe lens $3x10 e^-$ wells
OSCIR	Telesco @ U. Florida	128x128 Si:As BIBIB Rockwell	IRPTF 3m 0.24"/pix	5mJy pt src t=1hr N band	all refl 5-2μm $3x10 e^-$ wells
MIRCAM	Garden @ UC Irvine	128x128 Si:As BIBIB Rockwell	UKIRT 0.25"/pix IRTF 0.35"/pix	opnl 1995	all refl
camera	Rank/USCS	128x128 amber Si:Ga	NASA MtL 1.5m		ZnSe camera

TABLE 3. Mid-IR Array Imagers (USA).

2.2. *Mid-IR camera design considerations*

Let us assume that a "camera" consists of the instrumentation that accepts the radiation from a telescope and delivers this radiation to the detector array to form an image of the focal plane, and the hardware and software that read out the chip and display an image. Thus the "camera" includes the vacuum and cooling systems, as well as the lenses, mirrors, filters, and the detector array with its associated electronics to drive the chip and receive, record and display the signals. Since signal photons only come in a beam defined by the telescope, but background radiation is isotropic, it is very important to restrict the area and solid angle (i.e. the $A\Omega$) of the beam reaching the detector to that containing the signal photons delivered by the telescope. This is usually done by placing a cold input aperture at the telescope focus, and a cold aperture at the "Lyot stop" or pupil, where the telescope's secondary mirror is imaged. Additional internal baffling may also be needed to block background radiation from scattering into the beam. Figures 4

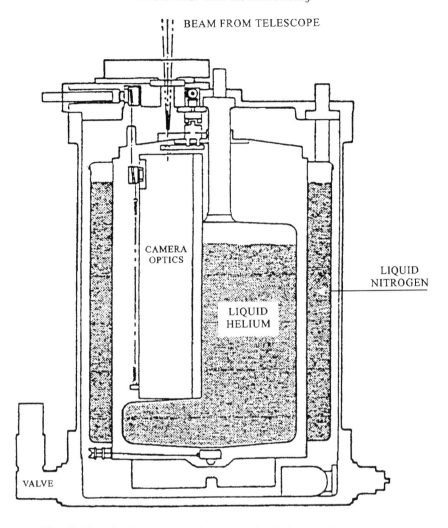

FIGURE 4. The Golden Gopher camera dewar, an "uplooker". The telescope beam enters through the window at the top. The liquid nitrogen and helium containers are shown and the volume at 4 K for the optics.

and 5 show the layout for the "Golden Gopher", a camera operating at the 1.5-m Mt. Lemmon telescope (Piña et al. 1993).

Mid-IR detector arrays usually need to be operated at temperatures from 6 to 12 K to reduce dark current (thermally generated electrons). This requires a dewar with liquid helium cooling, usually with a liquid nitrogen jacket to improve hold time, or a closed-cycle cooler which may also be aided or stabilized with cryogens. The chip is then either actively controlled to the correct temperature, or it may be partially isolated from the cryogen so that it self heats from the power dissipated by the readout. The detector enclosure radiates as a black body, and therefore must itself be cold. Temperatures less than 20 K are enough to reduce this background well below the background in the beam from the telescope. The optical elements and filters are also cold.

To reduce background, the number of warm optical elements between the detector and the source should be kept at a minimum. The primary and secondary telescope mirrors cannot be cooled below typical night-time temperatures, and the other essential element that cannot be cooled is the dewar window. Some window materials allow for an anti-reflection coating which helps reduce background introduced by multiple reflections. Good design should ensure that most of the radiation reflected into the beam at the window surfaces comes from cold baffles inside the camera. "Uplooking" dewars have no extra warm optical elements, at the expense of some internal design constraints. "Sidelooking" dewars use a steerable warm 45° tertiary mirror which simplifies the alignment of the camera at the expense of another warm reflection to add to the background. Uplookers need a more massive mechanism to be able to tilt the entire dewar through small angles for precise alignment.

A common optical arrangement is to use a collimating lens or mirror after the input aperture to form an image of the secondary (the Lyot stop). A cold aperture is placed at the pupil to restrict the background radiation to the solid angle defined by the telescope. This aperture can be made slightly smaller than the secondary image to mask out supporting structure immediately surrounding the secondary mirror. Figure 5 shows the optical layout of the "Golden Gopher" as an example. The bandpass filters are usually placed close to the pupil, at a slight angle to the beam to reduce ghost images. For narrow-band imaging a circularly variable filter (CVF) is often used. This requires the pupil size to be small (2–3 mm) to avoid significant variation in bandpass across the pupil, which would result in a degradation of both spectral resolution and throughput. Spectral resolutions up to $R = 100$ can be achieved. Finally a camera lens or mirror is used to re-image the telescope focal plane onto the detector array. The overall magnification (in this case the ratio of the camera-to-collimator focal lengths) is chosen to give the required plate scale. This is usually selected to match the pixel size to about half the diffraction-limited point source image size.

Because of the excellent atmospheric seeing in the thermal IR, it is essential that the optics of both the telescope and camera do not degrade the image quality. Many cameras use all reflecting optics which has the advantage of being totally achromatic. If the beams are slow, fairly simple arrangements can give good image quality (e.g. Figure 5). For wider fields and faster beams folded systems and aspherics are required. A gold coated mirror reflects close to 99% of the incident mid-infrared radiation, thus folding the beam multiple times does only minor damage to the system throughput. Although there are many materials available for lens construction in the mid-IR, if a camera is required to function over the large wavelength range from 5 to 25 μm, significant chromatic aberrations are impossible to avoid in a single optical system. Several different lens systems are needed, each optimized for a separate wavelength region. This also requires precise mechanisms for moving the selected lenses into the beam, and space to store the extra optics.

All instruments need a transmissive window which must also be able to support a vacuum. For 5 and 10 μm ZnSe is an excellent choice, it can be anti-reflection (AR) coated to reduce reflection losses and extra background. Unfortunately it does not transmit beyond 17 μm. For use at 20 μm KRS5 and KBr are the most common materials, but unfortunately both have problems. KRS5 has a high refractive index ($n = 2.4$) and AR coatings are not stable. KBr has a low refractive index ($n = 1.5$), but it is hygroscopic and degrades quickly, especially if it is close to the dew point (as windows frequently are). Precautions to extend the life of KBr include using a sealed cover (except when observing), and actively warming the window. Many IR materials are difficult to work with. Problems include hygroscopic materials (salts), poisonous substances (KRS5), and a lack of optical transmission (silicon and germanium). The *Infrared Handbook* provides an extensive list of the optical and mechanical properties of materials.

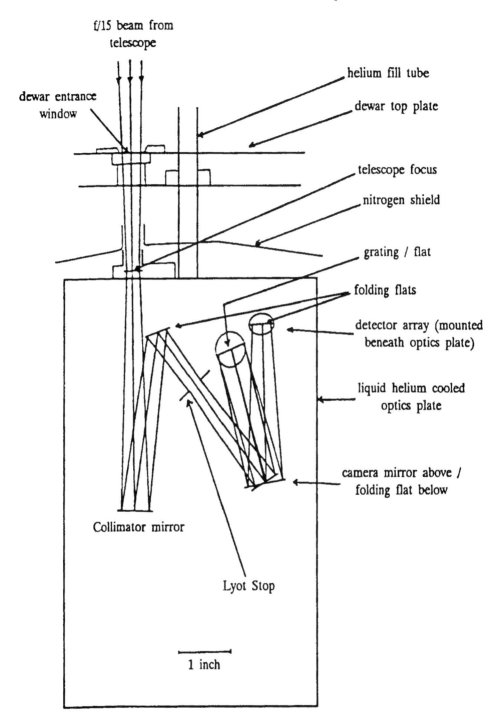

FIGURE 5. The optical layout of the Golden Gopher. All optical elements are gold-coated mirrors. The array is beneath the plane of the page, looking up. The filter assembly, located at the Lyot stop, is not shown.

Instrument Name	PI/Org	Array	Telescope & FOV	Sensitivity	Comments
TIRCAM	Persi IAS Frascati	10x64 Si:As Hughes	TIRGO 1.5m 1.23"/pix	0.8mJy/sq" t=1m	2 ZnSe lenses
TCMIRC	Robberto Torino	58x62 Si:Ga Hughes 58x62 InSb	TIRGO	0.13mJy/sq" t=1m	cryocooler
TIMMI	Lagage Saclay	192x128 Si:Ga 3x10 e⁻ wells	ESO3.6m 0.48"/pix	0.01Jy NEFD t=1m, 1pix	5-14μm
CAMIRAS	Lagage Saclay	192x128 Si:Ga 3x10 e⁻ wells	CFHT 0.3"/pix NOT 0.41"/pix	0.1Jy/pix t=1s	5-17μm CVF
C10μm	Lagage Saclay	64x64 Si:Ga 3x10 e⁻ wells	CFHT 0.4"/pix &	first obs	cryocooler
MIRACLE	Genzel MPE/ROE	58x62 As:Si Hughes SBRC	UKIRT 0.17"/pix	330mJy 10σ t=10m 1μm@10μm	7-22μm CVF
MAX	Beckwith MPIA/ROE	128x128 As:Si BIBIB Rocwell 3x10 e⁻ wells	UKIRT & Calar Alto 3.5m	first obs Nov 95 0.26"/pix UKIRT	cryocooler all refl R=100 grism
NIMPOL	Smith ADFA CanberraOZ	128x128 Si:Ga Amber	AAT 0.25"/pix	80mJy/sq" t=1m prelim	lenses CVF polarimeter

TABLE 4. Mid-IR Array Imagers (non USA).

Filters for the 5 and 10 μm atmospheric windows are available off the shelf in sizes up to about 25 mm from several manufacturers. However they are not standard (or inexpensive) items for wavelengths beyond about 13 μm. Mid-IR filters are usually made with multilayer coatings on a germanium substrate. CVFs are made as an annulus with wedge shaped coatings. The transmitted wavelength depends on position. If a filter must be cut there is an increased risk that its layers may delaminate on subsequent cooling, resulting in "leaks" outside the desired bandpass. Since this process is not always visually apparent it is prudent to check filter characteristics spectroscopically, preferably when cold.

2.3. Camera sensitivity calculations

The expected sensitivity of a camera can be calculated given relevant parameters of the telescope and site, instrument transmission, detector quantum efficiency, and pixel size on the sky. The sensitivity is usually expressed as the noise equivalent flux density (NEFD). This is the minimum signal flux required to generate the same number of electrons in the detector as the system noise. A fairly common choice for plate scale is to set the pixel size to match the diffraction limited beam. Selecting 2 pixels across the FWHM of the diffraction limited beam (i.e. 2 pixels in λ/D) matches a natural scale size. We will calculate the background power, signal power, and system performance for diffraction limited cameras as functions of telescope size, and detector and instrument parameters.

$$D = \text{telescope diameter (m)}$$
$$A = \text{telescope area } (\pi D^2/4) \text{ (m}^2)$$
$$\epsilon = \text{(atmos + tel + window) effective emissivity (assuming atmos trans=1)}$$
$$T = \text{effective temperature of background emitters (K)}$$
$$\lambda = \text{wavelength of observation } (\mu m)$$
$$\nu = \text{frquency of observation (Hz)}$$
$$\delta\lambda = \text{bandwidth } (\mu m)$$
$$\delta\nu = \text{frquency bandwidth (Hz)}$$
$$B(\lambda, T) = \text{Planck function (W m}^{-2}\text{ ster}^{-1}\text{ } \mu m^{-1})$$
$$tr = \text{instrument transmission}$$
$$\eta = \text{quantum efficiency of detector}$$
$$G = \text{photoconductive gain of detector}$$
$$\Omega = \text{solid angle subtended per pixel } (\lambda/2 \cdot D \cdot 10^6)^2 \text{ (ster pix}^{-1})$$
$$S = \text{source strength in Jy per pixel (1 Jy= } 10^{-26} \text{ W m}^{-2}\text{ Hz}^{-1})$$
$$P_B = \text{background power incident per pixel } \epsilon B(\lambda, T)\delta\lambda\Omega tr \text{ (W pix}^{-1})$$
$$P_S = \text{signal power incident per pixel } A \cdot tr \cdot S \cdot 10^{-26}(\nu/\lambda)\delta\lambda \text{ (W pix}^{-1})$$

Taking some typical values to assess the order of magnitude for background and signal: let $\epsilon = 0.1$, $T = 273$ K, $\lambda = 10$ μm, $\delta\lambda = 1$ μm, $tr = 0.5$, then

$$B(\lambda, T) = 6.16 \text{ W m}^{-2}\text{ ster}^{-1}\text{ } \mu m^{-1}$$

and

$$P_B = 6.04 \times 10^{-12} \text{ W pix}^{-1} \text{ or } 3.02 \times 10^8 \text{ photon s}^{-1}\text{ pix}^{-1},$$

while

$$P_S = 1.18 \times 10^{-14} SD^2 \text{ W pix}^{-1} \text{ or } 5.93 \times 10^5 SD^2 \text{ photon s}^{-1}\text{ pix}^{-1}.$$

Note that because the pixel size is locked to the telescope diffraction size, the background per pixel is not a function of telescope diameter. For a typical Si:As BIB detector $(\eta G)=0.6$ and $(\beta G)=2.2$ (see "Detectors" chapter), and the electron generation rates are

$$n_B = \eta G P_B = 1.8310^8 \text{ e}^-\text{s}^{-1}\text{pix}^{-1}$$

for the background and

$$n_S = \eta G P_S = 3.5610^5 SD^2 \text{ e}^-\text{s}^{-1}\text{pix}^{-1}$$

for the source.

The electrons generated from the background will be larger than the source generated electrons for all but the few brightest sources on the largest telescopes. When the background flux dominates it will also limit the noise performance (Other noise sources should be carefully constrained to be smaller than background noise, or performance will be degraded.) For n_B incident photons and the detector parameters above the noise per pixel in 1 second is given by (see "Detectors" chapter):

$$\text{noise pix}^{-1}\text{ s}^{-1} = \sqrt{(\frac{\eta}{\beta})(\beta G)^2 n_B} = 1.5510^4 \text{ e}^- \text{ pix}^{-1}\text{ s}^{-1}.$$

Signal to noise per pixel will be unity in 1 second when the number of signal generated electrons equals the number of noise generated electrons. This happens when

$$n_S = 3.5610^5 SD^2 = 1.5510^4 \text{ e}^- \text{ pix}^{-1}\text{ s}^{-1}$$

or

$$S_{NEFD} = 4.3610^{-2}/D^2 \text{ Jy pix}^{-1}$$

and gives the condition for S/N = 1 in 1 s for 1 pixel for the detector parameters chosen. The S/N in background limited conditions should increase with $\sqrt{\text{time}}$.

More useful numbers for system performance are the "point source" NEFD, and the NEFD per square arcsec (for extended sources). With the pixel size defined by the diffraction limited image it should be possible to place most of the energy from a point source on 3×3 pixels (the detailed structure of the low level wings of the image will be determined by seeing, scattering, and the exact profile of the diffraction pattern from the telescope). If we assume that a fraction f of the point source flux is contained in the 3×3 pixels, and that these pixels are summed together, then the noise will be increased by a factor of 3, ($3 = \sqrt{\text{number of summed pixels}}$) and the NEFD degraded by a factor of $3/f$. Under these conditions the NEFD is $S_{\text{NEFD}} = 0.13/(fD^2)$ Jy (for a point source).

For an extended source and 2 pixels in λ/D at 10 μm, the number of pixels per square arcsec is given by p where

$$p = (\frac{2D \times 10^6}{\lambda})^2 \cdot 2.3510^{-11} = 0.94d^2 \text{ pix arcsec}^{-2}.$$

The total electrons generated per square arcsec by the background is then given by

$$\text{BKGD}(e^- \ s^{-1} \ \text{arcsec}^{-2}) = n_B(e^- \ s^{-1} \ \text{pix}^{-1}) \cdot p(\text{pix arcsec}^{-2}),$$

so that

$$\text{BKGD}(e^- \ s^{-1} \ \text{arcsec}^{-2}) = 1.7210^8 D^2 = (1.3110^4 D)^2.$$

If S_{EXT} is the extended flux in Jy per square arcsec then the total electrons generated by the signal is

$$\text{SIGNAL}(e^- \ s^{-1} \ \text{arcsec}^{-2}) = 3.56 \times 10^5 \cdot S_{\text{EXT}} \cdot D^2.$$

For a S/N = 1 in 1 sec, equating the signal to the square root of the background

$$3.56 \times 10^5 S_{\text{EXT}} D^2 = 1.31 \times 10^4 D$$

so that the NEFD is

$$S_{\text{NEFD}} = 0.037/D \text{Jy}.$$

The above calculations assume that the source can be observed all the time, i.e. that the 1 second is both the time spent observing the object and the elapsed time. In reality most observations use both chopping and nodding. This reduces the efficiency by at least a factor of 2, since the source is in the beam only half the total time, and there are unavoidable overheads to allow for the chop settle and nod settle times. (Note that the factor of 2 can be reclaimed if the source is small enough, or the array large enough, for the entire chop/nod pattern to fit on the array). Assuming a factor of α loss in efficiency, such that elapsed time is a factor of α larger than the time spent integrating on source, and referring to elapsed time for the calculations, the NEFDs are:

$$S_{\text{NEFD}} = (0.13\sqrt{\alpha})/(fD^2) \text{ Jy}$$

for a point source S/N = 1 in 1 s elapsed time, and

$$S_{\text{NEFD}} = (0.037\sqrt{\alpha})/D \text{ Jy}$$

for an extended source S/N = 1 in 1 s elapsed time.

For example, for a 3-m telescope, 0.34 arcsec pixels, $\alpha = 3$, summing 3×3 pixels to get 60% of the total point source energy, with the detector and conditions described above, this gives NEFD = 42 mJy for a point source and NEFD = 21 mJy arcsec^{-2} for an extended source where the NEFDs are measured in elapsed time.

To assess the relative merits of telescopes of different sizes for different astronomical applications, several factors need to be considered. Selecting the pixel size by the size of the diffraction-limited image implies a small field of view for the larger telescopes. Using this criterion the field decreases and the spatial resolution increases with increasing telescope diameter.

The time required to reach a given signal to noise ratio for a given point source strength, S (Jy), is proportional to D^{-4},

$$t = \frac{0.0169\alpha(S/N)^2}{S^2 f^2 D^4} sec.$$

For an extended source of strength $S_{EXT}(Jy/arcsec^2)$, the time required to reach a given S/N ratio is proportional to D^{-2},

$$t = \frac{1.37 \times 10^{-3}\alpha(S/N)^2}{(S_{EXT})^2 D^2} s.$$

Clearly, measuring the faintest point sources is much better done with large apertures.

It is also interesting to calculate which telescope is superior in mapping large regions of emission. We will compare sensitivities assuming that the same area of sky must be covered in a given, fixed observing time, the only variable being the telescope diameter. If the region contains randomly distributed point sources, the D^4 time advantage is partly offset by the D^{-2} disadvantage of the larger number of fields to cover. The point source sensitivity (in the time available per field), is a factor of D better. The positional information also improves with increasing diameter. If the field is expected to contain extended emission then the D^{-2} disadvantage (of the number of fields) is offset exactly by a D^2 advantage in time, i.e. the extended source sensitivity is not a function of telescope diameter. Again, however, the spatial information content increases with diameter.

To compare the performances of the imagers listed in Table 3, the NEFDs reported must be normalized to account for the differences between the individual instruments and the assumptions made above. Integration times can be normalized using S/N $\approx \sqrt{\text{time}}$. The signal to noise changes as $\sqrt{\delta\lambda \cdot tr \cdot \epsilon\eta}$, allowing for variations in bandwidth, transmission, emissivity, and quantum efficiency. The ratio of the pixel size to the diffraction limited beam size is more complex to normalize. The signal to noise per arcsec^2 is independent of pixel size. The point source sensitivity is also independent of pixel size when the pixel scale is smaller than the diffraction image. If all other parameters are held fixed, as the pixel scale becomes larger than the diffraction-limited image, the signal remains constant but the background increases as the solid angle. The sensitivity then drops with linear pixel scale. This explains the increase in sensitivity when the transition was made from single detector instruments with several arcsec per pixel to arrays with typically a few tenths of arcsec per pixel.

2.4. Data acquisition and calibration

Mid-IR observational techniques have many aspects in common with standard optical procedures, and some significant differences. The frame time is set by the time taken for the background level to fill the wells to the desired amount. Sequential frames are then coadded into a frame buffer (the *object frame*), rather than stored individually. (This avoids storing excessively large amounts of data). This is repeated until the end of one half of a chop cycle, when the secondary chopper moves to its new position. After a delay for the chop to settle, data is coadded into a new buffer (the *sky frame*) for the second half of the chop. This whole procedure then repeats, using the same buffers, for as many chops are needed for half the nod time. Then the telescope nods, data is rejected until the telescope settles, and 2 new *object* and *sky* buffers are used for the coadded data in the second half of the nod cycle. When the nod has completed the 4 buffers can be permanently stored by the host machine, and appropriately added together for display and instrument status checks. The sequence is repeated for the number of nods required for the integration.

Chop and nod frequencies are set by experience with the instrument and site. Usually the lowest frequencies that do not adversely affect the background subtraction (by introducing extra noise) are chosen. (Lower frequencies are more efficient, as less time is lost to settling.) Piña *et al.* (1993) have noted that a faster chop can help when weather conditions are marginal, and a faster nod is needed when local temperatures are changing, for example at the beginning of the night just after the dome is opened. The need to chop and nod the telescope also requires the capability of guiding on a chopped beam (usually the *middle* beam, i.e. the overlap of the 2 chop *dog-bones*). Although near- IR or mid-IR guiders would often be optimum, usually an optical guider is used on a star outside the IR field.

Calibration is done with standard reference stars and an extinction curve to extrapolate to zero airmass for each wavelength. Intermediate calibration can be done with reference to a sector chopper as described in the spectroscopy section. Several references give fluxes (or magnitudes) for bright reference stars (e.g. Rieke *et al.* 1985).

Monitoring the amount by which the wells fill in each readout provides a useful diagnostic of instrument problems and changing sky conditions, and may be used on a short timescale to correct for changing transparency. A small subset of the array can be selected for this, to reduce calculation overhead, with values reported every nod cycle. Some form of "quick-look" software to monitor the incoming image is essential, both to confirm the telescope pointing and guiding, and to watch the build up of signal to noise in the image. For accurate photometry frequent calibration measurements should be made, especially at the edges of atmospheric windows, and in the poorer 5- and 20-μm windows.

Flatfielding is considerably different from optical work with CCDs. The chop/nod technique cancels background radiation to the system noise level. The resulting image has a zero background, i.e. off source pixels (the "sky level") are zero. (If this is not the case the data is likely to be severely compromised and the problem should be addressed at once.) The residual non- flatness in the data comes from variations in pixel to pixel sensitivity and changes in instrument throughput over the field. Two techniques have been successfully used to form flat fields, both use sky emission. The first subtracts a measure of the zenith sky from a measure of sky at large airmass. The second makes a difference in sky background at a given telescope position, measured with slightly different frame times (Piña *et al.* 1993). It is important to note that both these techniques use a differencing procedure (as does the data acquisition), and both fill the wells to about the same amount as the regular data acquisition. This avoids problems with nonlinearity in the chip, and with reset pedestals in non differenced data.

2.5. *Astronomical images—a sample of the science*

A fairly recent survey of mid-IR science is presented in (McLean 1994). A sample of other results includes a high spatial resolution 11.7-μm image (from SpectroCam-10) of NGC 7469 showing a 3-arcsec diameter ring around an unresolved nucleus (Miles *et al.* 1994). Aitken *et al.* (1991) present polarimetric images of the galactic center, with discussion of the magnetic field structure. Fomenkova *et al.* (1995) describe changes in the jet emission in the comet Swift-Tuttle over a 3-week period and give a measure of the size of the nucleus.

Emission Process	Species and Wavelengths (μm)
dust grains	silicates (10, 20), SiC (11.5), MgS (24-44), graphite (30)
PAH (polycyclic aromatic hydrocarbons)	misc. C-H, C=C bending & stretching modes at 3.3, 3.4, 6.2, 7.7, 8.7, 11.3, 12.7μm
molecular emission	PAHs, H_2 (5.0, 12.3, 17.0μm) CO (2.3, 4.7)
molecular absorption	H_2O, CH, NH_3, CH_3OH, CO,CH_4, C_2H_2, H_2CO, etc
forbidden line emission	AIII 9.0, SIV 10.5, NeII 12.8, SIII 18.7, AIII 21.8, NeV 24.3
recombination lines	H (parts of series n=6 to 12) & parts of HeI and HeII series.

TABLE 5. Some Mid Infrared Spectral Features

3. Mid-infrared spectroscopy

The scientific motivation for spectroscopic information is as great for IR spectral regions as for any other energy band. Mid-infrared spectrometer design for ground-based observations is discussed in this section, including different instrument types with a variety of elements for light dispersion. Observing techniques, and wavelength and flux calibration problems are covered. As for imagers, a list of currently operational instruments has been assembled from the published literature.

3.1. *Introduction*

Mid-infrared astronomical emission is rich in spectral features. Table 5 gives a partial list illustrating the diversity of the processes involved. As examples, one of the most frequently seen dust features is the silicate feature. This is broad (almost 5 μm) and usually smooth, indicating amorphous grains. It is seen in both emission from small warm silicate particles, and in absorption towards sources surrounded by cooler silicate grains (Knacke 1989; Blanco et al. 1993). Polycyclic aromatic hydrocarbon (PAH) features originate from complex molecules forming very small grains (see reviews by Muizon 1993, and Roche 1989). PAH emission is often observed near the interface regions between ionized and neutral species. In some situations PAH emission can dominate other thermal emission. Spectrally PAH features are typically about 0.25 μm wide. Molecular emission can arise from symmetric molecules if their symmetry is broken during vibration (e.g. CH_4), or from molecules absorbed on dust grains, where transitions in vibrational but not rotational states are seen (e.g. Parmer 1994). Molecular hydrogen emission gives information on shock excited regions. Molecular absorption probes the chemistry near newly forming stars. Atomic fine structure lines give data on a variety of phenomena, including abundances, ionization states and the hardness of the ionizing spectrum, bulk gas velocities, etc. With very little extinction, mid-IR spectroscopy is ideal for probing star formation regions, both local and extragalactic, and dust obscured galactic nuclei. The most comprehensive set of low resolution astronomical spectra was obtained by the 8 to 23 μm slitless spectrometer (LRS) on *IRAS* (Olnon & Raymond 1986; Volk 1993).

Spectral resolutions from 10 to 30 000 have been used to study these features. The instruments use a variety of dispersing elements including circular variable filters (CVF), prisms, low-order gratings, echelles and Fabry-Perot interferometers (FP). Table 6 lists active instruments (from recent literature) and Table 7 some of those under construction.

The CVFs and FPs differ from the grating instruments in that they sample a two-dimensional image at a given wavelength region. Several wavelength settings must be used to cover the required spectral range. Most grating instruments use a long slit, and obtain many wavelengths simultaneously for the section of the image under the slit. To gain complete wavelength coverage of a complex extended source the grating instrument must be repositioned repeatedly to cover the source with the slit. If both spectral and spatial coverage is needed on an extended source, the two types of instruments can be considered to complete the "data cube" in different ways. For sources significantly smaller than the field of the detector array, grating instruments are more efficient, making use of more of the pixels more of the time.

3.2. *General design constraints*

The optimum combination of spectral resolution and wavelength coverage varies with the intended science goals. To study dust emission or the PAH features, resolutions of a few hundred are adequate as the features are smooth, and do not break up at higher resolution. A large simultaneous wavelength coverage is an advantage. The choice of $R = 100$ for several currently operational grating instruments is dictated by the requirement to cover the entire 8–13 μm window, at one grating setting, on an array of about 100 pixels (sampling at 2 pixels per spectral resolution element). As arrays become larger, higher spectral resolutions can be achieved while still covering the entire window with one setting. As long as the instrument is background-noise limited, there is no disadvantage to this. For continuum spectra these can be convolved to lower resolutions to improve signal to noise. For an unresolved line in a dispersed spectrum for a background-limited instrument, the signal to noise increases as the square root of the resolution, until the line becomes resolved. For observations of atomic emission lines in many galactic sources a resolution of a few thousand is optimum. At this point some velocity information is also available. For extragalactic objects (and novae) the lines are generally broader. A better compromise for these sources is R around 1000. For observations of molecular absorption features much higher resolutions ($R \simeq 10^5$) are appropriate. A large simultaneous wavelength coverage is not such a high priority for line studies since the lines are widely spaced. The capability to retune to other ranges is important.

Both gratings and Fabry-Perot instruments can be designed to achieve resolutions from a few to around $R \simeq 10^5$. The choice between them is mostly made based on the angular extent of the typical object to be observed, as discussed above (see e.g. Lacy 1989, 1993). The normal "scaling" rules for grating spectrograph design (see e.g. Schroeder, p. 226) are simpler for the mid-IR wavelength regime, where it is both highly desirable and usually practical to set the slit width to the diffraction-limited image size from the telescope. The maximum resolution obtainable with a diffraction-limited slit and a properly sampled focal plane, is determined approximately by the grating width compared to the wavelength (again refer to Lacy 1993). Thus for $R = 3000$ at 10 μm, a grating width of 3 cm is needed, for $R = 30\ 000$ a 30-cm grating is required. Since the grating size determines the size of the instrument, and the optics must be cooled, $R = 30\ 000$ is a reasonable upper limit for mid-IR grating spectrometers.

Spectrometers have lower background fluxes than imagers. For grating instruments the background decreases as the resolution increases. Fabry-Perots must use a cold predisperser in addition to a cold Fabry-Perot interferometer to reduce background. With lower backgrounds, longer integrations are needed to fill detector wells. If the wells are not

Instrument Name	PI/Org	Array	Telescope dispersing elt	Spectral Resolution	Comments
Irshell	Lacy @ Texas	10x64 Si:As Hughes	IRTF et al. echelle	R=1000 to 30000	all refl (ex mag lens) 5-26μm
Spectro-Cam-10	Houck @ Cornell	128x128 As:Si BIBIB Rockwell	Hale 5m low & med res gratings	R=200 & 2000	all refl 8-13μm also camera
LWS	Jones & Puetter @ UCSD	96x96 Si:As MC2 (BIB) AESC	Keck 10m low & med res gratings	R=100 & 1400	all refl 3-26μm also camera
CGS3	Towlson @ UCL/ROE	32x1mm linear discrete Si:As	UKIRT 3.8m gratings	R=60-80	8-13 μm & 16-26μm
OSCIR	Telesco @ U. Florida	Si:As 128x128 BIBIB Rockwell	IRTF 3m low & med res gratings	R=100 & 1000	all refl 5-26μm also camera
BASS	Hackwell Aerospace	2 arrays 1x64 Si:As BIB	IRTF & 1.5m MtL 2 fixed prisms	R=80	2-13μm
GLADYS	LeVan @ AFGL	58x62 Si:Ga	WIRO 2.3m	R<100	NaCl prism 9-14μm
MAX	Beckwith @ MPIA	128x128 As:Si BIBIB Rockwell	UKIRT & Calar Alto grism		also camera first use 11/95
UCLS	Smith @ ADFA, OZ	30x1mm Si:As discrete	AAT/UKIRT 2 gratings	R=50-150 & R=300	8-25μm

TABLE 6. Mid-IR ($\lambda > 5\mu$m) Array Spectrometers. Spectrometers in use (up to 1995)

filled, much tighter constraints must be placed on other noise sources in order to remain background-noise limited. In order to avoid temporal "staggering" of the integrations on different parts of the chip, it is preferable to read the entire chip quickly (in a burst) at the end of the frame time. For the example given in the introduction to this series (3-m telescope, 0.35 arcsec pixels, 1×10^6 e$^-$ well depth, $\epsilon = 10\%$, $R = 10$ ($\Delta\lambda = 1.0$ μm) $\lambda = 10$ μm), the read rate required is 360 Hz. In spectroscopic mode at $R = 1000$, this would be 3.6 Hz. For a 3.6-Hz frame rate, a 1.8-Hz chop would allow one frame per half chop cycle. For higher-resolution instruments Lacy (1993) reports successful background subtraction under good sky conditions with nodding only.

The field of simultaneous polarimetry and spectroscopy is technically challenging in the IR as it is in the optical, but scientifically very rewarding. The most productive spectro- polarimetry measurements have been made by Smith and Aitken (e.g. Smith *et al.* 1994; Aitken *et al.* 1985, 1993).

3.3. *Low- and moderate-resolution grating (and prism) spectrometers*

Several instruments listed in Table 6 and 7 have similar characteristics, i.e. a combination of a low resolution mode ($R \simeq 100$), a moderate resolution mode ($R \simeq 1000$), and imaging. Since the cost of building these instruments is high, it is desirable to combine functions whenever reasonably convenient. The spectroscopic modes drive the design, and the inclusion of imaging is usually not a major modification. Since it is frequently necessary to image an object to make the best decision about where to place the spec-

Instrument Name	PI/Org	Array	Telescope dispersing elt	Spectral Resolution	Comments
Michelle	Glasse @ ROE	Si:As 128x128 BIBIB Rockwell	UKIRT & Gemini 4 grat & echelle	100<R<30000	8-25μm also camera 1997 delivery
VISIR	Lagage @ Saclay	not chosen	VLT		delivery 2000 also camera
MIRAS	Smith @ ADFA, OZ	Amber Si:Ga 128x128	AAT 3 gratings	R=150 & 800	8-25μm 1996 delivery
Fabry-Perot	Garden @ UC Irvine	Si:As 128x128 BIBIB Rockwell	UKIRT 0.25"/pix IRTF 0.35"/pix^2 tandem FPs	R up to to 256^2	all refl upgrade Rocwell chip
Echelon	Lacy @ Texas	20x64 Hughes Si:As BIB	Texas & Hawaii disp echelle	R=150,000 (2km/s)	ecehlon 1m long

TABLE 7. Mid-IR ($\lambda >5\mu$m) Array Spectrometers. Spectrometers in design or construction (1995)

trometer slit, having those capabilities available in a single instrument also makes efficient use of telescope time. An $R = 1000$ resolution mode can be added to a modest sized instrument at little extra expense, as it requires a beam size of less than an inch at the grating. Designs typically use a single grating turret carrying at least two gratings and a mirror. Precise control of angle is needed to set the wavelength needed, with fairly rapid rotation to change observing modes. As for imaging instruments, all-reflecting optics is a common choice to avoid chromatic aberrations over the large wavelength range.

The simplest optical arrangement (shown in Figure 6) is to place the input aperture (slit or rectangular mask) at the telescope focus. The collimator is placed at its focal length beyond the aperture, and forms a parallel beam, and an image of the secondary. A cold aperture is placed at this image (Lyot stop) with the filters close to this. The grating is also usually fairly close to the pupil. The camera reimages the input aperture on the array. The ratio of camera-to-collimator focal lengths is determined by the telescope diffraction-limited image size and the desired plate scale. The absolute value of the collimator focal length (and the grating size) is determined by the desired spectral resolution.

This arrangement has the potential disadvantage of additional diffraction at the slit, which can be physically narrow. A cleaner optical system is to use "fore-optics" to image the pupil first, before the spectrometer optics. The disadvantage of this approach is the increase in complexity, and in the number of optical components. SpectroCam-10 uses this system, and uses the fore-optics to increase the speed of the beam. Michelle also will use this technique. LWS does not use fore-optics. Its smallest slit is about $30 \times \lambda$ across at 10 μm. LWS uses oversized optics, and delivers good throughput and resolution.

All these instruments use blazed, plane reflection gratings, which are very efficient (see discussion by Schroeder, p. 245). Replicas, usually made on aluminum blanks and gold coated, appear to be able to withstand thermal cycling. However for the relatively small size of these gratings (compared to optical instruments), it is not unrealistic to use a master ruling, custom made for the exact requirements. It is not necessary to cool the entire optical chain to the detector temperature as long as the detector housing and filters are cold enough to avoid excess internal background emission, and the detector field of

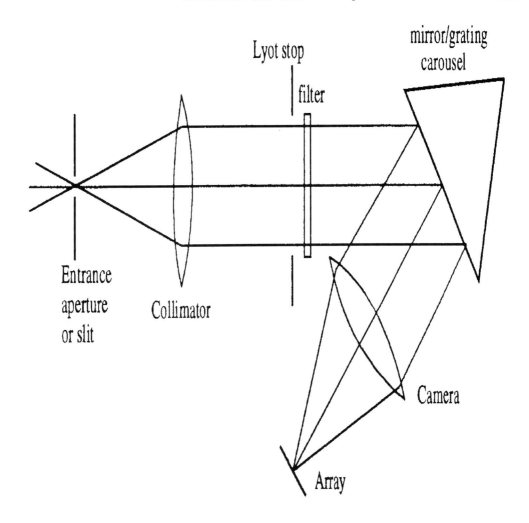

FIGURE 6. Schematic layout of simple grating imaging spectrometer requiring three mechanisms (aperture, filter and grating)

view through the instrument is well baffled. SpectroCam-10 (Hayward 1993; Miles 1993) cools the grating and fore-optics to 77 K.

A variety of optical configurations have been used. Since the focal ratio of the beam from the telescope is usually slow (chopping secondaries are small), and the arrays are (unfortunately) also small, many fairly simple solutions are available to deliver diffraction limited imaging quality. As larger arrays become available the pixel sizes will be smaller (e.g. Rockwell's 256×256 chip has 50-μm pixels) and the optical designers will face more of a challenge.

3.4. Echelle spectrometers

Because of the much larger grating sizes needed for high-resolution echelle instruments it is hard to combine an echelle with low- and moderate-resolution instruments discussed above. However the UK instrument now in construction for UKIRT and Gemini (Michelle) combines $R \cong 100$, $R \cong 1000$ and an $R \cong 15000$ to 30000 echelle mode.

Michelle also has an imaging option through a separate optical path, a field rotator, and a selection of 5 gratings (see e.g. Glasse *et al.* 1993)

One of the first, and most successful array spectrometers is IRSHELL, which has now been in operation since 1986 (Lacy *et al.* 1989). This instrument uses a 22 cm wide R2 echelle (63° blaze) with a typical resolving power of 10 000. An optional lens, placed in front of the detector, can increase both dispersion and sampling of the focal plane to give $R = 24\ 000$. Replacing the echelle with a first order grating can decrease the resolution to $R = 1000$. IRSHELL uses 2 warm mirrors external to the dewar window. The first reimages the telescope's secondary onto the second warm mirror which can act as a chopper, in lieu of a chopping secondary, when needed. A cold baffle just inside the dewar window forms a cold stop very close to this pupil. The dewar window is a positive lens which converts the beam to $f/7.5$, the slit is at a telescope focus made by the first external mirror. An off-axis paraboloid, used for both collimator and camera, makes a 10-cm diameter beam for the echelle. Since the array is small (10×64) no cross disperser is used. Unwanted orders are blocked by filters. This highly successful instrument has been used on several telescopes.

Lacy (1993) has found from experience that at high resolution he does not need to chop when atmospheric conditions are good. He scans the telescope across the object and coadds frames for about 1 arcsec of telescope motion. He takes a measure of sky emission from the ends (off source) of these scans. He reports that about 50% of the time at Mauna Kea is good enough so that this technique does not introduce extra noise.

Lacy is currently constructing a new instrument to achieve resolutions up to $R = 150\ 000$ with a 1-m long cross dispersed R-10 echelon.

3.5. *Fabry-Perot interferometers*

Fabry-Perot instruments are attractive for applications requiring high-resolution multi-spectral imaging. Applications include scanning across emission lines in H II regions and planetary nebulae, and scanning molecular features in complex star formation regions. The two plates of the FP need to be cooled, and a predisperser (CVF, bandpass filter, or second FP) is needed to reject both signal and background from unwanted orders. To change the wavelength transmitted, the plate spacing must be precisely controlled. Piezo-electric translators provide adequate motion, although some care needs to be taken using high voltages in the same enclosure as an expensive, static sensitive array. The etalons can be made of ZnSe (Watarai 1994), or at longer wavelengths of free standing metal mesh (Stacey 1993).

3.6. *Data reduction and calibration*

The 1-D spectral data on an object must first be extracted from the 2-D chop/nod (or other) difference frames. Since the background subtracts to zero, this is simpler than the extraction of optical spectral data on CCDs. If the spectrum is straight and aligned with the pixels, fairly simple routines can be used to extract the optimum signal as a function of position in the cross-dispersion axis. The individual coadded buffers accumulated in the chop/nod cycle can be manipulated to give the spectrum of the telescope imbalance signal (which should be uniform along the length of the slit), and the spectrum of the "raw" telescope + sky emission (which may also include a pedestal). The telescope imbalance signal should look like a blackbody at the telescope temperature convolved with the instrumental response. It can be used to correct for instrumental transmission, but not atmospheric transmission. For Fabry-Perot data the spectrum for the desired spatial position must be extracted from the data cube.

The most straightforward technique for calibrating spectra, both to unfold atmospheric and system transmission, and to provide an absolute flux reference, is to observe calibration stars with the same instrument settings, and at a similar air mass. Wavelength calibration, at low resolution, can be done with reference to known filter edges and atmospheric features (e.g. O_3), at higher resolution atmospheric features are more useful. Objects with bright line emission, such as planetary nebulae, can be used for absolute wavelength checks.

Another useful technique used for IRSHELL (see Lacy 1993) is to measure spectra for sky, dark (by closing the slit), and a blackbody at telescope temperature. The data are flatfielded by division with (sky-dark) or (blackbody-dark). Then the data are divided by (blackbody-sky) which has also been flat fielded. The (blackbody-sky) spectrum is, to first order, proportional to atmospheric transmission. This technique corrects well for water vapor features (which are usually low in the atmosphere), and surprisingly well for species well mixed in the atmosphere. It fails for O_3, which is higher in the atmosphere, and much cooler than telescope temperature.

LWS uses a similar technique. It has a sector chopper mounted above the window. Each blade on the sector has a carefully selected emissivity, from close to 100% (black paint), to close to 0% (highly reflecting). Between each blade the dewar views an unobstructed beam from the telescope. The sector wheel can then be "chopped" (back and forward) to give alternate views of a variety of emissivities at telescope temperature, or sky. This can be used to get differenced data corresponding to (blackbody- sky) above. These frames can be used to correct for instrument and atmospheric transmission, and sensitivity variations. Since it takes very little time for a sector measurement, they can be frequently interleaved with data acquisition to monitor system function as well as provide calibration.

4. Mid-infrared detector arrays

This section summarizes some of the parameters important in the selection and operation of detector arrays for ground-based astronomical use between 5 and 33 μm. The atmosphere restricts observations beyond around 30 μm until 300 μm where submillimeter techniques are used. Single detectors will only be discussed to illustrate the semiconductor physics. All the chips now available use extrinsic silicon for the detector elements, indium bump bonded to a MOSFET readout. Except for some high spectral resolution instruments, a severe design constraint is the large photon arrival rate which requires fast frame times.

4.1. *Extrinsic photoconductors*

Mid-IR photon energies are small $E = h\nu = \frac{hc}{\lambda}$ that is $E \cong 0.1$ eV for $\lambda \cong 10\mu$m. The bandgaps of intrinsic photoconductors are too large for use with photons with $\lambda > 5\mu$m so dopants are used to create a material (an extrinsic photoconductor) with the appropriate excitation energy. For a general summary of the operation of photoconductors see Bratt (1977) or the text "Detection of Light: from the Ultraviolet to the Submillimeter" by G.H. Rieke. Currently the most commonly used materials are arsenic doped silicon (Si:As) which responds out to 23 μm (27 μm for BIB operation, see below), or antimony-doped silicon (Si:Sb) which goes to slightly longer wavelengths (Lucas et al. 1994). With these small excitation energies it is critical to cool the detector to reduce thermal excitation into the conduction band (usually known as dark current). Typical operating temperatures are around 10 K which is usually achieved with a liquid helium cooled cryostat and some array self heating.

The photon absorption coefficient increases as the dopant concentration is increased.

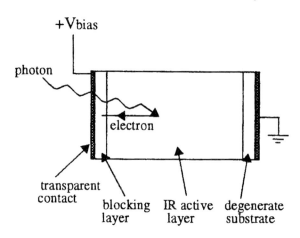

FIGURE 7. Cross section of BIB detector

The ultimate limit on dopant level is the solubility of the dopant. However, when the dopant reaches a high enough concentration so that the electron wave functions from neighboring impurity atoms overlap, undesirable conductivity modes develop. One of these, known as "hopping", allows an electron to move from one impurity site to another, contributing to the dark current in a way that cannot be controlled by cooling. To avoid "hopping", dopant levels must be limited to about 10^{16} cm^{-3}. (For reference, silicon has 5×10^{22} atom cm^{-3}, highly pure silicon has 10^{13} cm^{-3} of impurities). The resulting absorption coefficients are much smaller than for intrinsic materials, requiring much larger volumes to achieve reasonable efficiency. With typical dimensions of about 1 mm, individual detectors are large and array construction is difficult. Large detectors are also susceptible to particle hits. This is a critical problem for space based observations. Fortunately there is another technique, known as "blocked impurity band" devices (BIBs), in which a detector's electrical and optical properties can be separately optimized.

4.2. *Blocked impurity band detectors*

A blocked impurity band material consists of 2 layers, a heavily doped (about $10^{17.5}$ cm^{-3} for silicon) IR-absorbing layer and a thin, high purity, high-resistance blocking layer (Sclar 1984; Szmulowicz 1987; Bharat 1994; Lum et al. 1994; Le Pham et al. 1994). Figure 7 shows a typical geometry. One electrical contact is made to the front of the blocking layer (this must be transparent). The other contact is to the back of the IR-active layer. The heavy doping provides efficient absorption of photons in a small volume, making array construction much easier and, because of the much smaller volume, reducing problems from particle hits. Quantum efficiencies of 40–80% can be achieved. Figure 7 shows (for simplicity) a "front-illuminated" device. In practice, back-illuminated BIBs (BIBIBs) are generally used. In these the photons pass through a transparent degenerately doped layer to reach the active layer. All the electrical contact can be made to one side of the device. The filling factor is 100% with a few percent "optical" cross-talk at pixel boundaries.

In BIB material the impurity band has a finite width compared to the impurity level in normally doped material, reducing the excitation energy and extending the wavelength cutoff to slightly higher values, (e.g. Si:As goes from 24 μm to 27 μm cutoff). In addition generation-recombination noise (G-R noise) is reduced by $\sqrt{2}$ compared with non-BIB materials since the electrons are swept into the high-purity blocking layer before they can recombine. The blocking layer also provides the high electrical resistance required to overcome Johnson noise for low light level operation, and blocks the unwanted conduction modes (hopping) in the valence band.

The transparent contact on the blocking layer is held at a positive potential compared to the back contact. This pulls electrons from the IR-active layer towards the blocking layer. An electron excited to the conduction band passes through the blocking layer to the contact, but thermal electrons moving in the impurity band in unwanted conductivity modes are stopped at the blocking layer as this band is not continuous. The region in the IR active layer near the blocking layer sustains a high electric field and is depleted of charge carriers. This depleted region is responsible for most of the photogenerated charge collected at the electrodes. With a 4-μm thick blocking layer, and a bias voltage of 4 V, a thickness of about 20 μm is typical for the depleted region of the IR-active layer. (This assumes a donor concentration of 5×10^{17} cm^{-3} and a density of 10^{13} cm^{-3} for ionized acceptors.) A typical mean free path for electrons close to the blocking layer is 0.2 μm. Since the electric field close to the blocking layer is high, electrons can gain enough energy between collisions to ionize neutral arsenic impurity atoms. These secondary electrons can also produce ionization, giving a cascade of electrons. The multiplication stops inside the blocking layer, where much more energy is needed to ionize the intrinsic material. This process is called photoconductive gain. The bias voltage controls the magnitude of the gain which is typically 2–3, but may be made as high as 5–10 to overcome amplifier noise. Since the gain process is noisy it increases the overall noise as well as the signal.

The signal and noise can now be calculated. Let $\beta = \frac{\langle G^2 \rangle}{\langle G \rangle^2}$ be the gain dispersion. If N photons hit the detector, the noise is given by

$$\text{noise}^2 = \langle \sum_{i=1}^{\eta N} G_i^2 \rangle = \sum_{i=1}^{\eta N} \langle G_i^2 \rangle = \langle G^2 \rangle \eta N = \eta \frac{\langle G^2 \rangle}{G^2} G^2 N = \eta \beta G^2 N = \frac{\eta}{\beta} (\beta G)^2 N.$$

The noise is written in this way since ηG, and βG (hence $\frac{\eta}{\beta}$) are measurable quantities. The signal is given by

$$\text{signal} = \langle \sum_{i=1}^{\eta N} G_i \rangle = \sum_{i=1}^{\eta N} \langle G_i \rangle = G \eta N.$$

Thus the signal to noise is given by

$$\frac{S}{N} = \frac{G \eta N}{\sqrt{\eta \beta G^2 N}} = \sqrt{(\frac{\eta}{\beta}) N}.$$

Hence the detective quantum efficiency is η/β (Herter 1989, 1994). For comparison with non-BIB devices G^2 must be replaced by $(\beta G)^2$ in expressions for G-R noise (also remember the $\sqrt{2}$ improvement). Typically $\beta < 2$. Performance improvements over non-BIB devices result from higher η, and higher G which is partially offset by β. Smaller volumes enable construction of first class arrays.

Some BIB devices show variations in temperature sensitivity. Telesco (private communication) reports (for a 128^2 Rockwell deep well chip) a 40% increase from 6 to 10 K, then a fairly flat response to 12 K where dark current starts to dominate. For the Aerojet chips we have noticed similar changes. However Garden (private communication) reports little change with temperature (measured for Rockwell chips).

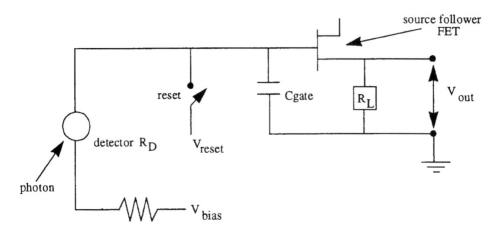

FIGURE 8. Schematic diagram of source follower FET: the integrating amplifier.

4.3. *Detector array readout*

4.3.1. *Array construction—indium bump bonding*

All mid-IR arrays in astronomical use consist of 2 components. The detecting layer, consisting of many elements of extrinsic silicon, is bonded to a separately constructed readout array by indium bumps. This technique for attaching the layers together has many difficulties, including the need to make both parts to exactly the same dimensions, and to squeeze the parts together hard enough to make good contact but not break either fragile piece. The final product must be physically robust enough to be temperature-cycled from 300 K to 10 K. The low yield from the indium bump bonding process helps explain why IR chips are expensive. However the ability to optimize the readout on a separate component from the detecting elements provides a tremendous advantage. Currently, the maximum physical size of wafer that can survive repeated thermal cycling is about 1 cm². The lower bound to the size of a pixel is constrained by the number of components needed per pixel in the unit cell. It is fortunate that the typical pixel sizes (from 50 to 100 μm, square pixels) are a reasonable match for astronomical instruments.

4.3.2. *The Source-follower FET, integration and reset*

The function of the readout is to receive and integrate the weak electrical signals generated by the detector, and to multiplex these signals to the output lines for subsequent amplification and digitization. MOSFETs are used for this, both as source followers for signal amplification and impedance transformation, and as switches for multiplexing. Operating as a source follower, a MOSFET has almost ideal properties, in that a charge Q on the gate capacitance C_g, establishes a voltage $V = Q/C_g$ which controls the current in the channel of the FET from source to drain. This current can be monitored at any time without disturbing the charge on the gate, providing the option of multiple non-destructive reads. Figure 8 shows a typical source follower circuit. The FET also provides a low-impedance output as required by subsequent electronics. Typical gate capacitances are 0.3–3 pF. If this is allowed to integrate up to 1 V the charge accumulated is $2 \times 10^6 - 2 \times 10^7$ electrons. The "well depth" is the charge (in electrons) that can accumulate while the detector is still operating in a linear regime.

The unit cell contains the source follower and other switching FETs which allow for readout via the multiplexer and reset of the integrating nodes. In a typical circuit (see Figures 8 & 10) the charge integrating on the gate capacitance controls the current

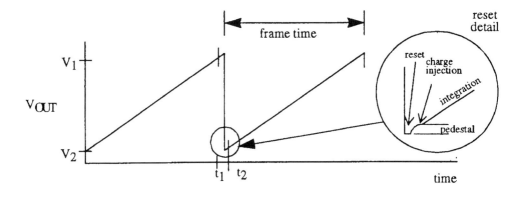

FIGURE 9. Reset and read timing in δ-reset sampling $Q = (V_1 - V_2)C_g$

through the FET and hence V_{OUT}. As long as the total integration time is much shorter than $R_d C_g$ the voltage at the output is proportional to the charge that has passed through the detector. The system may also be nonlinear because as charge builds up on the capacitor the effective bias across the detector is reduced, changing the responsivity. Normally the voltage swing at the gate is restricted to a value significantly less than V_{BIAS} to minimize this effect. The frame rate for reading the chip is set by the flux level incident on the detector, its quantum efficiency (η) and photoconductive gain (G), and restrictions on the amount of charge which may be allowed to accumulate.

There are several procedures for reading out the cell. The commonly used "destructive read" samples the voltage V_{OUT} after the integration, then shorts the capacitor using the reset switch. Usually V_{OUT} is sampled again after the reset, and the difference in voltages is proportional to the accumulated charge. This technique, shown in Figure 9, is called delta-reset sampling, and is adequate for most ground-based applications where the wells can be filled. For full wells, with a well depth of say 10^6 electrons, the noise should be dominated by the counting statistics of the photons, i.e. the noise is

$$\sigma = \sqrt{\frac{\eta}{\beta}(\beta G)^2 N} \approx 10^3.$$

This is the background-noise limited condition. Other noise sources need to be kept at a level low enough so that, when they are added in quadrature with the background, no significant excess contribution is observed. These noise sources include kTC noise ($(1/2)C\langle V^2 \rangle = \langle Q^2 \rangle = 1/2kT$, the random noise on a capacitor), FET noise (noise associated with fluctuating gate voltage) and amplifier noise (noise associated with following electronics). If the flux levels and timing are such that the wells cannot be filled, it is much harder to achieve background-limited performance. This may happen in high-resolution spectroscopy. However usually the inverse is more of a constraint - i.e. building a system that responds fast enough to readout the whole chip before saturation!

Non-destructive reads sample V_{OUT} without following with a reset. The timing of the reads can be adjusted for maximum benefit, e.g. reads can be made many times as the signal voltage ramps up, and a best fit computed. Although there can be some noise benefits to this type of procedure, a significant disadvantage is the extra time required in a regime where achieving a short frame time is critical. It is more often used in lower background environments, e.g. for near-IR arrays.

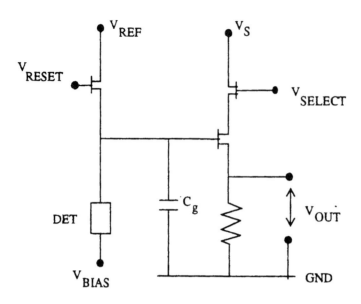

FIGURE 10. A typical Unit Cell

4.4. *Multiplexing*

The balance of the components in the unit cell perform the multiplexing operation. The number of outputs and architecture vary from chip to chip. For example the 20 × 64 element chips have 20 outputs, one for each row. The Aerojet 96×96 element chip used in the Keck LWS instrument had originally 1 output line, a custom modification connected 96 output lines, again 1 per row. Rockwell 128 × 128 chips have 4 or 16 outputs with a more complex pattern for addressing pixels. There are speed advantages in having more output lines as long as resources are available to run multiple parallel amplifiers and A/D converters. (The faster the frame time, the larger the bandwidth that can be tolerated before saturation. Since S/N increases as the square root of the bandwidth, this is particularly important for optimizing performance.)

Some multiplexers (although not many currently available) offer the capability to address any pixel on the chip randomly, without going through the entire sequence. This can be a very useful feature, both for debugging a system (where controlling the timing and watching the function of a single pixel simplifies operation), and for subarraying, where to gain speed, only a portion of the full array is used.

The unit cell shown in Figure 10 omits some of the multiplexing FETs but shows the source follower, select and reset FETs. When the integration is complete the V_{SELECT} line is toggled to turn on the select FET. The output voltage V_{OUT} is then proportional to the charge that has accumulated on C_g, which represents the combined capacitances of the gate of the source follower FET, the detector, and the reset FET. After time to sample V_{OUT} the V_{RESET} line is toggled to reset the capacitor to V_{REF}, which is also sampled. The difference between these two samples across a reset is called delta reset sampling. The voltages actually correspond to different integrations, although they are closer in time to each other than the reset is to V_{OUT} at the end of its particular integration (which is true correlated double sampling). When the reset and select lines are deasserted there is a small amount of charge injection onto C_g, as shown in Figure 9. This type of "pedestal" can vary from pixel to pixel in a different way from the charge accumulated as a result of the signal. The chopping technique assures that the pedestal is subtracted out from any observations. Flatfielding techniques also need to be differential for the same reasons.

4.5. *Control and readout electronics*

The signal lines from the array are usually brought out of the dewar and to the pream-plifier by the shortest route to avoid pickup and signal losses. The amplifiers must have adequate bandwidth to follow the fast signals, yet low enough noise not to contribute significantly compared to background noise. The A/Ds should be chosen with potentially conflicting requirements, i.e. speed, and resolution adequate to sample the noise. The fastest systems use individual preamps and A/Ds for each channel, but with care, some multiplexing can save on hardware costs. It is highly desirable to run the chip continuously, whether data is being recorded or not, to avoid drifts caused by self heating from power dissipated in the readout. Breaks in readout continuity can cause temperature and sensitivity fluctuations in the chip, which may appear as a fake signal, destroying the background subtraction from the chop/nod procedure. The control lines into the chip should be isolated from the signal lines, and the fast digital electronics should be contained in its Faraday cage for as complete isolation as possible.

The data rate from a mid-IR array is high and can rapidly fill any bulk storage medium. For most observations it is adequate to "coadd" sequential frames for the duration of a given chop position. Most data shows only systematic effects until the coadded data from the other chop position is subtracted. One common procedure is to coadd into 4 buffers, for the 2 chop and 2 nod positions, and to "download" the coadded data during a nod cycle. This is also the natural time to "coadd" those 4 buffers (with appropriate signs) for a preview of the data quality (Jones, et al. 1994).

4.6. *Measured quantities for a 20 × 64 array*

The UCSD IR group has operated a 20×64 element BIB from Aerojet ElectroSystems in a camera for several years. We have accumulated a significant data base on the operation of this chip. In many ways it is comparable to other BIB devices, so a discussion of its properties is useful.

This chip has 100-μm square pixels and has a half-pixel shift between the upper and lower 10×64 sections. It has 20 output lines, although for historical reasons we only sample 16 of these. We operate the chip in imaging mode with 0.83 arcsec per pixel at the 1.5-m Mt. Lemmon Observatory near Tucson AZ. With a 1-μm bandwidth at 10 μm it achieves a background-limited NEFD of 25 mJy min$^{-1/2}$ arcsec^{-2}. The source follower gate capacitance is 0.3 pF. The minimum frame time is 2.75 ms, limited by a single controller and A/D converter. Samples are digitized with a 12-bit A/D with a 1-μs convert time. The wells fill to about 10^6 electrons in a single frame (approx 50% of the well depth), corresponding to an effective emissivity of 30% for the overall system (telescope+sky+instrument window).

The chip has excellent cosmetic properties. The sensitivity variation from pixel to pixel is less than 5% with only 4 unusable pixels. The source follower gain is 0.65. The dark current is a strong function of both detector bias and temperature. At 1 V bias and 9 K the dark current is insignificant at $10^{-13.5}$ A. The quantities η, β, and G are difficult to determine individually, however direct measures of ηG and βG have been made, giving $\eta G=0.57$ and $\beta G=2.2$ (to about 10% uncertainty). Figure 11 shows the relative spectral response of a typical detector.

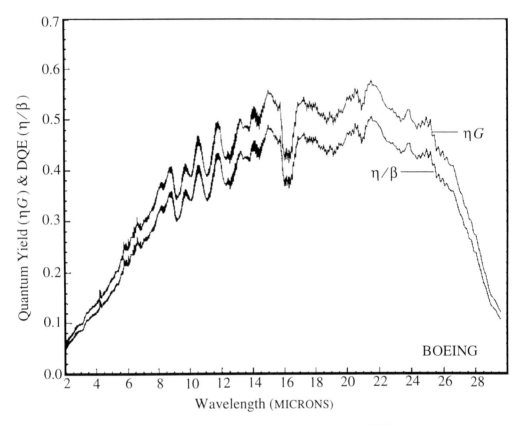

FIGURE 11. Relative spectral response of typical BIB array

4.7. *Future prospects*

Rockwell currently offers 128×128 chips of Si:As BIB with 75-μm pixels and 4 or 16 read lines. Well depths of a few $\times 10^6$ and about 3×10^7 are available. The larger well depths allow a slower frame rate without saturating the pixels, which is desirable because of the larger number of pixels to read per frame. These are ideal for high background imaging applications, but the smaller wells are more appropriate for spectroscopic instruments, where the background is lower. Typically the read noise for the deep well devices is larger than for the shallow well chips. If the wells can be filled, the read noise is not significant. However if the need to chop (to cancel a fluctuating background) limits the integration time and if the background is low, the larger wells may not have time to fill, resulting in higher noise levels.

The next generation of chips beyond the 128×128 devices is about to appear. Rockwell has a 256×256 chip, with 50-μm pixel size, using Si:As BIB for the detecting material. It is possible, now the IR astronomy community has a larger influence on the market for these new chips, that some dedicated devices may be made for astronomical use. The trade offs among well depth, read noise, pixel size, number of read lines is a complex one, and will be different for each type of instrument

5. The Long Wavelength Spectrometer for Keck

The Long Wavelength Spectrometer (LWS) for the Keck Telescope is an imaging spectrometer with diffraction-limited imaging and low- and high-resolution spectroscopy modes. In both imaging and spectroscopic modes it has a plate scale of 0.12 arcsec per pixel. The detector employed is a direct readout Si:As impurity band conduction array, with 96 by 96 pixels, manufactured by GenCorp Aerojet Electronic Systems Division. Direct imaging can be accomplished through a selection of 16 filters. In spectroscopy mode LWS can obtain spectra at resolutions of 100 and 1400 at 5, 10, and 20 μm. Details of the design and operation are discussed here.

5.1. *Introduction*

The Keck telescopes have been designed for good thermal IR performance. The IR instruments are placed at a forward Cassegrain focus situated 4 m in front of the primary mirror vertex. A chopping secondary of 0.527-m diameter (measured from extreme corners) delivers an $f/25$ beam. The secondary is gold coated and 5% undersized. It is supported by lightweight thin spiders from the larger secondary cage which supports the optical secondary mirror, and has a central hole to the sky to mask out the main secondary support structure. The optical baffles are retractable. The thermal background from the telescope is low and stable. The primary mirror segments are phased, typically to about 30 nm, and produce a mid-IR image dominated by diffraction. All the IR instruments are mounted in the forward Cass module which has four available slots. Three of these currently house the NIRC (Near Infrared Camera), LWS, and an optical guider. This design was adopted for rapid instrument interchange and flexibility in scheduling observations. All instruments are cryogen cooled, and the module is rolled out of the telescope each day for cryogen servicing. The electronics are situated immediately behind the instruments in the same module. Heat from the electronics is removed from the telescope by circulating coolants. There is no physical access to any of the Cassegrain instruments during observing.

There are some interesting consequences of this telescope and instrument geometry. The telescope with its 36 segments does not have circular symmetry. Its azimuth-altitude drive requires rotation of the instrument (or field of view) as an object moves across the sky. This is accomplished by rotating the module. The instruments requires either a hexagonal pupil to match the overall shape of the primary and which rotates counter to the module to remain fixed with respect to the telescope, or a non-rotating circular pupil, either circumscribed or inscribed on the overall hexagonal symmetry. LWS uses a non rotating inscribed circular pupil. The instruments are all off axis (2 arcmin for LWS, which is an "uplooker", less for NIRC, which is a "sidelooker"). The telescope focal plane is relatively steeply curved. Instrument design must take account of these factors.

From the earliest days of planning for the large telescopes it was clear that thermal IR instrumentation would be important. At these wavelengths the diffraction-limited spatial resolution scales inversely with primary mirror diameter, and the time required to observe a point source of a given flux level scales inversely as the fourth power of the diameter. Many scientific programs which would otherwise be too time consuming can be done with appropriate instrumentation given these advantages. Planning for LWS started as soon as good mid-IR arrays were available and their astronomical performance was demonstrated. After the usual preliminary design phases and fund-raising activities, fabrication started in 1991. First light in the laboratory was in July 1993. LWS was delivered to Keck in August 1995, first light at the telescope was September 1995. Several runs since then have allowed us to measure some of its performance characteristics.

5.2. *Instrument design*

The LWS was designed to give diffraction -limited imaging at 10 and 20 μm, and long-slit spectroscopy at resolutions of 100 and 1400. A rotating grating turret contains the low- and high-resolution gratings and a mirror for imaging mode. A plate scale of 0.12 arcsec pixel^{-1} was selected for adequate sampling of the 0.25 arcsec FWHM of the 10-μm diffraction pattern. LWS also functions at 3 and 5 μm, but with lower sensitivities because of a lower detector quantum efficiency at shorter wavelengths. For minimum background and optimum performance, LWS is an uplooking dewar. The largest array available when the instrument design was frozen was the Aerojet 96 × 96 chip which is currently in use. The optical system was constructed to accept a larger field of view for future upgrades. An imaging option was included in the spectrometer design to provide a more versatile instrument. This allows for deep imaging on the faintest sources and preliminary imaging to optimize spectrometer slit positions on complex objects not yet studied at such high spatial resolutions. The overall dewar layout is shown in Figure 12.

Significant constraints were placed on the overall design by the size and geometry required to fit into the forward Cass module, matching the focal plane of NIRC and the optical guider. Many aspects of the electronics are common to both LWS and NIRC, although each instrument has its own set of electronics mounted in the module.

5.3. *The Optical design*

A schematic view of the optics is shown in Figure 13. The telescope focal plane is imaged immediately behind the dewar window. A cold aperture driven by an external mechanism limits the field of view as is appropriate for imaging or spectroscopy. A 4-position aperture wheel selects either a square aperture for imaging or 2-, 4-, or 6-pixel wide slits for spectroscopy. The collimator delivers a parallel beam to the grating and forms an image of the telescope secondary between the collimator and grating. A circular baffle at this point matches an inscribed circle within the overall hexagonal secondary image shape. This provides good background radiation control, and avoids problems as the instrument rotates with respect to the telescope. A 16-position filter wheel is placed immediately behind the baffle.

The ratio of camera-to-collimator focal lengths was chosen to give the desired plate scale. The value of the camera focal length was chosen for the appropriate spectral resolution given the detector parameters and the physical size available in the cryogenic volume. Gold coated mirrors are used throughout, and the optical path is folded many times to pack the system into a small volume to reduce cryogen vessel sizes and cryogen consumption.

The full detector array of 96 × 96 pixels would give a field of 11 × 11 arcsec in imaging mode. At $R = 100$ ($\lambda/\Delta\lambda$), the full 8- to 13-μm atmospheric window is covered by 96 pixels at one grating setting, with a slit length of 11 arcsec. At $R = 1400$ the 96-pixel coverage for a single setting is approximately 0.35 μm, and 17 settings are needed to cover the full window with some overlap. Unfortunately the array is not correctly bump bonded over the full 96 × 96 pixels. About 1/3 of the chip is unusable, reducing the wavelength coverage in spectroscopy, and the field of view for imaging. This defect is not correctable without a change of chip, but appears to be stable over time.

The angle between the input and output beams on the grating is 15°. A custom order,

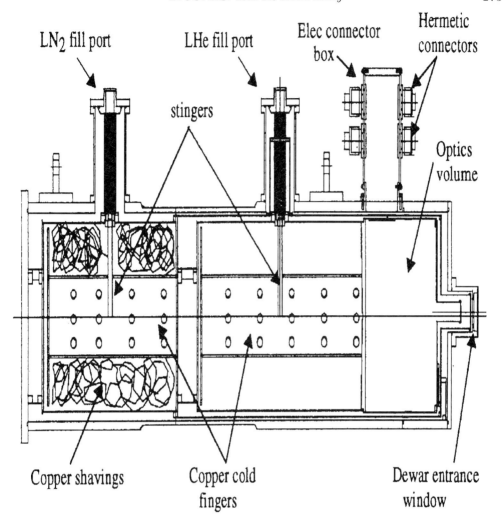

LN$_2$ fill port

LHe fill port

Elec connector box

Hermetic connectors

stingers

Optics volume

Copper shavings

Copper cold fingers

Dewar entrance window

FIGURE 12. The LWS dewar shown on its side. The cryogen volumes are normally only half filled through the "stingers" shown, so that no cryogens are spilled whatever angle the dewar is used at. The copper cold fingers provide good conduction to the cryogen at any angle.

gold-coated 8 groove mm^{-1} first-order grating blazed for 10 μm is used for low resolution. The spectral resolution (for 2 pixels) is approx $R = 100$ and does not vary strongly across the array. The high-resolution grating has 50 groove mm^{-1}, and is blazed for first order 19.5 μm at 29.2°. The spectral resolution (2 pixels) will vary with wavelength from a low value of $R = 1000$ at 8 μm through $R = 1400$ at mid range. At longer wavelengths the resolution is limited to approx 1400 by the number of grooves illuminated.

The complete assembly is packaged in a liquid helium cooled volume 10 inches in diameter and 6.25 inches high. The optics are divided into 2 levels, with the upper level components supported by a "deck" and the lower level by the optics plate. Figure 14 shows this 2 level layout. The upper level of components collimates the beam. The filter wheel is parallel to and just beneath the deck. The lower-level components are the grating, camera and detector array.

At 2 arcmin off axis, a geometric ray trace of the collimated telescope shows approx-

input aperture

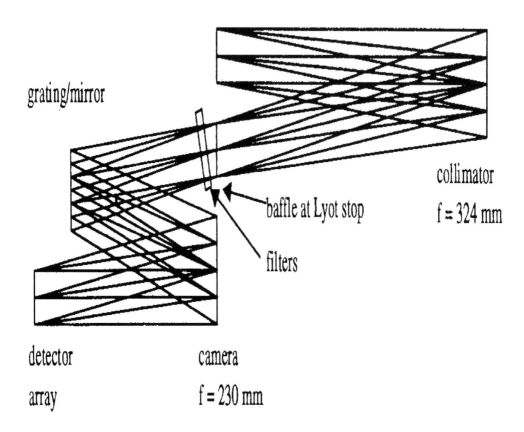

grating/mirror

collimator

f ≈ 324 mm

baffle at Lyot stop

filters

detector

array

camera

f = 230 mm

FIGURE 13. Schematic of optical layout. For clarity the folding flats are not shown.

imately 0.15 arcsec RMS of coma, too large to tolerate given a 0.25 arcsec diffraction-limited beam. Tilting the secondary out of true collimation is used to induce coma of the opposite sign from that caused by the off-axis position. The resulting geometric image size is 0.05 arcsec RMS across the LWS field. This secondary tilt must rotate as the instrument rotates. A geometric optical ray tracing was done to assess the image quality in the different modes of operation. Rays were traced through the off-axis telescope to the focal plane. All the geometrical images are at (or below) pixel size (0.12 arcsec). Laboratory testing confirmed the design field of view, plate scale and image quality.

The LWS has two entrance windows to support both 10- and 20-μm operation. A fixed window of KBr is used for 20 μm operation. This window is transparent from optical wavelengths to roughly 30 μm. Unfortunately this window material is hygroscopic. To protect this window when not required for 20-μm operation, a second movable ZnSe window is flipped into place. This protects the KBr window from the atmosphere by forming a sealed volume containing desiccant. The movable window is anti-reflection coated and transparent from 0.5 μm to 17 μm.

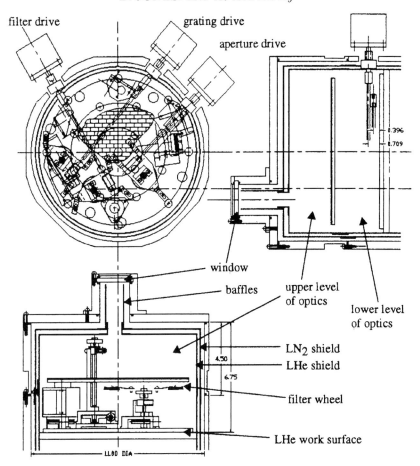

FIGURE 14. Schematic diagram of LWS cryogenic optical volume. The KBr window froms a vacuum seal above the cooled baffles in the snout. The upper optics layer houses the entrance aperture, collimator and folding flats. The lower layer supports the grating carousel, camera and array.

Figure 15 shows the dewar mounted into its one fourth section of the IR module. A sector wheel mounted above the window has 6 blades and is driven by a stepper motor to create calibration signals. Three of the blades are used, each alternating with a direct view to the sky between the blades. The blades are coated (on the dewar side) to radiate as a black body at telescope temperature, a grey body, and a good reflector which creates a low background by reflecting the field of view back into the dewar.

The entrance window was placed as close to the edge of the dewar as possible to minimize the amount by which the incoming beam is off axis in the telescope. The beam center is at 150 mm from the telescope axis (120 arcsec off axis). At this position effects of image quality, focal plane curvature, instrument tip, and optical and mechanical clearances are all important. The radius of curvature of the $f/25$ focal plane is 846 mm. At a position 150 mm off axis this results in a 9.5° tilt in the focal plane in the sense that more off-axis images focus closer to the secondary. Each of the 4 positions in the LWS aperture wheel has this 9.5° tilt incorporated to match the telescope focal plane.

LWS is also tipped at 0.7° (toe-in at the top) to point to the center of the secondary. This is achieved by adjusting the tilter at the base of the dewar. Correct pointing was

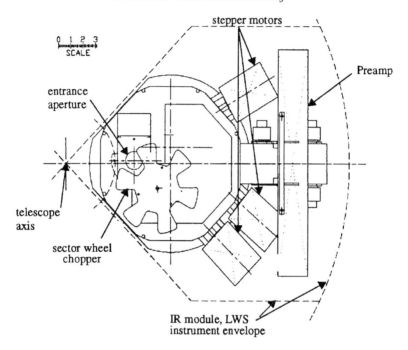

FIGURE 15. LWS dewar in the Keck IR module as seen from above. The outer dashed lines show the edge of the LWS quadrant of the IR module. The sector wheel chopper and window flipper mechanisms are controlled by stepper motors mounted to the dewar top plate.

established by using a capability of the primary mirror that exploits its segmented design. The primary can be made to separate the image of a star into 36 images, one from each segment. This is accomplished by "fanning out" the segments, effectively increasing the primary's effective focal length a small amount. These 36 images should fall into a regular pattern, the spacing of which can be selected. When the instrument is accurately pointed at the secondary all 36 images are visible. Inverting the positions of the images (by "fanning in" the segments by the same amount, enables an accurate measure of the best focus. Distortions in the array of images can also be analyzed for astigmatism and coma.

5.4. *The detector array*

The LWS focal plane employs a 96×96 element Si:As impurity band conduction detector manufactured by GenCorp Aerojet Electronics Systems Division. The detector is mounted on a custom sapphire PC board (designed and built by Aerojet) which allows the 120 signal and control lines to be fanned-out to a 120-pin Isotronics package.

The special packaging of the LWS detector allows a direct readout the 96 parallel data lines available on the chip. Each unit cell has its own source follower circuit and columns of detectors are readout simultaneously. Detector timing chooses which of the 96 "columns" of data and reset pixel values are selected. One aspect of this chip is that the pixel data voltage levels and pixel reset voltage levels are not available at the same time for each pixel, but are interspersed with the voltage levels for other pixels. This means that delta-reset sampling must be performed in the digital domain and that $2 \times 96 \times 96$ digital values must be obtained per data frame. The readout noise at a frame rate of 600 frame s^{-1} is approximately 200 e$^-$ RMS.

The LWS detector is an impurity band conduction (IBC) device. To ensure IBC operation and to ensure stable detector responsivity, the detector temperature must be maintained relatively constant. The array clocking rate determines the electrical power dissipated on the chip. To ensure stability in temperature and sensitivity clocking is continuous whether data is being acquired or not. The largest swings in detector temperature occur because of changes in dewar angle to the vertical changing thermal impedance to the cryogens. This causes slow sensitivity changes of about 30% which must be removed from the data by calibration. The sector wheel mounted above the dewar window is used for this function.

5.5. *The mechanical and cryogenic design*

The dewar was designed to fit into a 90° segment of the Keck IR module, with the window and optics at the top. It is 40 inches long, supported from below by a tilter which allows for fine pointing of the field of view at the secondary mirror. Five hermetic connectors carry the 96 output lines, bias and clocking signals, and housekeeping data. The preamp is mounted on the dewar close to the connectors to minimize cable length. The level shifter, which generates clocking and bias voltages, is mounted on the preamp. All other electronics is mounted in the back part of the module, behind the dewar.

There are 3 mechanisms active inside the dewar, controlling the aperture wheel, the filter wheel, and the grating turret. All mechanisms are identical, 100:1 antibacklash worms coupled to 3:1 step-down pulley driven by stepper motors. A step size of 2 arcsec is achieved by this mechanism for a 200 steps/rev motor substepped 10:1. Penetrations through the dewar wall are ferrofluidic seals. An internal encoder (microswitch) provides a contact once per revolution of the internal mechanism. A similar external encoder identifies a unique position on the motor shaft. These encoders together provide a "home" signal which is used for initial setup. The drives and encoders give sub-pixel precision and repeatability on all mechanisms. The drives run in either direction with a small backlash correction.

The outer vacuum jacket of the dewar is aluminum. The inner vessels are stainless cylinders with copper end plates. Since the dewar must operate in any orientation the cryogens will not always be in contact with the top plates (see Figure 12). A reasonable thermal path from the "top" copper plates to the cryogens is provided by internal copper fins. In spite of using high-purity copper for this, there is a resulting temperature rise of about 0.5 K when the dewar is close to vertical resulting from the typical heat load over the longer thermal path. This requires more care in calibration of data, as the detector sensitivity is temperature dependent. The radiative surfaces are soft aluminum foil wrapped, giving an effective emissivity of about 4%.

To prevent the dewar dumping cryogens during observing the vessels are filled only to half capacity through "stingers" which extend the fill tubes to the center of each cryogen volume. The cryogenic full volumes are 19 liters of liquid helium and 15 liters of liquid nitrogen. The total heat loads including radiation, conduction, and power dissipated on the focal plane are about 10 W for the LN2, and 200 mW for the LHe, this gives expected hold times of about 35 hours for a half-volume cryogen fill.

An unwelcome phenomenon was observed when the dewar was first operated in a vertical (window up) position. The helium boil-off rates increased to higher than 10 times their value when the dewar was horizontal. This effect occurred independently of cryogen fill level (always less than half), and was found to be caused by oscillations in the cold gas in the dewar neck pumping energy into the liquid helium. These oscillations have been reduced, but not eliminated, by using a spiral baffle in the helium neck, and by running the pressure of the helium vessel about 4 psi above ambient. The hold time achieved by this is close to 24 hours.

5.6. *LWS electronics*

Figure 16 shows the analog electronics stream. The 96 output lines from the array are amplified and multiplexed 4:1 in the preamp. The 24 analog lines from the preamp pass to the "coadders" in the electronics rack. After a programmable gain and offset stage they are digitized and added into storage—this is the "coadder" function. In order for LWS output signals to settle to 12-bit accuracy in a reasonable time, it was essential to minimize the output capacitance driven by the on-focal-plane MOSFET source-follower circuits. This requires keeping the detector cables short (12 inches inside the dewar and 6 inches outside the dewar) and to have receiver circuits for each of the 96 parallel data lines.

The control electronics hardware includes a Force 5C control computer residing in the electronics rack in the IR Instrument frame below the instrument volume, a S56X controller card, the timing generator, the co-adders and the level shifter.

The electronics is controlled from a Force 5C board that resides in a VME card cage in the electronics racks below the instrument volume in the IR instrument frame. This computer, called "Coyote", runs all of the LWS "host" software. It communicates to the timing generator, and the co-adders through the S56X interface card. All of the timing generator and co-adder software is loaded through Coyote and the data collected by the co-adders is passed through Coyote for recording on disk. Coyote controls the LWS stepper motors through an RS-232 port and picks up housekeeping data (e.g. detector and dewar temperatures) residing in the VME card cage from the VME bus.

The S56X card provides the communication between Coyote and the timing generator and co-adders, and runs the program that controls the Keck chopping secondary mirror and the sector-wheel chopper. Having the chopping control software run on the S56X card allows the timing generator card to concentrate solely on providing steady clocking signals to the detector. This is crucial since electronics self-heating of the detector (which is readout rate sensitive) is important to the detector temperature and hence its sensitivity and dark current level.

The level shifter receives the TTL clocks from the timing generator and conditions the signals, changing their voltage rails, noise properties, etc., and then passes them on to the LWS detector. It also provides for optical isolation between the digital signals from the timing generator and the low noise clocking environment needed for the detector. It has pot setable bias voltages for running the detector, e.g. detector common voltage, substrate voltage, unit-cell source-follower circuit power, etc.

5.7. *LWS software*

The LWS software runs on three basic platforms: (1) the control room computer, (2) Coyote, and (3) the Motorola M56001 DSPs. It communicates with the telescope computers for housekeeping information, e.g. time, telescope position, etc., and to control or learn of telescope operations, e.g. secondary chopper control and status, telescope pointing, etc. The control room computer provides the astronomer interface and provides control and display of LWS control GUI's, and Quick-Look data displays. It also provides keyword value updating, and the FITS file writing facility. The Force 5C VME-based computer, Coyote, runs the stepper motor control, housekeeping, and data acquisition. There are 3 basic M56001 based processors in the IRE. The first, on the S56X controller card, controls the communications link between Coyote and the other processors, and controls chopping. The other M56001 based processors are the timing generator and the

Detector Electronics Readout Chain

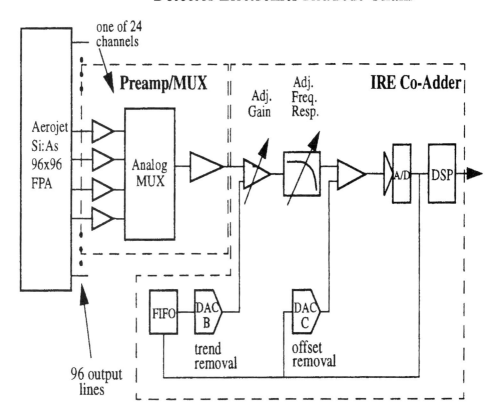

FIGURE 16. LWS electronic readout. Schematic diagram of the chain of readout electroics for the LWS, showing the focal plane array (FPA) with its 96 output lines, one of 24 channels in the PolyCom Preamp/MUX, and one IRE Co-Adder card. As can be seen, the Co-Adder cards provide programmable gain (in the range 1 to 10), frequency response roll-off (knee at 2 kHz, 20 kHz, 200 kHz, or 2 MHz), and offset and trend removal. The first stage gain of the Preamp/MUX is 2.0; the second stage gain is 3.7.

Co-Adder cards. These provide the detector clock timing and co-add the data coming off of the detector before passing the data to Coyote. The acquire program is responsible for collecting the co-added data from the Co-Adder cards via the pass-through processor (S56X card) and writing the data to disk in FITS format via the FITS RPC utility. Each new observation that requires data archiving to disk creates a new acquire task and at the end of the observation the acquire task dies.

All aspects of the real-time configuration of the LWS instrument are controlled through keyword values. However, astronomer control of the LWS is made routinely through the instrument control GUIs (written in DataViews). These GUIs "protect" the observer from setting incompatible keyword values and also provide natural instrument set-ups.

The Quick-Look software is written in IDL and is accessible from any UNIX shell prompt line. Quick-Look provides real-time views of the data as it is transferred from the Co-Adder cards to Coyote for archiving through a memory mapped file. A variety of displays are possible, including sky frames and object frames in both beam positions as well as difference, signal, and balance frames. In addition, it provides data frame buffer arithmetic that can be used to flat-field and dark frame subtract the displayed data in real time as well as tools that can select a particular frame, save and/or load frames, zoom the display, read the cursor, make plots of cuts through the data, make contour plots, perform frame statistics, perform area flux measurements, and extract and calibrate spectra along with wavelength calibration and sky reference spectra. This capability to assess image quality is crucial to efficient use of telescope time.

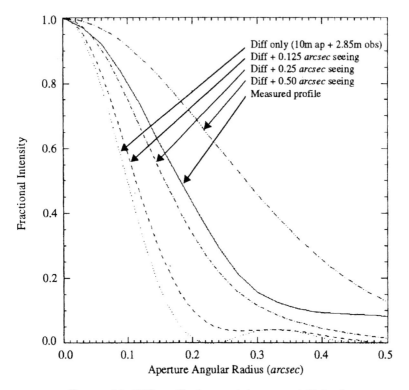

FIGURE 17. PSF profile for a point source at 11.7 microns

5.8. *Instrument commissioning*

The first critical question to be addressed with LWS at Keck was that of image quality, since the instrument design was motivated by the goal of achieving diffraction limit at 10 μm. This is closely followed by sensitivity, then by the usual list of parameters that make an instrument useful, user friendly and efficient. These tests are still ongoing, this section reports the initial results.

5.8.1. *Image quality*

The primary mirror is typically phased to 30 nm, far better than needed for 10-μm observing. The dome is actively cooled to typical night temperatures during daytime, and appropriate care is taken by the Keck facility not to degrade the excellent seeing at the site. LWS image profiles of a point source observed at 11.7 μm are shown in Figures 17 and 18. The calculated point spread functions and enclosed energy distributions were made for a circular 10m diameter aperture with a 2.85-m circular central obscuration. No attempt was made to allow for diffracted energy from the gaps between the segments, or from the overall hexagonal structure. Seeing was assumed to have a Gaussian profile with the given full width at half maximum, and was added in quadrature.

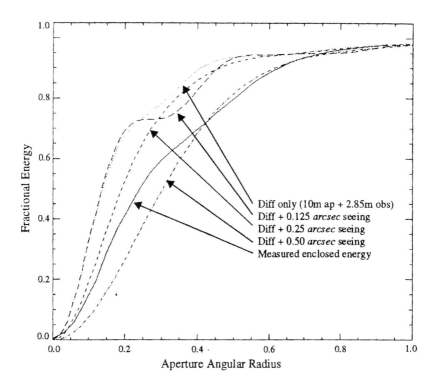

FIGURE 18. Enclosed energy for a point source at 11.7 microns within the angular radius shown

The observed profiles are for an integration of about a minute and therefore include effects of image wander. Many of the standard star images show hexagonal diffraction spikes. While none of the computed profiles matches the observed exactly, the best fit is diffraction with a seeing component of just over 0.25 arcsec FWHM, giving an observed FWHM of 0.36 arcsec, which contains about 40% of the energy. With this type of image on LWS, simultaneous measures of seeing made with the red sensitive CCD guide camera may be just under an arcsec to about 0.6 arcsec. Clearly there is a substantial improvement in seeing at longer wavelengths, even for telescopes with large apertures.

5.8.2. *Sensitivity*

Referring to lecture 3 on imaging $S_{\mathrm{NEFD}} = (0.13\sqrt{\alpha})/(fD^2)$ Jy for a point source, S/N=1 in 1 second of elapsed time. From the above LWS images $f = 0.4$, and for the Keck telescope $\alpha = 4$. This predicts $S_{\mathrm{NEFD}} = 8$ mJy (in 1 second). Early measurements of sensitivity have yielded S_{NEFD} closer to 28 mJy in 1 second from integrations of 30 minutes. However on the first few runs on the telescope the noise has had a non-Gaussian character, and been larger than measured in the lab before delivery. Recently a disconnect on the array bias line inside the dewar became intermittent, then complete. After repair the noise appears to be at previous lab levels. We are optimistic that LWS can achieve the 8-mJy level, and that in time Keck will improve its efficiency in the chop/nod mode, enabling another sensitivity gain.

5.9. *Acknowledgments*

This series of lectures results from experience gained over the years with several projects. The "Golden Gopher" at the Mt. Lemmon Observatory was constructed and optimized by Robert Piña. LWS was a joint effort with Richard Puetter. Aerojet provided arrays free of charge. Along the way grants from NASA, NSF, the California Space Institute, and CARA (California Association for Research in Astrophysics) have helped fund new instruments and observations.

REFERENCES

AITKEN, D. K. ET AL. 1985 *Mon. Not. R. Astron. Soc.* **215**, 813.

AITKEN, D. K. ET AL. 1989 *Mon. Not. R. Astron. Soc.* **262**, 456.

AITKEN, D. K. ET AL. 1991 *Astrophys. J.* **380**, 419.

BHARAT, R. 1994 in *Infrared astronomy with arrays* (ed. I. McLean). Astrophys. & Sp. Sci. Lib., vol. 190, p. 376. Kluwer.

BLANCO, A. ET AL. 1993 in *Astronomical infrared spectroscopy: future observational directions* (ed. S. Kwok). A. S. P. Conf. Ser., vol. 41, p. 267. A. S. P.

BRATT, P. R. 1977 *Semiconductors and Semimetals* **12**, 39.

CABRITT, S. ET AL. 1994 in *IR Astronomy with Arrays* (ed. I. S. McLean). Astrophys. & Sp. Sci. Lib., vol. 190, p. 311. Kluwer.

CAMERON, M. ET AL. 1992 in *Progress in Telescope and instrumentation technologies* (ed. M-H. Ulrich). ESO Conf. & Workshop Proc., vol. 42, p. 705.

CAMPINS, H. ET AL. 1985, *Astron.J.* **90**, 896.

DENNEFELD, M. ET AL. 1994 in *Infrared Astronomy with Arrays: The next generation.* (ed. I. S. McLean). Astrophys. & Sp. Sci. Lib., vol. 190, p. 315. Kluwer

FOMENKOVA, M. N. ET AL. 1995 *Astron. J.* **110**, 1566.

GARDEN, R. P. 1994 *Proc. SPIE.* **2198**, 487.

GEZARI, D. Y. ET AL. 1988 *Proc. SPIE* **973**, 287.

GEZARI, D. Y. & BACKMANN, D. E. 1994, in *IR Astronomy with Arrays* (ed. I. S. McLean). Astrophys. & Sp. Sci. Lib., vol. 190, p. 195. Kluwer.

GLASSE, A. C. H. & ATAD, E. E. 1993 *Proc. SPIE* **1946**, 629.

HAYWARD, T. L. ET AL. 1993 *Proc. SPIE* **1946**, 334.

HERTER, T. ET AL. 1989 in *Proc. 3rd. IR Det. Wkshp.* NASA Tech. Memo 102209, p. 427.

HERTER, T. 1994 in *Infrared astronomy with arrays* (ed. I. McLean). Astrophys. & Sp. Sci. Lib., vol. 190, 409.

HOFFMANN, W. F. ET AL. 1993 *Proc. SPIE* **1946**, 449.

JONES, B. & PUETTER, R. C. 1993 *Proc. SPIE* **1946**, 610.

JONES, B. ET AL. 1994 in *Infrared astronomy with arrays* (ed. I. McLean). Astrophys. & Sp. Sci. Lib., vol. 190, 187.

KNACKE, R. F. 1989 in *Interstellar Dust* (ed. L. J. Allamandola & A. G. G. M. Tielens). IAU Symp. 135, p. 415. Kluwer.

KORFF, D. 1973 *J. Opt. Soc. Am.* **63**, 971.

KWOK, S. ET AL. 1993 in *Astronomical infrared spectroscopy: future observational directions* (ed. S. Kwok). A. S. P. Conf. Ser., vol. 41, p. 123. A. S. P.

LACY, J. H. ET AL. 1989 *Publ. Astron. Soc. Pacific* **101**, 1166.

LACY, J. 1993 in *Astronomical infrared spectroscopy: future observational directions* (ed. S. Kwok). A. S. P. Conf. Ser., vol. 41, p. 357. A. S. P.

LAGAGE, P. O. ET AL. 1993 *Proc. SPIE* **1946**, 655.

LAGAGE, P. O. ET AL. 1994 in *Infrared Astronomy with Arrays* (ed. I. S. McLean). Astrophys. & Sp. Sci. Lib., vol. 190, p. 323. Kluwer.

LENA, P. 1986 *Observational Astrophysics.* Springer.

LE PHAM ET AL. 1994 in *Infrared astronomy with arrays* (ed. I. McLean). Astrophys. & Sp. Sci. Lib., vol. 190, 383.

LEVAN, P. D. ET AL. 1989 *Publ. Astron. Soc. Pacific* **101**, 1140.

LORD, S.D. 1992 *NASA Technical Memorandum 103957.*

LOW, F. J. & RIEKE, G. H. 1974 in *Astrophysics* (ed. N. Carleton). Methods of Exp. Physics, vol. 12A, pp. 415–462. Academic Press.

LUCAS, C. ET AL. 1994 in *Infrared astronomy with arrays* (ed. I. McLean). Astrophys. & Sp. Sci. Lib., vol. 190, 425.

LUM, N. ET AL. 1993 *Proc. SPIE* **1946**, 100. bibitem[]MCLEAN, I. S. (ED.) 1994 *Infrared Astronomy with Arrays.* Astrophys. & Sp. Sci. Lib., vol. 190. Kluwer.

MEIXNER, M. ET AL. 1994 in *Infrared Astronomy with Arrays* (ed. I. S. McLean). Astrophys. & Sp. Sci. Lib., vol. 190, p. 207. Kluwer.

MILES, J. W. 1993 in *Astronomical infrared spectroscopy: future observational directions* (ed. S. Kwok). A. S. P. Conf. Ser., vol. 41, p. 407. A. S. P.

MILES, J. W. ET AL. 1994 *Astrophys. J.* **425**, L37.

MILONE, E. F. & STAGG, C. R. 1993 in *Astronomical infrared spectroscopy: future observational directions* (ed. S. Kwok). A .S. P. Conf. Ser., fol. 41, p. 395. A. S. P.

MUIZON, M. J. 1993 in *Astronomical infrared spectroscopy: future observational directions* (ed. S. Kwok). A. S. P. Conf. Ser., vol. 41, p. 79. A. S. P.

OLNON, F. & RAYMOND, E. 1986 *Astrophys. J. Supp.* **65**, 607.

PARMAR, P. S. ET AL. 1994 *Astrophys. J.* **430**, 786.

PERSI, P. ET AL. 1994 in *Infrared Astronomy with Arrays* (ed. I. S. McLean). Astrophys. & Sp. Sci. Lib., vol. 190, p. 331. Kluwer.

PIÑA, R. K. ET AL. 1993 *Proc. SPIE,* **1946**, 640.

RANK, D. M. & TEMI, P. 1993 *Proc. SPIE* **1946**, 417.

RESSLER, M. E. 1994 in *Infrared Astronomy with Arrays* (ed. I. S. McLean). Astrophys. & Sp. Sci. Lib., vol. 190, p. 429. Kluwer.

RIEKE, G. H. ET AL. 1985 *Astron.J.* **90**, 900.

RIEKE, G. H. 1994 *Detection of Light: from the Ultraviolet to the Submillimeter.* Cambridge University Press.

ROBBERTO, M. ET AL. 1994 *Proc. SPIE* **2198**, 446.

ROCHE, P. F. 1989 in *Infrared spectroscopy in astronomy* (ed. B. H. Kaldeich). Proc. 22nd. Eslab Symp., p. 79.

ROELLIG, T. L. ET AL. 1994 in *Infrared Astronomy with Arrays* (ed. I. S. McLean). Astrophys. & Sp. Sci. Lib., vol. 190, p. 333. Kluwer.

SCHROEDER, D. J. 1987 *Astronomical Optics.* Academic Press.

SCLAR, N. 1984 *Prog. in Quantum Electronics* **9**, 149.

SELBY, M. J., WADE, R. & SANCHEZ-MAGRO, C. 1979 *Mon. Not. R. Astron. Soc.* **187**, 553.

SIBILLE, F. ET AL. 1994 in *Infrared Astronomy with Arrays* (ed. I. S. McLean). Astrophys. & Sp. Sci. Lib., vol. 190, p. 335. Kluwer.

SMITH, C. H. ET AL. 1994 *Proc. SPIE* **2198**, 736.

WATARAI, H. ET AL. 1994 in *Infrared Astronomy with Arrays* (ed. I. S. McLean). Astrophys. & Sp. Sci. Lib., vol. 190, p. 339. Kluwer.

STACEY, G. J. ET AL. 1993 *Proc. SPIE* **1946**, 238.

STOCK, J. & KELLER,G. 1960 *Telescopes* (ed. G. P. Kuiper & B. M. Middlehurst). Stars & Stellar Systems, vol. 1, p. 38. Univ. Chicago Press.

SZMULOWICZ, F. & MADARSZ, F. L. 1987 *J. Applied Physics* **62**, 2533.

VOLK, K. 1993 in *Astronomical infrared spectroscopy: future observational directions* (ed. S. Kwok). A. S. P. Conf. Ser., vol. 41, p. 63. A. S. P.

WALKER, G. 1987 *Astronomical Observations.* Cambridge Univ. Press.

WATERAI, H. ET AL. 1994, *IR Astron with Arrays* (ed. I. McLean). Astrophys. & Sp. Sci. Lib., vol. 190, p. 339. Kluwer.

WOLFE, W. L. & ZIZZIS, G. J. 1978 *The Infrared Handbook.* Office of Naval Research.

Polarimetry with large telescopes

By SPERELLO DI SEREGO ALIGHIERI

Osservatorio Astrofisico di Arcetri, Largo E. Fermi 5, I–50125 Firenze, Italy

The new generation of 8-10m telescopes is opening up important possibilities for polarimetry of astrophysically interesting sources, mainly because the large collecting area is particularly advantageous in this technique, which requires high S/N ratio. This course starts by emphasizing the importance of polarimetry in astronomy and giving some examples of polarizing phenomena in everyday life. Then an introduction to the Stokes parameters and to Mueller calculus is given, with examples on how to describe the most common polarizing optical components, and the main mechanisms producing polarized light in astrophysics are reviewed. The section devoted to instruments starts with a brief overview of the classical photopolarimeter, follows with a description of an imaging polarimeter, with examples of data obtained and an analysis of the sources of errors, and ends with a discussion of modern spectropolarimetry. The following section is devoted to an analysis of the gains of large 8–10 m telescopes for polarimetry and to a review of the polarimeters planned for them. The course ends with a discussion of polarimetry of AGN, as an example of a field of research, where polarimetry has provided important results, by disentangling unresolved geometries and mixed spectral components.

1. The beauty of polarimetry

Astronomy is an observational science, not an experimental one in the usual sense, since for the understanding of the objects in the Universe we cannot perform controlled experiments, but have to relay on observations of what these objects do, independently of us †. Apart from a few disturbing cosmic rays, several elusive neutrinos, and, maybe, some still undetected gravitational waves, the astronomical observations are concerned with electromagnetic radiation at all wavelengths, which can be fully described by the 4 Stokes parameters: I, the total intensity, Q and U describing linear polarization, and V describing circular polarization (see section 2.1). All these parameters depend on the wavelength λ. Therefore the task of observational astronomy is to study I, Q, U and V, as a function of wavelength, position in the sky, and time.

All the other courses at this School deal with the detection of just one of these parameters, i.e. the total intensity I. It is clear that, if you limit yourself to this, you would be throwing away aspects of the information that is carried to us by electromagnetic radiation. Polarimetry is the technique which studies all 4 Stokes parameters. In the following my aim is to describe how polarization can be measured, in particular with a large optical telescope, and how useful it is in astrophysics. Although in the past polarimetry has been regarded as a very specialized technique of limited use, and so complicated that only a few specialists could use it, I hope to be able to demonstrate to you that with modern instruments it becomes relatively easy and that with large telescopes there are good reasons and enough photons for recording all 4 Stokes parameters for almost every observation, changing the habit acquired with smaller telescopes to mostly record just one of them, i.e. the intensity.

† There are, of course, a few exceptions, like planetology, for objects near enough that we can reach them, and laboratory astrophysics for those few cases that we can reproduce on Earth. These exceptions however are rare enough not to change the observational spirit of astronomy.

1.1. *Some every–day polarizing phenomena*

Our eyes are nearly blind to polarization (see however section 1.1.1), probably because the Sun emits mostly unpolarized light, which for this reason is called "natural". Nevertheless Nature provides us with very beautiful examples of polarized light, like the rainbow. In the following short sections I try to show you a couple of these examples.

1.1.1. *Can we see polarization?*

One of the reasons why polarization is regarded as a strange phenomenon is that our eyes are insensitive to it. However this is not completely true, as the trained human eye can detect linear polarization on a smooth and diffuse source like the blue sky, by seeing the so called Haidinger's brush (e.g. Minnaert, 1954). This is a bow–tie–shaped yellow object about 4° long, which appears in the centre of the visual field when linearly polarized light is viewed (Fig. 1). Sometimes a blue "brush" is seen perpendicular to the yellow one. Training is required because the colour sensitivity of our eye locally adjusts to make the brushes disappear within a second or two. However this natural eye adjustment can be used to enhance the sensitivity to the brushes, by changing the viewing direction every couple of seconds, or by slightly rotating a linear polarizer in front of the eye once per second or so. In this case the brushes rotate in our eyes and this will bring the yellow part of the brush to a portion of the retina which has just been adapted to be more sensitive to yellow light. Therefore the rotation can enhance the colour contrast of the brushes.

Once one has seen the Haidinger's brushes, it becomes easier to detect them even against a coloured background like the blue sky, e.g. by looking at 90° from the Sun near sunrise or sunset. In this case the yellow brush should appear horizontal.

The human eye can also see circularly polarized light, since this also produces the brushes. In particular, for right–circularly polarized light the yellow brush appears rotated clockwise from the vertical, while it appears rotated counterclockwise for left–circularly polarized light. Therefore an observer can tell the handedness of polarization merely by the eye (Schurcliff, 1955).

Some insects, like bees, have a much better ability to see polarization than we do, and they use it for navigation purposes.

1.1.2. *How do Polaroid sunglasses work?*

Polaroid sunglasses are known to be particularly good in selectively eliminating annoying reflections in the outdoors. The reason for it is simple and is related to polarization. In fact reflections are heavily polarized in the direction perpendicular to the plane of reflection. In the open air reflections mostly occur from horizontal surfaces, like water or ground, illuminated from above. In these cases the plane of reflection is vertical and the polarization horizontal. Polaroid sunglasses are filters which effectively cut horizontally polarized light and let the vertical one through.

Another application of Polaroid glasses is in tridimensional cinematography. The visual perception of tridimensionality is achieved because our two eyes receive a slightly different image. In 3–D movies two different pictures are projected on the screen, one of which is polarized horizontally, the other one vertically. Equipped with special glasses in which one lens is a horizontal polaroid and the other one is a vertical one, we can selectively detect the two pictures with each of our eyes and obtain a 3–D effect. 3–D television could work the same way, in particular by using liquid crystal displays, whose light is strongly polarized.

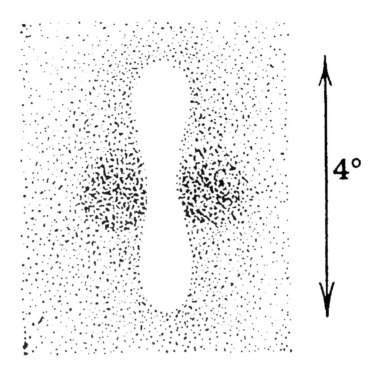

FIGURE 1. The so–called Haidinger's brushes, which the trained human eye can see on a uniform visual field which is strongly polarized, like the blue sky at about 90° from the Sun. The light brush is yellow, while the perpendicular one is blue. The yellow brush is perpendicular to the direction of oscillation of the **E** vector (from Minnaert, 1954).

2. A formal description of polarization

2.1. *The Stokes parameters*

In order to take into account its polarization properties, electromagnetic radiation is best described with the four Stokes parameters. In the following we shall define them, examine their properties and their relationship with the more customary degree and direction of polarization, and give the foundations of Mueller calculus.

2.1.1. *Definitions*

Given a direction of observation in the sky, let us consider an orthogonal reference frame linked to the equatorial coordinate frame, in which the z axis points to the observer, the xy plane is in the plane of the sky with the x axis pointing to the North and the y axis pointing to the East (Fig. 2). For a simple electromagnetic wave (i.e. a monochromatic and completely polarized one) propagating along the z axis towards the observer the **E** vector vibrates in the xy plane and its components are (apart from an inessential phase factor that depends on the choice of $t = 0$):

$$E_x = a_x cos(2\pi\nu t)$$

$$E_y = a_y cos(2\pi\nu t + \delta)$$

As viewed from the z axis the tip of the **E** vector describes an ellipse with major axis a, minor axis b and position angle θ (Fig. 3):

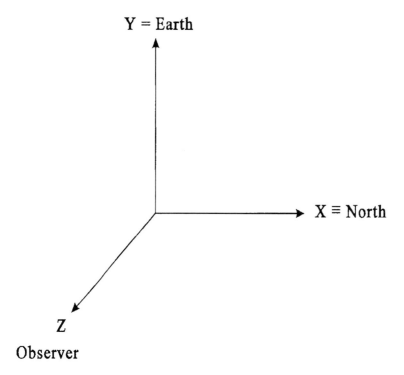

Y = Earth

X ≡ North

Z
Observer

FIGURE 2. The orthogonal reference frame linked to the equatorial coordinate frame, which is used to describe an electromagnetic wave coming to the observer along the z axis and vibrating in the xy plane.

$$a = \sqrt{\frac{a_x^2 + a_y^2}{1 + tan^2\left\{\frac{1}{2}arcsin\left[(2a_x a_y/\left(a_x^2 + a_y^2\right))\,sin\delta\right]\right\}}}$$

$$b = \left|a\,tan\left\{\frac{1}{2}arcsin\left[(2a_x a_y/\left(a_x^2 + a_y^2\right))\,sin\delta\right]\right\}\right|$$

$$\theta = \frac{1}{2}arctan\left[(2a_x a_y/\left(a_x^2 - a_y^2\right))\,cos\delta\right]$$

$$a^2 + b^2 = a_x^2 + a_y^2.$$

If the components E_x and E_y have the same phase ($\delta = 0$), then the **E** vector vibrates in a plane and the light is called linearly polarized at the position angle θ (if $a_y = 0$ then $\theta = 0°$ and if $a_x = 0$ then $\theta = 90°$).

If E_x and E_y have equal amplitudes ($a_x = a_y$) and are out of phase by 90°, then the tip of the **E** vector describes a circle and the light is called circularly polarized. In such case, if $\delta = +90°$, then the **E** vector proceeds like a right handed screw producing right circularly polarized light; on the other hand, if $\delta = -90°$, one has left circularly polarized light. Note that, just as circularly polarized light can be viewed as being made up of linearly polarized components that are out of phase with each other, linearly polarized light can be viewed as being composed of equal quantities of right and left circularly polarized components.

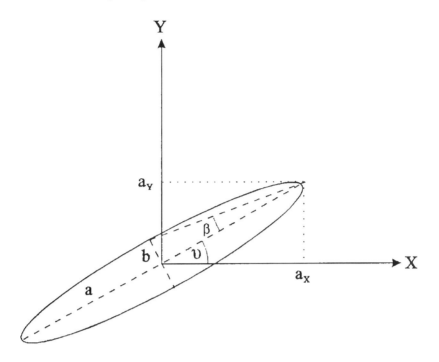

FIGURE 3. The ellipse — projected in the xy plane — described by the tip of the **E** vector for a monochromatic and completely polarized electromagnetic wave propagating along the z axis.

For the purpose of polarimetry it is convenient to describe the electromagnetic radiation in terms of the Stokes vector, made up of 4 real parameters:

$$\mathbf{S} = \{I, \ Q, \ U, \ V\}.$$

The Stokes parameters are defined as follows:

I is the total intensity;

$Q = I_0 - I_{90}$ is the difference in intensities between the horizontal (or North–South) and vertical (or East–West) linearly polarized components;

$U = I_{+45} - I_{-45}$ is the difference in intensities between linearly polarized components oriented at $+45°$ and $-45°$ (from the North)†;

$V = I_{rcp} - I_{lcp}$ is the difference in intensities between right and left circularly polarized components.

The first parameter I is clearly a positive number, while the latter three can be described as the "preference" of the radiation for horizontal, $+45°$ linear, and right circular polarization, respectively. If the radiation is **completely polarized** and is described as at the beginning of this section, then the Stokes parameters are:

$$I = a_x^2 + a_y^2$$

$$Q = a_x^2 - a_y^2 = I\cos2\beta\cos2\theta$$

$$U = 2a_x a_y \cos\delta = I\cos2\beta\sin2\theta$$

† Position angle in the sky is measured counterclockwise (i.e. from North to East), starting from the North.

$$V = 2a_x a_y \sin\delta = I\sin 2\beta$$

where: $\beta = \arctan\frac{b}{a}$. For this completely polarized light $I^2 = Q^2 + U^2 + V^2$, and there is some redundance among the 4 parameters.

One of the advantages of the Stokes vector is that it can also describe unpolarized or partially polarized light, since unpolarized light can be described by an electric vector that, at any instant in time, corresponds to a well defined polarization state, however fluctuates randomly between different polarization forms on a time scale that is large compared with the period of the light, but small with respect to the time interval over which the measurement is performed. Therefore, over a relatively extended period of time, all polarization biases are averaged out and the beam appears to be unpolarized. Similarly for partially polarized light, which can be regarded as the sum of a completely polarized component and an unpolarized one. This is represented in the Stokes vector by taking a time average of the 4 parameters:

$$I = \langle I \rangle = \langle a_x^2 + a_y^2 \rangle$$

$$Q = \langle I_0 - I_{90} \rangle = \langle a_x^2 - a_y^2 \rangle$$

$$U = \langle I_{+45} - I_{-45} \rangle = \langle 2a_x a_y \cos\delta \rangle$$

$$V = \langle I_{rcp} - I_{lcp} \rangle = \langle 2a_x a_y \sin\delta \rangle.$$

This is a more rigorous and general definition of the Stokes parameters, since it applies to all types of radiation, completely–, partially–, or un–polarized. For unpolarized light the parameters Q, U, and V are zero, while for partially polarized light: $0 < (Q^2+U^2+V^2) < I^2$.

It is customary to describe polarization in terms of its degree and position angle. In general, the degree of polarization is defined as:

$$P = \frac{\sqrt{Q^2 + U^2 + V^2}}{I}.$$

Clearly the degree of polarization is equal to 1 (or 100%) for completely polarized radiation. The degree of linear polarization and the position angle of the **E** vector† are:

$$P_l = \frac{\sqrt{Q^2 + U^2}}{I}$$

$$\theta = \frac{1}{2}\arctan\frac{U}{Q}.$$

Finally the degree of ellipticity is:

$$P_e = \frac{V}{I}.$$

If $V = 0$ the radiation is linearly polarized; then:

$$P = P_l = \frac{I_{max} - I_{min}}{I_{max} + I_{min}}$$

where I_{max} and I_{min} are the maximum and minimum intensities transmitted by a 'perfect' linear polarizer when it is rotated.

Some authors use the normalized Stokes vector, for which the first parameter I remains the same and the others are defined as: $Q_n = Q/I$, $U_n = U/I$, $V_n = V/I$. These normalized parameters are more easily related to the degrees of polarization (e.g. $P_l = \sqrt{Q_n^2 + U_n^2}$).

---

† θ, being the position angle of the direction of vibration of the **E** vector, is only defined between $0°$ and $180°$.

2.1.2. *An operational definition*

For those of you who like to visualize physical phenomena I offer the following operational definition of the Stokes parameters. Let us take four filters, all with a transmission of 50% for unpolarized light, which have the following characteristics:

F_1 always has the same transmission of 50%, independently of the state of polarization of the incoming light.

F_2 is completely opaque for linearly polarized light with $\theta = 90°$.

F_3 is completely opaque for linearly polarized light with $\theta = -45°$.

F_4 is completely opaque for left circularly polarized light.

In order to measure the Stokes parameters of some incoming light, we analyze it with the four filters in turn and we measure the radiation transmitted after the filters with intensities J_1, J_2, J_3, and J_4. Then we simply obtain the Stokes parameters as follows:

$$I = 2J_1$$

$$Q = 2(J_2 - J_1)$$

$$U = 2(J_3 - J_1)$$

$$V = 2(J_4 - J_1)$$

2.1.3. *Examples of Stokes parameters*

We give here some examples of Stokes vectors for the most usual types of radiation. For simplicity we always take a total intensity $I = 1$. The angle θ is the position angle in the sky of the plane of vibration of the **E** vector.

$\mathbf{S} = \{1,\ 1,\ 0,\ 0\}$ for light completely linearly polarized North–South ($\theta = 0°$).

$\mathbf{S} = \{1,\ -1,\ 0,\ 0\}$ for light completely linearly polarized East–West ($\theta = 90°$).

$\mathbf{S} = \{1,\ 0,\ 1,\ 0\}$ for light completely linearly polarized with $\theta = 45°$.

$\mathbf{S} = \{1,\ 0,\ -1,\ 0\}$ for light completely linearly polarized with $\theta = -45°$.

$\mathbf{S} = \{1,\ cos2\theta,\ sin2\theta,\ 0\}$ for light completely linearly polarized at an arbitrary θ.

$\mathbf{S} = \{1,\ 0,\ 0,\ 1\}$ for light completely right circularly polarized.

$\mathbf{S} = \{1,\ 0,\ 0,\ -1\}$ for light completely left circularly polarized.

$\mathbf{S} = \{1,\ 0,\ 0,\ 0\}$ for natural, unpolarized light.

2.1.4. *Properties of the Stokes parameters*

We recall that the four Stokes parameters are real numbers and that they have the dimensions of a light intensity. An important property is the additivity of the Stokes parameters:

If a beam of light results from the combination of two incoherent beams, then its Stokes parameters are the vector sum of the parameters of the single beams:

$I = I_1 + I_2$; $Q = Q_1 + Q_2$; $U = U_1 + U_2$; $V = V_1 + V_2$.

For example each beam of light described by the Stokes vector $\mathbf{S} = \{I, Q, U, V\}$ can be decomposed in the sum of an unpolarized beam $\mathbf{S}_u = \{I_u, 0, 0, 0\}$ and in a completely polarized beam $\mathbf{S}_p = \{I_p, Q, U, V\}$, where $I_p = \sqrt{Q^2 + U^2 + V^2}$ and $I_u = I - I_p$.

The importance of the Stokes parameters is emphasized by the principle of optical equivalence:

It is impossible by means of any instruments to distinguish between various incoherent sums of simple waves that may together form a beam with the same Stokes parameters.

The principle of optical equivalence shows that the Stokes parameters are not merely some interesting quantities, but the complete set of quantities that are needed to characterize the intensity and state of polarization of a beam of light, inasmuch as it is subject to practical analysis. Further distinctions are theoretically possible but do not correspond to measurable differences (van de Hulst, 1957).

2.2. *The foundations of Mueller calculus*

Any optical component, which changes the intensity and the state of polarization of a beam of radiation without introducing coherent effects, produces a linear transformation on the Stokes vector representing the beam, and can therefore be described as a 4×4 matrix, called the Mueller matrix. So if **S** is the Stokes vector describing the radiation beam incident on the optical component represented by the Mueller matrix **M**, then the Stokes vector of the emerging beam is given by:

$$S' = M \cdot S$$

The effect of several successive optical components (e.g.an instrument) can be represented by the product of the Mueller matrices of the single components. Mueller calculus is therefore very useful in designing complex instruments for polarimetry.

2.3. *Polarizing devices*

In the following I shall discuss the main kinds of optical components which change the state of polarization, and give some examples of them and of their Mueller matrices.

2.3.1. *Linear polarizers*

Linear polarizers are optical devices which extract from the incoming beam the component linearly polarized along an axis fixed with respect to the polarizer itself. A perfect linear polarizer with its axis oriented at the position angle θ can be described by the following Mueller matrix:

$$M = \frac{1}{2} \begin{pmatrix} 1 & cos2\theta & sin2\theta & 0 \\ cos2\theta & cos^2 2\theta & cos2\theta sin2\theta & 0 \\ sin2\theta & cos2\theta sin2\theta & sin^2 2\theta & 0 \\ 0 & 0 & 0 & 0 \end{pmatrix}$$

For example a perfect linear polarizer with $\theta = 0$ acts as follows:

$$\frac{1}{2} \begin{pmatrix} 1 & 1 & 0 & 0 \\ 1 & 1 & 0 & 0 \\ 0 & 0 & 0 & 0 \\ 0 & 0 & 0 & 0 \end{pmatrix} \cdot \begin{pmatrix} I \\ Q \\ U \\ V \end{pmatrix} = \begin{pmatrix} \frac{I+Q}{2} \\ \frac{I+Q}{2} \\ 0 \\ 0 \end{pmatrix}$$

Let $T_{\parallel}$ and $T_{\perp}$ be the transmission of a linear polarizer for linearly polarized light parallel and perpendicular, respectively, to its axis. A perfect linear polarizer has $T_{\parallel} = 1$ and $T_{\perp} = 0$. Any real polarizer, of course, is not so good: its quality is measured by the extinction ratio $R_e = \frac{T_{\perp}}{T_{\parallel}}$, which should be as small as possible.

The most common kind of linear polarizers are the dichroic sheet polarizers. In dichroic materials the molecules are oriented such that their dipole transition moments are aligned along a specific axis. The light polarized along this axis is absorbed, while that polarized perpendicularly is transmitted. The extinction ratio of dichroic polarizers is of the order of 10^{-4}. The so–called Polaroid sheets are commercially available and are good between 400 and 700 nm.

Considerably better are the polarizers made of birefringent materials. Birefringence is the property of some crystals, like calcite, which have a different index of refraction for light polarized parallel and perpendicular to their optical axis. Therefore after refraction a beam of unpolarized light is split into two beams which are completely polarized into

perpendicular directions and are deviated by different angles: the ordinary ray is polarized perpendicular to the axis of the crystal, while the extraordinary ray is polarized
parallel to it. Birefringent polarizers are better than the dichroic ones, because, if the
crystal is pure, then the polarization of the ordinary and extraordinary beams is rather
complete over a large wavelength range.

One kind of birefringent polarizers is the polarizing prisms, like the Glan–Taylor and
the Glan–Thompson prisms (see Figure 4), which use the total reflection of the ordinary
ray, since this has a higher index of refraction. The extraordinary ray is transmitted
undeviated. This type of polarizer has a very good extinction ratio, smaller than 10^{-5},
but they have a small acceptance angle of about $10°$.

Beamsplitting polarizers are also made of birefringent material, but they have the
useful property that both the extraordinary and the ordinary ray are transmitted separated by a small angle, which makes them both available for measurement. Examples
are the Rochon and the Wollaston prisms (see Figure 4). The Wollaston prism has the
important feature that the two beams emerge deviated by the same amount, providing
a good symmetry between them. The beamsplitting polarizers have a good extinction
ratio of about 10^{-5}. They do however suffer from some chromatism, in the sense that
the separation between the two output beams depends slightly on the wavelength. This
chromatism can be reduced by using appropriate materials like magnesium fluoride.

2.3.2. Retarders

Retarders introduce a phase shift δ between the two components of the input beam
with linear polarization parallel and perpendicular to a specific axis of the retarder, which
we assume to have a position angle ρ. The two most common types of retarders are the
half–wave plate and the quarter–wave plate, which we discuss briefly because they are
very important in polarimetric instruments.

For the half–wave plate the retardation is $\delta = 180°(\pm 2n\pi)$. Its Mueller matrix is (ρ is
the position angle of the plate optical axis):

$$M = \begin{pmatrix} 1 & 0 & 0 & 0 \\ 0 & cos4\rho & sin4\rho & 0 \\ 0 & sin4\rho & -cos4\rho & 0 \\ 0 & 0 & 0 & -1 \end{pmatrix}$$

The half–wave plate does not change the degree of polarization, but rotates the plane
of polarization of linearly polarized light to the symmetric direction with respect to its
optical axis and reverses the direction of circularly polarized light (Fig. 5). If the half–
wave plate is set to rotate, then the direction of linear polarization of the outcoming light
rotates at twice the angular speed of the plate.

The quarter–wave plate has a retardation $\delta = 90°(\pm 2n\pi)$ and its Mueller matrix is:

$$M = \begin{pmatrix} 1 & 0 & 0 & 0 \\ 0 & cos^2 2\rho & sin2\rho cos2\rho & -sin2\rho \\ 0 & sin2\rho cos2\rho & sin^2 2\rho & cos2\rho \\ 0 & sin2\rho & -cos2\rho & 0 \end{pmatrix}$$

Then the quarter–wave plate transforms circular polarization into linear one and vicev-
ersa (Fig. 6), and it therefore makes possible to measure circular polarization using a
linear polarizer.

The main problem of retarders is their chromatism, since the retardation depends on
the wavelength. Therefore special materials are used to build achromatic and superachro-
matic retarders, for which the deviations in phase shifts are limited to a few degrees over
a wavelength range from 400 to 800 nm.

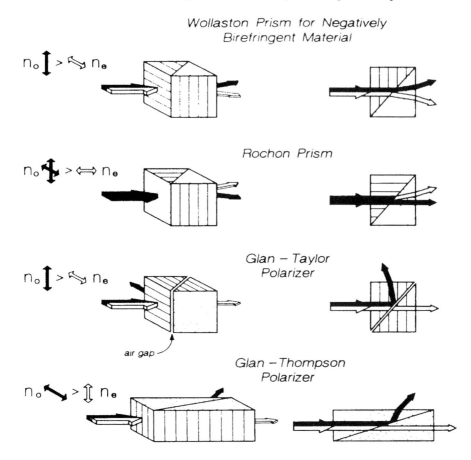

FIGURE 4. Some examples of polarizers made of two wedges of birefringent material such as calcite, which has an index of refraction for the ordinary ray (n_o) larger than that for the extraordinary one (n_e). The lines in the wedges show the direction of the optical axis of the crystal (from Kliger, Lewis, & Randall, 1990).

2.3.3. *Depolarizers*

The purpose of depolarizers is to transform any kind of polarized radiation to an unpolarized one. Obviously the Mueller matrix of a perfect depolarizer is:

$$M = \begin{pmatrix} 1 & 0 & 0 & 0 \\ 0 & 0 & 0 & 0 \\ 0 & 0 & 0 & 0 \\ 0 & 0 & 0 & 0 \end{pmatrix}$$

However real depolarizers do not exist, and can only be approximated by pseudo–depolarizers, for which the outcoming radiation is unpolarized only if it is averaged over some wavelength range, some period of time, or some area. For example the Lyot depolarizer averages over wavelength: it consists of two retarder plates with very large retardation, much larger than 360°, and with $\delta_2 = 2\delta_1$ and $\rho_2 = \rho_1 + 45°$. The polarization of the output beam varies very rapidly with wavelength, and by averaging over a few tens of nanometers the polarization is reduced to 1% of its input value. In astron-

INPUT **OUTPUT**

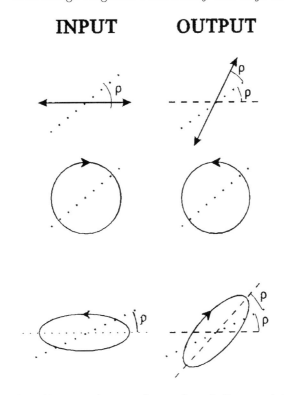

FIGURE 5. Examples of input and output beams for a half–wave plate. The dotted line shows the optical axis of the plate, while the thick continuous line draws the path of the tip of the **E** vector projected on the xy plane for an electromagnetic wave proceeding along the z axis.

omy, where long integrations are common, one can greatly reduce linear polarization by rapidly rotating a half–wave plate and averaging over time.

2.3.4. An example

Let us take two (perfect) linear polarizers and orient them so that their axes are perpendicular to each other ($\theta_1 = 90°$, $\theta_2 = 0°$). We all know that they are completely opaque. The fact that no light is transmitted is also shown by the relevant Mueller matrix:

$$M = \frac{1}{2}\begin{pmatrix} 1 & 1 & 0 & 0 \\ 1 & 1 & 0 & 0 \\ 0 & 0 & 0 & 0 \\ 0 & 0 & 0 & 0 \end{pmatrix} \cdot \frac{1}{2}\begin{pmatrix} 1 & -1 & 0 & 0 \\ -1 & 1 & 0 & 0 \\ 0 & 0 & 0 & 0 \\ 0 & 0 & 0 & 0 \end{pmatrix} = \frac{1}{4}\begin{pmatrix} 0 & 0 & 0 & 0 \\ 0 & 0 & 0 & 0 \\ 0 & 0 & 0 & 0 \\ 0 & 0 & 0 & 0 \end{pmatrix}$$

If we now insert between these two linear polarizers a third one with its axis oriented such that $\theta_3 = 45°$, some light is transmitted, as shown by the Mueller matrix:

$$M = \frac{1}{2}\begin{pmatrix} 1 & 1 & 0 & 0 \\ 1 & 1 & 0 & 0 \\ 0 & 0 & 0 & 0 \\ 0 & 0 & 0 & 0 \end{pmatrix} \cdot \frac{1}{2}\begin{pmatrix} 1 & 0 & 1 & 0 \\ 0 & 0 & 0 & 0 \\ 1 & 0 & 1 & 0 \\ 0 & 0 & 0 & 0 \end{pmatrix} \cdot \frac{1}{2}\begin{pmatrix} 1 & -1 & 0 & 0 \\ -1 & 1 & 0 & 0 \\ 0 & 0 & 0 & 0 \\ 0 & 0 & 0 & 0 \end{pmatrix} =$$

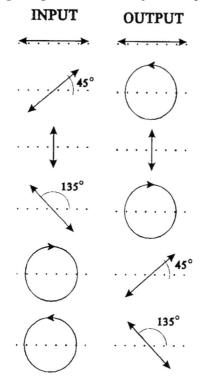

FIGURE 6. Examples of input and output beams for a quarter–wave plate. The symbols are
the same as in Figure 5.

$$\frac{1}{4}\begin{pmatrix} 1 & 0 & 1 & 0 \\ 1 & 0 & 1 & 0 \\ 0 & 0 & 0 & 0 \\ 0 & 0 & 0 & 0 \end{pmatrix} \cdot \frac{1}{2}\begin{pmatrix} 1 & -1 & 0 & 0 \\ -1 & 1 & 0 & 0 \\ 0 & 0 & 0 & 0 \\ 0 & 0 & 0 & 0 \end{pmatrix} = \frac{1}{8}\begin{pmatrix} 1 & -1 & 0 & 0 \\ 1 & -1 & 0 & 0 \\ 0 & 0 & 0 & 0 \\ 0 & 0 & 0 & 0 \end{pmatrix}$$

The transmitted light is linearly polarized at the position angle $0°$, like the orientation
of the **last** polarizer:

$$\frac{1}{8}\begin{pmatrix} 1 & -1 & 0 & 0 \\ 1 & -1 & 0 & 0 \\ 0 & 0 & 0 & 0 \\ 0 & 0 & 0 & 0 \end{pmatrix} \cdot \begin{pmatrix} I \\ Q \\ U \\ V \end{pmatrix} = \begin{pmatrix} \frac{I-Q}{8} \\ \frac{I-Q}{8} \\ 0 \\ 0 \end{pmatrix}$$

This simple experiment, apart from giving to the unaware the magic impression that
we are able to revive light, demonstrates two important facts. First that a linear polarizer
does not merely act statistically on the incoming photons by letting through only those
that are appropriately oriented, but does actually change the state of polarization of single
photons (the polarization of photons is discussed in Chapter 1 of Dirac, 1958). Second
that the polarizing components are not commutable: in fact, if the third polarizer is
inserted before or after the first two, no light is transmitted.

3. Polarizing Mechanisms in Astrophysics

This section is intended to provide a basis for understanding the variety of physical
mechanisms which produce polarization in the radiation of celestial objects. The purpose

is to draw your attention to the information on astronomical objects which can be gained by measuring polarization. However, because of its brevity, the following discussion is not necessarily complete and leaves much room for deeper studies to the interested researcher.

3.1. *Reflection*

We have already seen that the reflection from solid surfaces produces linear polarization in everyday life (section 1.1.2). The same is true in astrophysics: well known examples occur particularly in the Solar System, which has a variety of solid bodies reflecting the light of the Sun, like the Moon, the planets and the asteroids. The reflected light is linearly polarized with a degree that can approach unity, and depends on the detailed structure and composition of the surface. The **E** vector of the reflected light always vibrates perpendicularly to the plane of reflection, i.e. the plane which contains both the incident and the reflected rays. This property can provide useful information on the orientation of surfaces which are not easily resolved from the Earth.

3.2. *Scattering*

Large solid bodies are not easily seen in reflection outside the Solar System, because the reflected light is overwhelmed by the direct light emitted by the objects which the solid bodies should reflect. Much easier is to see diffuse particles of gas and dust which scatter light from other sources, probably because the high scattering efficiency per unit mass makes them visible at large distances from the original source of light. Therefore scattering is a very common phenomenon in astrophysics, and is present in different varieties which we examine in turn.

The polarizing properties of scattering by a distribution of spherical or randomly oriented particles can be described with the help of a Mueller matrix, called the "phase matrix", relating the incident Stokes vector S_i and the scattered one S_s (White, 1979):

$$\begin{pmatrix} I_s \\ Q_s \\ U_s \\ V_s \end{pmatrix} = \begin{pmatrix} P_1 & P_2 & 0 & 0 \\ P_2 & P_1 & 0 & 0 \\ 0 & 0 & P_3 & -P_4 \\ 0 & 0 & P_4 & P_3 \end{pmatrix} \cdot \begin{pmatrix} I_i \\ Q_i \\ U_i \\ V_i \end{pmatrix}$$

The quantities P_i are all functions of the scattering angle χ. P_1 is the so called "phase function", the quantity which is used to characterize scattering when one is concerned with intensity alone. P_2/P_1 is the degree of linear polarization of the scattered light for unpolarized incident light. Conventionally the scattering plane, i.e. the plane containing the incident and scattered rays, is along the x axis (the North–South direction), which means that P_2 is negative for light polarized perpendicular to the scattering plane, and positive for light polarized parallel to that plane. P_3 represents a term for scattering of incident light polarized at $45°$ from the scattering plane, while P_4/P_1 is the fraction of incident light polarized at $45°$ which is converted into circular polarization. In the following sections we will examine the elements of the phase matrix for some types of scattering.

3.2.1. *Electron Scattering*

Free electrons scatter radiation with a cross section which is independent of wavelength, called the Thomson cross section:

$$\sigma_T = \frac{8\pi}{3} \left(\frac{e^2}{mc^2} \right)^2 = 0.665 \times 10^{-24} cm^2$$

The scattered radiation is linearly polarized and, if χ is the scattering angle, then the elements of the phase matrix are:

$$P_1 = \frac{1}{2}(cos^2\chi + 1); \quad P_2 = \frac{1}{2}(cos^2\chi - 1); \quad P_3 = cos\chi; \quad P_4 = 0$$

Therefore the degree of linear polarization of the electron scattered light for unpolarized incident light is:

$$P_l = \frac{sin^2\chi}{1 + cos^2\chi}$$

which reaches 100% for 90° scattering. However in real situations scattering is produced by many electrons which do not all scatter exactly at 90°: therefore the observed polarization never reaches 100%. As in reflection, the direction of the scattered $\mathbf{E}$ vector is perpendicular to the scattering plane ($P_2 < 0$). Electron scattering does not produce circular polarization ($P_4 = 0$). It occurs whenever there are large quantities of ionized gas in the vicinity of a bright source, like in the solar corona, in the envelopes around early type stars, and in the narrow line region of active galactic nuclei.

As the cross section does not vary with wavelength, the scattered spectrum reproduces the incident one. However, if the temperature is high, the thermal motion of the electrons will produce a smearing of the incident spectrum. The Doppler–broadening produced by electrons at temperature T, scattering at an angle χ, is:

$$\Delta v_D = 41600\sqrt{(1 - cos\chi)\frac{T}{7 \times 10^7 K}} \; km \; s^{-1}$$

The Doppler–broadening can therefore be important in hot gases like those emitting the X–ray halos in cluster of galaxies.

3.2.2. *Rayleigh Scattering*

If the scattering particles are not electrons, but have a size which is much smaller than the wavelength ($a \ll \lambda$), like molecules and small dust grains, then one has the so–called Rayleigh scattering, which has the peculiarity that the cross section depends strongly on the wavelength:

$$\sigma_R = \frac{8\pi}{3}\left(\frac{2\pi}{\lambda}\right)^4 |\alpha^2|$$

where $|\alpha^2|$ is a geometric factor which depends on the size and shape of the particle. Therefore the scattered radiation F_d is much bluer than the incident one F_i. For example, if the incident radiation is described by a power law:

$$F_i(v) \propto v^{-\alpha} \quad \text{then} \quad F_d(v) \propto v^{-\alpha+4}$$

The phase matrix for Rayleigh scattering is equal to that for electron scattering (see Section 3.2.1): therefore the scattered radiation is linearly polarized and the degree of polarization depends on the scattering angle like in electron scattering. The $\mathbf{E}$ vector is again perpendicular to the scattering plane. Astrophysical examples of Rayleigh scattering occur in the Solar limb, in the Zodiacal light, in the diffuse Galactic light and in reflection nebulae, particularly in the IR.

3.2.3. *Mie Scattering*

When the size of the scattering particles becomes comparable with the wavelength, the approximations of Rayleigh are not valid and one has to apply the Mie theory. The scattered light becomes less blue, but it is still strongly polarized with the $\mathbf{E}$ vector perpendicular to the scattering plane. A very good discussion of Mie scattering is given by van de Hulst (1957) and cannot be reported here (see also White, 1979, for the phase matrix and section 8). Mie theory applies to scattering from dust grains with the normal size distribution for cosmic dust, which occurs in a variety of astrophysical situations, from the Solar system to distant galaxies and active Galactic nuclei (e.g. Manzini & di Serego Alighieri, 1996).

I would like to stress that scattering and reflection, whatever the detailed mechanisms are, always produce light which is strongly linearly polarized with the **E** vector perpendicular to the reflection/scattering plane. This characteristic makes it possible to safely diagnose scattering and to use it for disentangling unresolved geometries.

3.2.4. *Interstellar polarization*

It is well known that often the light of stars which has travelled across the interstellar medium, is polarized. This is due to the forward scattering of the radiation by non-spherical dust grains which are aligned by the Galactic magnetic field. The interstellar polarization is essentially linear ($V \lesssim 10^{-4}$) and is usually smaller than $\sim 5\%$. The **E** vector is parallel to the plane containing the Galactic magnetic field vector **B** and the line of sight.

The polarization varies with wavelength and reaches a maximum value P_{max} for a wavelength λ_{max} in the visible. The wavelength dependence has been found empirically to follow "Serkowski's law" (Fig. 7):

$$\frac{P(\lambda)}{P_{max}} = e^{-K \ln^2 \frac{\lambda_{max}}{\lambda}}.$$

The quantity K was originally taken to be 1.15 (Serkowski, Mathewson & Ford, 1975), but was later found to be better fitted by (Wilking, Lebofski, & Rieke 1982):

$$K = -0.10 + 1.86\lambda_{max}(\mu m).$$

Therefore Serkowski's curve becomes broader for smaller λ_{max}. P_{max} is associated with the dust column density and with the degree of alignment of the interstellar grains with the magnetic field, while λ_{max} is linked to the average dimension of the polarizing grains.

In more detail, λ_{max} can vary between 340 and 1000 nm and its average value is 550 nm. It has been found that λ_{max} is proportional to the total–to–selective extinction ratio $R_V = A_V/E_{B-V}$ and that:

$$P_{max}(\%) \leq 9.0 E_{B-V}.$$

This empirical relationship is useful to put upper limits to the contribution of Galactic interstellar polarization to the linear polarization of extragalactic objects. The study of the interstellar polarization gives important constraints on the interstellar medium and on the diffuse Galactic magnetic field (for more details see Whittet, 1992).

3.3. *Zeeman Effect*

This effect deals with the linear and circular polarization of spectral lines emitted in a magnetic field, as for example in Sunspots and in magnetic stars. When the line of sight is perpendicular to the magnetic field at the source (transverse Zeeman effect) an emission line is separated into three linearly polarized components. The **E** vector of the central one is parallel to the magnetic field, while that of the side ones is perpendicular. On the other hand, when the line of sight is parallel to the magnetic field (longitudinal Zeeman effect), an emission line is split into two circularly polarized components, while the central one is missing. The split in wavelength between the lateral components and the central one is proportional to the magnetic field B (Landi Degl'Innocenti, 1992):

$$\Delta\lambda_Z(\mathring{A}) = 4.67 \times 10^{-13} B(Gauss)\lambda_0^2(\mathring{A})$$

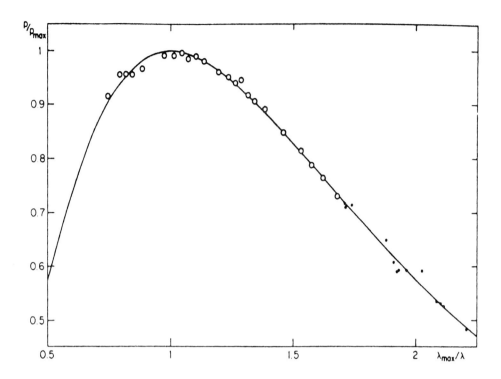

FIGURE 7. Serkowski's curve describing the wavelength dependence of interstellar polarization (from Serkowski et al., 1975).

where λ_0 is the wavelength of the central component. Therefore observation of the Zeeman effect and of the associated polarization provides important information on the strength and orientation of the magnetic field at the source.

3.4. Synchrotron Emission

Synchrotron radiation is emitted by charged particles (mostly electrons) accelerated by a magnetic field. This radiation is strongly linearly polarized ($P_l \lesssim 75\%$) with the **E** vector parallel to the particle orbital plane, and therefore perpendicular to the local magnetic field. Because of the normal energy distribution of electrons, synchrotron radiation is particularly strong at radio wavelengths. However its tail is often extended to the visible range (for more details see Rybicki & Lightman, 1979). This is observed in supernova remnants, pulsars, radio galaxies and quasars.

4. Photopolarimetry

The classical way of measuring polarization in astronomical objects is by means of a photopolarimeter. This instrument measures the polarization of the light falling over an aperture at the telescope focal plane by means of a polarization analyzer and a photomultiplier. The basic scheme of a photopolarimeter is shown in Figure 8. The light of the object (and the surrounding sky) passes through an aperture, then through a rotatable linear polarizer, which lets through only the linearly polarized component in a given direction, then through a band selecting filter, and finally is refocused on a photomultiplier. If the object is linearly polarized, by rotating the polarizer the light intensity recorded by the photomultiplier varies sinusoidally, going through a maximum (I_{max}) and a minimum (I_{min}). To first approximation the degree of linear polarization is given by:

$$P_l = \frac{I_{max} - I_{min}}{I_{max} + I_{min}}.$$

There are several possible improvements, like adding a depolarizer in front of the photomultiplier to avoid spurious effects due to the fact that the sensitivity of the photocathode may depend on the state of polarization, inserting a compensator for the instrumental polarization †, or adding a quarter–wave plate in front of the polarizer, to measure circular polarization.

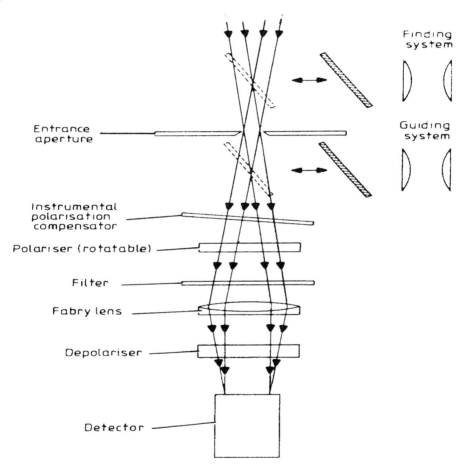

FIGURE 8. The basic scheme of a photopolarimeter (from Kitchin, 1984).

However there are two major problems connected with this simple scheme for a photopolarimeter. The first concerns the sky background, which must be subtracted. This subtraction is complicated by the fact that this background is strongly polarized, since it is mostly scattered light. This problem is solved by frequently switching to a pure sky measurement or by having a simultaneous sky measurement via a parallel channel. The

† Instrumental polarization results from all unwanted polarizing elements in the optical path, like the atmosphere when you are not observing at zenith, non–normal reflections, gratings, or even a non uniformly reflecting primary or secondary mirror.

second problem arises from the fact that the polarization results from intensity measurements taken at different times, and therefore is adversely affected by variations in the sky transparency and in the source itself. This difficulty is solved by using a beam splitting analyzer and measuring both output beams. In this way the polarization is obtained from *ratios* of intensities measured simultaneously and therefore independent of source and sky variations. When things are done properly a photopolarimeter can measure P with an accuracy of the order of 0.01%. However its use is limited to sources which are fairly bright compared to the sky falling over the aperture.

5. Imaging Polarimetry

As explained in the previous section, polarimetry results from ratios of very accurate photometric measurements. With modern large telescopes the most accurate photometric measurements, particularly on sources which are faint compared with the sky background, are made using CCD cameras, since the CCDs have a very high sensitivity and a very uniform response over an array of many pixels, which allows a very accurate a–posteriori sky subtraction. They also have a wide dynamic range and a large field, and therefore calibration sources can be found in the same frame as the object under study.

In analogy to modern CCD photometry, the best polarimetry of faint objects is now performed using CCD imaging polarimeters. The implementation of polarizers in front of a CCD is simplified by the recent availability at the major observatories of focal–reducer type CCD instruments, like for example EFOSC, the ESO Faint Object Spectrograph and Camera (Melnick, Dekker, & D'Odorico, 1989). These instruments often use lenses aligned in a single train, and therefore, if used at the Cassegrain focus, have very small instrumental polarization. Imaging polarimetry has also the advantage of easily providing polarization maps for extended sources (for an interesting application of polarization maps see Section 8.2.2). In the following I shall describe the imaging–polarimetry mode of EFOSC, as a very good example of a modern polarimeter.

5.1. *An example of imaging polarimeter*

EFOSC is essentially made of an aperture wheel at the telescope focal plane, a collimator producing a parallel beam, a filter and a grism wheel to put optical elements in the parallel beam, and a camera which refocuses the light on a CCD (Fig. 9). EFOSC becomes an imaging polarimeter by inserting a Wollaston prism in the grism wheel and a special mask in the focal plane. The prism splits the incoming parallel beam into two beams which are perpendicularly polarized. Therefore each object image at the focal plane of the telescope field produces two images with perpendicular polarizations on the output CCD image. Two Wollaston prisms are available in EFOSC, producing images split by 10 or 20 arcsec at the output. In order to avoid confusion and increased sky, one can insert at the telescope focal plane a mask with alternating transparent and opaque parallel strips with a width corresponding to the splitting. This mask produces the loss of half of the field of view. Therefore a polarization image of the whole field at one position angle requires two sets of exposures with the telescope moved by the image split (10 or 20 arcsec), in order to recover the part of the field occulted by the mask. Nevertheless, in many cases, the object under study, like for example a point source or a distant galaxy, is well contained in one strip and one exposure for each position angle is sufficient. A filter can be mounted in the filter wheel to select the band over which polarization is measured. One image with the instrument at the position angle ϕ gives the intensity of the linearly polarized components both with the E vector in the direction of ϕ and with the E vector in the direction of $\phi + 90°$. This is obtained for all the objects (and the sky)

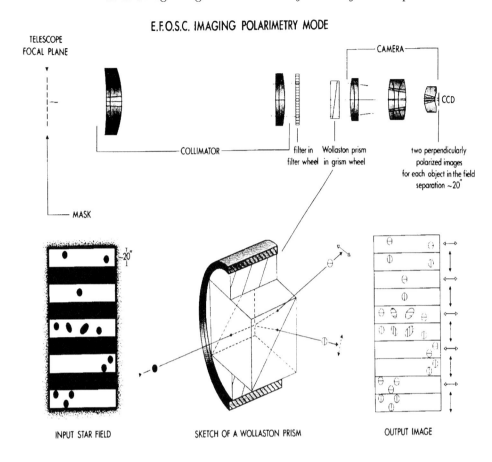

FIGURE 9. The optical configuration and an example of input field and output image for the imaging polarimetry mode of EFOSC

in the transparent parts of the mask. In order to obtain P and θ it is necessary to take at least two images with the instrument rotated† by 45°, for example with $\phi = 0°$ and $\phi = 45°$. Then the Stokes parameters and hence P and θ can be obtained simply from the intensities of the object with the E vector in the four position angles:

$$Q = i(0°) - i(90°) \ , \quad U = i(45°) - i(135°)$$

$$I = i(0°) + i(90°) = i(45°) + i(135°).$$

This simplified algorithm neglects the instrumental polarization and the possible imperfections in the Wollaston prism. More accurately, for each position angle ϕ, a parameter $S(\phi)$ is derived from the intensities of the object $i(\phi)$ and $i(\phi+90°)$ and from the average intensities of field stars, which are assumed to be unpolarized, $i^u(\phi)$ and $i^u(\phi + 90°)$:

$$S(\phi) = \left(\frac{i(\phi)/i(\phi + 90°)}{i^u(\phi)/i^u(\phi + 90°)} - 1 \right) \bigg/ \left(\frac{i(\phi)/i(\phi + 90°)}{i^u(\phi)/i^u(\phi + 90°)} + 1 \right).$$

† Actually what one really needs to rotate is the Wollaston prism; however since the mask has also to be rotated by the same amount to work properly, it is easier to rotate the whole of EFOSC using the telescope rotator/adapter, which already has that rotation capability.

$S(\phi)$ can be regarded as the component of the normalized Stokes vector describing linear polarization along the direction ϕ (for example $S(\phi = 0) = Q_n$). It is related to P and θ by:

$$S(\phi) = \sqrt{\frac{T_l - T_r}{T_l + T_r}} P cos2(\theta - \phi)$$

where T_l is the transmittance of unpolarized light for one of the two beams by two identical Wollaston prisms oriented parallel and T_r is the transmittance of unpolarized light by two identical Wollaston prisms oriented perpendicularly. Again, taking two frames with, say, $\phi = 0°$ and $\phi = 45°$ one gets:

$$P = \sqrt{\frac{T_l + T_r}{T_l - T_r}} \sqrt{[S(0°)]^2 + [S(45°)]^2} \quad , \quad \theta = \frac{1}{2}arctan\left(\frac{S(45°)}{S(0°)}\right).$$

Alternatively one can take frames at more than two position angles and fit a cosine curve to $S(\phi)$ (Fig. 10). This latter method has the advantage that the deviations of the single $S(\phi)$ values from the fitted cosine curve gives an indication of the accuracy achieved. The important point to stress here is that the parameter $S(\phi)$, from which P

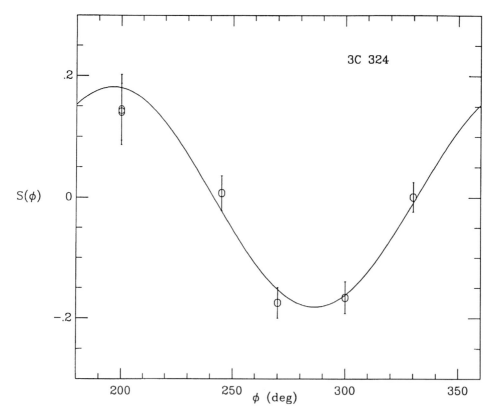

FIGURE 10. Example of the cosine curve fitted to the $S(\phi)$ data for 3C 324, a radio galaxy at $z = 1.206$ with $R = 21.7$. The fit provides the following polarization values in the R–band: $P_l = 18.0 \pm 2.6\%$ and $\theta = 16° \pm 5°$.

and θ are directly derived, is obtained from measurements made all on the same frame; therefore time dependent parameters like seeing and instrumental polarization have no effect and measurements of a polarized standard star are not necessary, provided one knows the direction of polarization for the two beams, which can be derived easily from the orientation at which the Wollaston prism is mounted and from the direction of the splitting of the two images.

For a perfect Wollaston prism $T_l = 1/2$ and $T_r = 0$. From catalogues of optical components one finds that for a good Wollaston prism $T_l \sim 0.40 - 0.47$ and $T_r \leq 10^{-5}$. Therefore $\sqrt{\frac{T_l + T_r}{T_l - T_r}}$ differs from 1 by less than 3×10^{-5} and the imperfections in the Wollaston can be neglected. The same is not true for the instrumental polarization which can be as large as $1 - 3\%$ and varies with position angle and zenith distance. The correction using field stars assumed to be unpolarized is therefore important. In some cases it is not possible to use field stars to correct for the instrumental polarization, either because there are no suitable stars in the field, or because the field stars are not necessarily unpolarized, like if one is working in fields with high extinction. In these cases the instrumental polarization has to be evaluated from observations of standard stars which are known to have a very small polarization ($P < 0.01\%$).

Because of the variation of the index of refraction with wavelength, the split produced by the Wollaston prism varies with wavelength. This produces a slight elongation of both images in the direction of the split. For EFOSC it has been computed that the elongations produced by the 20 arcsec Wollaston prism, made of calcite, are $0.58, 0.32$ and 0.17 arcsec (FWHM) for the B, V and R band respectively. The elongations produced by the 10 arcsec Wollaston, made of quartz, are $0.31, 0.11$ and 0.07 arcsec for the same bands. These elongations do not affect the measurement of $S(\phi)$ on whole objects since they are symmetric, but they may limit the angular resolution at which polarization changes can be noticed.

5.2. *Error Analysis*

The correct evaluation of the accuracy of a given polarization measurement is complicated by the fact that P, θ, Q, U, V do *not* follow a normal distribution even if they are derived from intensity measurements with a large number of photons, for which the photon statistics can be considered normal. A detailed discussion of the statistical behaviour of the polarization parameters and of the correct evaluation of the accuracy of polarization measurements is given by Clarke & Stewart (1986). Here I concentrate on the main practical aspects, particularly those connected with imaging polarimetry.

5.2.1. *The Polarization Bias*

Since the degree of polarization P is a positive definite quantity, in the presence of noise it is always overestimated, providing a biased value, and making the comparison of biased data very dangerous. This problem can be visualized as follows: let's take N measurements of the normalized Stokes parameters Q_{n_i} and U_{n_i}, which give N values of P:

$$P_i = \sqrt{Q_{n_i}^2 + U_{n_i}^2}.$$

Their average value $P_a = \frac{\Sigma P_i}{N}$ does not lie at the centre of the distribution of (Q_{n_i}, U_{n_i}) points (Fig. 11). Even an estimate of the centre (Q_{n_a}, U_{n_a}) of the distribution gives a biased value $P_c = \sqrt{Q_{n_a}^2 + U_{n_a}^2}$, since the circle with radius P_c does not divide the data in equal portions. The polarization bias can be corrected with the following formula:

$$P \sim P_{obs} \sqrt{1 - \left(\frac{\sigma_P}{P_{obs}}\right)^2}$$

where P_{obs} is the observed degree of polarization, P is the bias–corrected one, and σ_P is the r.m.s. error on the polarization (Fig. 12). For example, if $P_{obs} >> \sigma_P$ then the

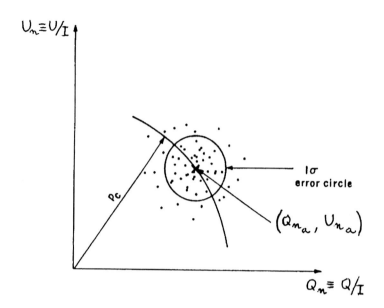

FIGURE 11. The distribution of repeated measurements of the normalized Stokes parameters Q_{n_i} and U_{n_i} (from Clarke & Stewart, 1986).

corrected P becomes equal to the observed one, while when the corrected P approaches zero the observed value does not go to zero but to σ_P.

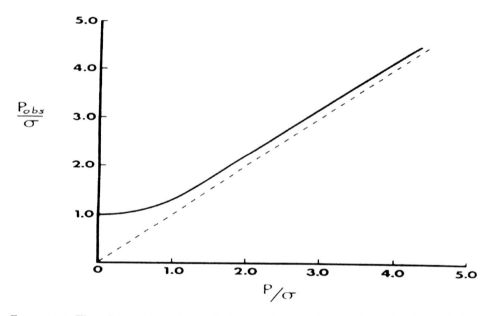

FIGURE 12. The relationship between the biased observed degree of polarization and the true one.

Since P and θ derive from the measurements in a combined way, the error on θ (σ_θ)

is related to the error on P. As a rule of thumb:

$$\sigma_\theta \sim 52^\circ \quad \text{for } P \sim 0$$

$$\sigma_\theta \sim 29^\circ \frac{\sigma_P}{P} \quad \text{for } P \gg \sigma_P$$

However the correct derivation of the error on θ should be done as explained in the next sections.

5.2.2. *The Analytic Estimate of σ_P and σ_θ*

As described in section 5.1, in the case of imaging polarimetry we first have to evaluate the errors on the intensity measurements i. All intensities are measured in total numbers of electrons counted in the CCD. The intensity of the object (i_{o+s}) is measured on an aperture containing N_{pix} pixels which includes the contribution from the sky background (i_s), hence for the object:

$$i_o = i_{o+s} - i_s.$$

For large intensities the photon statistics is normal:

$$\sigma_{i_o} = \sqrt{\sigma_{i_{o+s}}^2 + \sigma_{i_s}^2}$$

where:

$$\sigma_{i_{o+s}} = \sqrt{i_o + s N_{pix}}$$

$$\sigma_{i_s} = \frac{\sqrt{s N_{pix}}}{\sqrt{N_{ap}}}.$$

N_{ap} is the number of apertures equal in size to the object aperture, which have been used to evaluate the sky background per pixel s. Once we have obtained the errors on the intensity σ_i, we can propagate them to the errors on $S(\phi)$:

$$\sigma_{S(\phi)}^2 = \frac{4}{i^2(\phi + 90^\circ)} \left[\frac{i(\phi)}{i(\phi + 90^\circ)} + 1 \right]^{-4} \left[\sigma_{i(\phi)}^2 + \sigma_{i(\phi+90^\circ)}^2 \left(\frac{i(\phi)}{i(\phi + 90^\circ)} \right)^2 \right].$$

If $\sigma_{i(\phi)} \sim \sigma_{i(\phi+90^\circ)} = \sigma_i$, as is commonly the case, then:

$$\sigma_{S(\phi)}^2 \sim \frac{4\sigma_i^2}{i^2(\phi + 90^\circ)} \left[\frac{i(\phi)}{i(\phi + 90^\circ)} + 1 \right]^{-4} \left[1 + \left(\frac{i(\phi)}{i(\phi + 90^\circ)} \right)^2 \right].$$

These formulae are valid if the intensity of the field stars, which are used to correct for the instrumental polarization (see Sectiuon 5.1), is much larger than that of the object, so that the error on their measurement is negligible. When the values of $S(\phi)$ and of its error are known, P and θ are estimated fitting the cosine curve described in the previous section (see also Figure 10). Their errors σ_P and σ_θ are derived by means of standard least squares techniques for the non linear case.

5.2.3. *The Numerical Estimate of σ_P and σ_θ*

For non linear error propagation problems on quantities that do not follow a normal distribution, as in our case, it is convenient to build a stochastic model for the error analysis. Starting from the intensity measurements i_{o+s} and i_s and from their poissonian error distribution, and using a model of the polarimeter — giving the transformations from $i(\phi_i)$ to $S(\phi_i)$, then to P and θ — a Montecarlo simulation can be used to provide the distribution function of P and θ, and therefore a direct estimate of their errors and confidence intervals (Fosbury, Cimatti & di Serego Alighieri, 1993) (Fig. 13). In addition the simulation can be used to optimize the use of a given amount of observing time among a sequence of exposures at different angles ϕ and to estimate what can be gained using a large telescope.

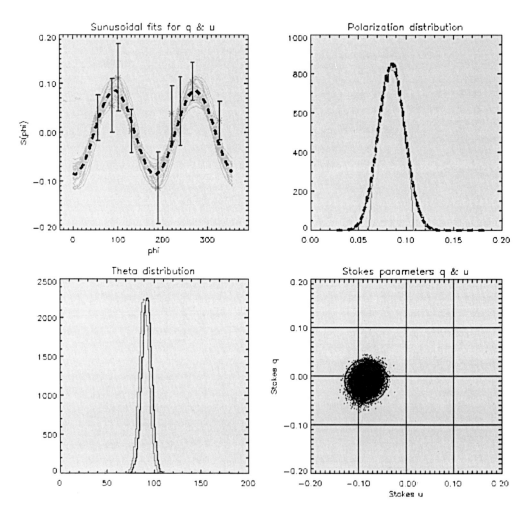

FIGURE 13. A representation of the stochastic model of imaging polarimetry. The top left panel shows the measured data for the distant radio galaxy MRC 2025-218 ($z = 2.16$) with their 1σ error bars and the fitted sinusoid (dashed line). The continuous curves represent every thousandth fit to 30000 simulated data sets with poisson noise added to sky and object counts. The top right panel shows the polarization distribution resulting from the simulations. The measurement corresponds to a $P/\sigma_P = 3.8$ and so the distribution function differs only slightly from a normal curve. The distribution function for the position angle of the **E** vector is in the bottom left panel, while the last panel shows the distribution of simulated points in the plane of the normalized Stokes parameters.

6. Spectropolarimetry

Spectropolarimetry is the technique to obtain the polarization as a function of wavelength at a spectral resolution and wavelength coverage better than with filters and imaging polarimetry, which we have seen in the previous section. The need to get the spectral variations in the polarization parameters has essentially three reasons. First, some physical mechanisms produce polarization, which has a degree changing with wavelength, like for example the Zeeman effect and the interstellar polarization (see Section 3.2.4 and 3.3). While for studying the wavelength dependence of the interstellar polarization, which has a very smooth spectral shape, imaging polarimetry with broad band filters is mostly sufficient, a much higher spectral resolution is necessary to resolve the Zeeman split of emission lines.

Another important justification for spectropolarimetry is that, even in those cases in which the original polarizing mechanism produces a polarization constant with wavelength, like electron or Rayleigh scattering, the measured Stokes parameters can vary with wavelength, if the polarization is diluted by another component with a spectral shape different from that of the polarized component. In order to understand what information can be gained by doing spectropolarimetry in this case let us consider a source which has two components in its spectrum: a polarized one $I_p(\lambda)$ with a degree of intrinsic polarization $P_p(\lambda)$, and an unpolarized one $I_u(\lambda)$. An example of this case is a Seyfert 2 galaxy, in which the bright active nucleus and the broad line region are hidden from direct view by some obscuring torus, according to the Unified Scheme for AGN. In this case the nuclear radiation and the broad lines can be seen through scattering by electrons located at the poles of the obscuring torus. Because the scattering geometry is asymmetric, the scattered light is polarized. Its spectrum (I_p) reproduces the direct nuclear spectrum that is seen in Seyfert 1 galaxies. The degree and angle of polarization $(P_p$ and $\theta_D)$ of the scattered component are constant with wavelength in this example. The observed spectrum also has a component (I_u) which comes from the stars and from the narrow line region in the galaxy and which is inevitably mixed with the scattered one and dilutes the observed polarization (Fig. 14).

If the spectrum of the unpolarized component is different from that of the direct one as in our example, the amount of dilution changes with wavelength and so does the observed degree of polarization P_{obs}:

$$P_{obs}(\lambda) = P_p(\lambda)\frac{I_p(\lambda)}{I_{tot}(\lambda)}.$$

Where $I_{tot}(\lambda) = I_p(\lambda) + I_u(\lambda)$ is the total observed spectrum. It is important to notice that one can reconstruct the spectral shape of the polarized component from the total spectrum and the observed polarization, if one has some information on the intrinsic polarization of the polarized component and on its wavelength dependence:

$$I_p(\lambda) = \frac{P_{obs}(\lambda)I_{tot}(\lambda)}{P_p(\lambda)}.$$

For example, if P_p is constant with wavelength, like for electron and Rayleigh scattering, even if its exact value is not known, the spectral shape of the scattered spectrum can be obtained just from the product of the total spectrum and the observed degree of polarization, i.e. the so called polarized flux.

As a third case of application of spectropolarimetry, let us now suppose that the observed spectrum has two polarized components $I_1(\lambda)$ and $I_2(\lambda)$ with a different degree and, in particular, a different angle of polarization, like for example a scattered and a synchrotron component. Then the observed polarization would be a combination of the

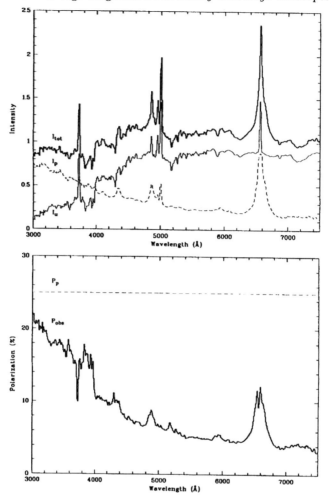

FIGURE 14. The top panel shows a spectrum (I_{tot}) which is the sum of two components: a polarized one (I_p), which has a constant degree of linear polarization P_p, and an unpolarized one (I_u). This example is a realistic representation of a Seyfert 2 or a radio galaxy, where unpolarized stellar radiation and narrow line emission dilute the polarization of a scattered type 1 spectrum. The bottom panel shows the wealth of information contained in the polarization spectrum: broad scattered lines are more evident than in the total spectrum I_{tot}, and the unpolarized narrow component causes the double peak in H_α; the stellar 4000Å break causes an obvious drop in the polarization; narrow unpolarized emission line produce an "absorption" in the polarization, except if they are also present in the scattered spectrum; stellar absorptions, like the MgI band around 5180Å, are seen in emission in the polarization. From observations of the total spectrum I_{tot} and of the polarization spectrum P_{obs}, is is possible to reconstruct the two components I_p and I_u separately (see text).

synchrotron and of the scattered polarization. The best way to quantitatively study this combination is by using the additivity property of the Stokes vectors, i.e., in case of linear polarization, plotting the two components in the Q, U plane (Fig. 15).

What I would like to stress here is that the combination of two linearly polarized components with a different spectrum produces a rotation in the polarization angle: suppose the scattered component is brighter in the blue, and the synchrotron one is brighter in the red, then the position angle would be closer to the one of the scattered component in

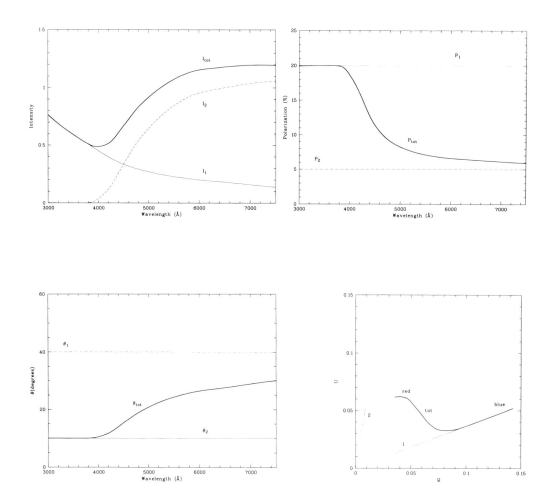

FIGURE 15. An example of spectropolarimetry of an object with two components ($I_1(\lambda)$ and $I_2(\lambda)$), which are both linearly polarized with degree of polarization P_1 and P_2 and position angle of the **E** vector θ_1 and θ_2. For simplicity we assume that the polarization degrees and angles are constant with wavelength. In the Q, U plane it is easy to see how the polarization of the two components combines, because of the additivity property of the Stokes vector (see the bottom panel): the lines marked "1", "2", and "tot" show how the tip of the Stokes vector changes with wavelength for the two components and their sum. The angles that the lines "1", "2" make with the horizontal are $2\theta_1$ and $2\theta_2$ respectively. The blue and red end of the sum are marked. The different spectral shape of the two components produces a rotation of the observed polarization angle.

the blue and would rotate closer to that of synchrotron in the red. Therefore the rotation in position angle is a clear sign of the presence of two polarized components.

In summary it is important to keep in mind that spectropolarimetry can disentangle mixed spectral components.

6.1. *An example of spectropolarimeter*

In order to illustrate the working principles of a spectropolarimeter I shall describe the spectropolarimetric mode of EFOSC. In section 5.1 we have already seen the main characteristics of EFOSC and how it can be used for imaging polarimetry.

Spectropolarimetry is implemented in EFOSC by inserting a Wollaston prism in the filter wheel, to work in combination with a grism on the grism wheel (Fig. 16). The Wollaston prism works in a similar way as for imaging polarimetry: it splits the incoming radiation into two perpendicularly polarized beams separated by a small angle. Therefore two spectra are projected on the CCD for each object in the slit and the intensity ratio between these two spectra gives the component of the Stokes vector for linear polarization in the direction of the split. The components in other directions can be obtained by rotating the whole instrument as for imaging polarimetry. However this procedure, while it may be acceptable for point sources, is certainly not acceptable for even slightly extended objects, since by rotating the slits different parts of the object are selected, and cannot be combined together to give a polarization measurement. Therefore recently a rotating half–wave plate (HWP) has been added to EFOSC as the first element in the parallel beam after the collimator. With such a device it is possible to take a full polarization measurement without rotating the slit. The usual practice is to obtain 4 spectra with the HWP at 0°, 22.5°, 45° and 67.5°, which allow the Stokes parameters Q and U to be derived.

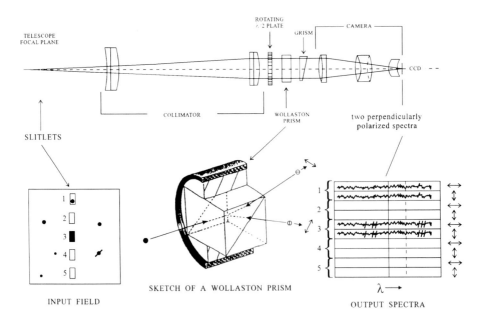

E.F.O.S.C.1 SPECTROPOLARIMETRY MODE

FIGURE 16. The spectropolarimetry mode of EFOSC1: at the top is a schematic drawing of the instrument; while at the bottom there is a sketch of the output spectra on the CCD for a given input field projected on the slitlets.

In order to avoid overlapping object and sky spectra, the slit is transparent only in slots which are as long as the separation produced by the Wollaston prism (about 20 arcsec) and are separated by opaque masks which are equally long. Therefore one obtains spectra only for sections of the sky corresponding to about half of the length of the slit. One of these sections contains the object of interest, the others are devoted to the sky, and possibly to a field star, which, if unpolarized (or assumed to be unpolarized), can be used to check the instrumental polarization. This configuration is well suited to extended objects whose dimensions are less that 20 arcsec. For larger objects only sections can be observed in a single set of 4 spectra.

The frames taken at a single HWP position angle contain two spectra of the object, referred to as the O (Ordinary) and E (Extraordinary) ray spectra, together with at least one O and E ray set of sky spectra (depending on the number of apertures along the slit and the CCD size). A determination of one Stokes parameter requires frames taken at two positions of the HWP separated by 45°: for determination of the Q Stokes parameter, from the exposure with the HWP at 0° one gets the ordinary and extraordinary spectra:

$$I_0^o(\lambda) = \frac{1}{2}(I + Q)G_o T_0$$

$$I_0^e(\lambda) = \frac{1}{2}(I - Q)G_e T_0$$

where G_o and G_e are the gains of the instrument after the Wollaston prism for the ordinary and for the extraordinary beam respectively, while T_0 is the transmission of the instrument before the Wollaston prism for the HWP set at 0°. The exposure with the HWP at 45° gives the two spectra:

$$I_{45}^o(\lambda) = \frac{1}{2}(I - Q)G_o T_{45}$$

$$I_{45}^e(\lambda) = \frac{1}{2}(I + Q)G_e T_{45}.$$

Then the normalized Q parameter can be obtained:

$$Q_n(\lambda) = \frac{R - 1}{R + 1}; \qquad R^2 = \frac{I_0^o/I_0^e}{I_{45}^o/I_{45}^e}.$$

Therefore Q_n does not depend on the gains nor on the transmissions. The other normalized Stokes parameter $U_n(\lambda)$ is similarly obtained from the exposures with the HWP at 22.5° and 67.5°. The zero point offset of the polarization position angle has to be calibrated with a laboratory measurement, or, better, with a polarization standard star. It depends somewhat on wavelength because of the chromatism of the HWP (Fig. 17).

The instrumental polarization must be calibrated using a zero polarization standard star. However our observations of zero polarization standard stars with EFOSC give: $P_{obs}(\lambda) < 0.1\%$ for $4000\text{Å} < \lambda < 9000\text{Å}$, because most of the instrumental polarization is removed by the observing procedure, combining spectra at the 4 HWP positions.

In order to assess the quality of the data and to be able to distinguish meaningful polarization measurements, depending as they do on small differences between signals, a realistic error analysis is required. Errors can be computed from the input data provided that the CCD Analog–to–Digital Units can be converted to electrons (to allow Poissonian statistics to be applied), also allowing a correct consideration of the readout noise to be included. This error handling requirement demands that dedicated software for spectropolarimetry data reduction is available.

Figure 18 shows the flow diagramme for spectropolarimetric data reduction. Flat fields must be taken with the Wollaston prism, since the CCD sensitivity is a function of polarization. In order to ensure that the input light before the Wollaston is unpolarized, a useful technique is to set the HWP continuously rotating and to have exposure times corresponding to many rotation periods: this effectively acts as a depolarizer (see section

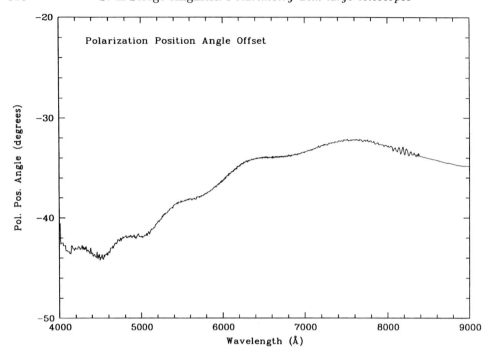

FIGURE 17. The zero point offset of the position angle of polarization, obtained with the half–wave plate in EFOSC at the ESO 3.6m telescope. It has been obtained with a blue and a red grism separately. The data for the two grisms overlap very precisely in the common region between 6000Å and 6900Å.

2.3.3). Most of the reduction steps are very similar to what is normally done for long slit spectroscopy. However I would like to stress a few points. First, it is very important that statistical errors derived from poissonian noise are propagated along the procedure: an efficient way to achieve this goal is to carry "error spectra" along with the normal signal spectra. Second, it is often necessary to rebin the data in wavelength in order to achieve a meaningful S/N ratio. However, since the Q and U Stokes parameters have non–normal error distributions (see Section 5.2), the rebinning must be done **before** computing them. Finally, since it is impossible to judge the results of spectropolarimetric observations by simply looking at the raw (or even flat–fielded) spectra, it is very important to be able to reduce the data on–line at the telescope, in order to have any significant feedback, which can positively influence subsequent observations. This reduction procedure has been implemented in the main reduction packages, like MIDAS and IRAF (Walsh, 1992).

    As an example of astronomical observations we show the results of the spectropo-larimetry of the radio galaxy 3C 195 at z=0.110 and V=17.8, obtained from 4 individual spectra of 20 minutes each (see Figure 19), taken with the ISIS spectrograph on the 4.2m William Herschel Telescope. We have also observed much fainter objects: for example we have obtained the degree of linear polarization, with an accuracy of about 2% over spectral bins about 100Å wide on the continuum, for a galaxy with V=21.3 from 8 spectra of 45 minutes each, taken with EFOSC on the 3.6m ESO telescope.

7. Polarimetry with 8–10m telescopes

    In this section I shall discuss the advantages for polarimetry of the large 8–10m telescopes. These advantages are demonstrated by the fact that all of the large telescopes

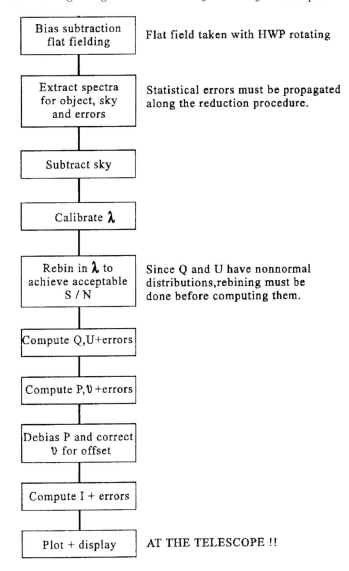

FIGURE 18. The flow diagramme of the reduction of spectropolarimetric data.

built in the last decade of this millennium have, or plan to have, polarimetric instruments. I will also describe some of these instruments.

7.1. *The gains of larger telescopes*

It is very instructive to understand in general, independently of polarimetry, the gains in limiting magnitude to be expected, i.e. in the faintest flux which is observable with a given signal–to–noise ratio (S/N), when one increases the diameter D of the telescope. For CCD observations of point sources (or of unresolved parts of extended objects) the S/N is given by:

$$S/N = \frac{(\pi/4)D^2 F_o E \Delta \lambda t}{\sqrt{(\pi/4)D^2 F_o E \Delta \lambda t + (\pi/4)D^2 S_s (\pi/4)\theta^2 E \Delta \lambda t + (r_n^2 + Tt)n}},$$

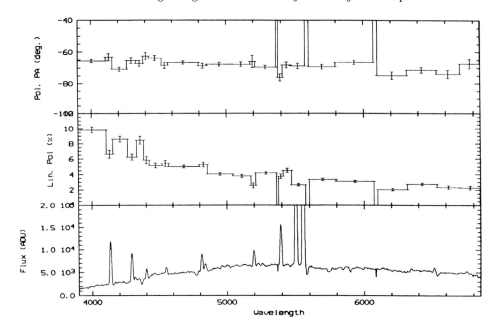

FIGURE 19. Results of spectropolarimetry of the radio galaxy 3C 195. It also shows the output format of the on–line reduction at the telescope.

where F_o is the photon flux density from the point source, S_s is the photon surface brightness (density) of the sky background, E is the total throughput (in counts per photon) including atmosphere, telescope, instrument and detector, $\Delta\lambda$ is the bandwidth, t is the exposure time, θ is the diameter (FWHM) of the seeing disk, r_n is the CCD readout noise, T is the rate of production of thermal electrons per pixel and n is the number of pixels illuminated by the PSF.

If the readout noise and the thermal noise of the CCD are negligible, as is the case for imaging with modern CCDs, particularly on a large telescope, then:

$$S/N = \frac{DF_o\sqrt{(\pi/4)E\Delta\lambda t}}{\sqrt{F_o + S_s(\pi/4)\theta^2}},$$

which shows that S/N increases linearly with telescope diameter and with the square root of exposure time. More interesting is to examine the dependence of the limiting flux on the observational parameters. This can be done easily in two cases:

(1) if the flux from the source is much smaller than the flux from the sky within the PSF ($F_o \ll S_s(\pi/4)\theta^2$, the sky–limited condition), then:

$$F_{lim} \propto \frac{(S/N)\theta\sqrt{S_s}}{D\sqrt{E\Delta\lambda t}};$$

(2) if the flux from the source is much larger than the flux from the sky within the PSF (photon–limited condition, which is the case for example when a very high S/N is needed), then:

$$F_{lim} \propto \frac{(S/N)^2}{D^2 E\Delta\lambda t}.$$

First we note that the source flux separating the two cases does not depend on telescope diameter. In the sky–limited condition the limiting flux decreases inversely with telescope diameter, while it decreases inversely with the square of the diameter in the photon–limited condition. For example a single 8m VLT† unit and the whole VLT array should reach 0.87 and 1.62 mag. fainter than the 3.6m respectively in the sky limited condition, and should reach 1.73 and 3.24 mag. fainter in the photon limited case. Therefore the gains of a larger telescope are higher for those applications that require a high S/N, like high resolution spectroscopy and polarimetry. Many scientists discuss advantages of large telescopes for the IR and, most appropriately, all large telescope projects foresee the use in the IR. However we should not forget that in the IR one is mostly working in the sky–limited condition, where the gains of larger telescopes are smallest. These gains are much larger for polarimetry, which must therefore be given adequate possibilities on the large telescopes.

It is also clear that in the sky–limited conditions good seeing is as important as a large telescope: for example a 4m telescope with 0.5 arcsec seeing is as good as an 8m telescope with 1.0 arcsec seeing (see Fig. 20). If you think about all the money and efforts that are spent on increasing the size of our telescopes, you would immediately agree that some of that money and effort should be spent in improving the image quality.

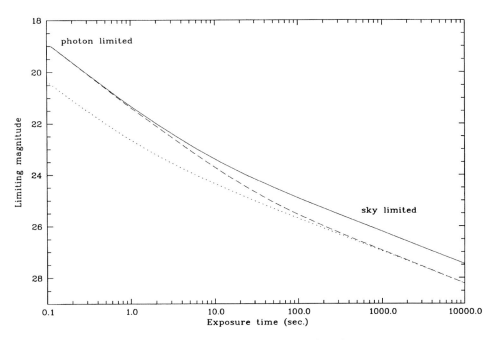

FIGURE 20. The limiting magnitudes to be reached (for S/N=5) as a function of exposure time for some seeing/telescope diameter combinations: the continuous line is for a telescope diameter D=4m and for a seeing FWHM $\theta = 1arcsec$, the dotted line is for D=8m and $\theta = 1arcsec$, and the dashed line is for D=4m and $\theta = 0.5$. The other parameters (see the formula in the text) are: E=0.4, $\Delta\lambda = 800\mathring{A}$, $r_n = 7e/pix$, $n = 9pix$, the sky background is $21.6m_V/arcsec^2$, applicable for broad band imaging. The plot can also be read with the exposure time in units of hours instead of seconds, for E=0.2, $\Delta\lambda = 0.5\mathring{A}$, as in medium resolution spectroscopy.

From the equations in this section you could gain the impression that a sufficiently long exposure time could compensate for a smaller telescope and that, given the rapid increase of total project cost with telescope size, this might even be a cost–effective

† The Very Large Telescope (VLT) is an array of 4 independent 8.2m telescopes, which is being constructed by ESO on Mount Paranal in Chile.

solution. However we all know that this is not true, mainly because there is a practical limit to the exposure time which is set by the density of cosmic ray signatures that one is prepared to accept and by the duration of the periods of best observing conditions (i.e. seeing, sky transparency and brightness, zenith distance). Many of us have experienced the fact that it is often worthless to combine exposures taken with different conditions: one is better off by taking just the data obtained with the best conditions and throwing away all the rest. Therefore there is a limit to compensating for a smaller telescope by increasing the exposure time, and some observational results can only be obtained with a large telescope.

7.2. *Polarimetric instruments for large telescopes*

In the previous section we have seen that the large increase in collecting area provided by the new generation of 8–10m telescopes will bring particularly large gains for polarimetry, which is a technique that requires a large number of detected photons to achieve an acceptable S/N ratio †. It is important to realize that these gains will not only allow to reach fainter sources, but also to improve the accuracy of the polarization measurements on relatively bright sources. Therefore it becomes possible to make meaningful polarization measurements also on sources that have a very low polarization level (say below 1%), thereby greatly increasing the number of objects good for polarimetry. In fact, given the large number of photons available and the very efficient polarimetry modules on the very large telescopes, it might well become common practice to measure all 4 Stokes parameters in *every* observation: in a few years we will all be used to polarimetry!

7.2.1. *Spectropolarimetry at the Keck 10m telescope*

The large gains expected with 8-10m telescopes for polarimetry have been met so far (end of 1995) by the only optical polarimeter in use on a large telescope, the Low Resolution Imaging Spectrometer (LRIS) on the Keck 10m telescope. LRIS is a spectrograph with imaging capabilities, which uses a blue and a red camera to cover the whole range from 3100Å to 10000Å (Oke et al. 1995). The blue camera has not been implemented yet and LRIS works so far only down to about 4000Å with a scale of 0.21 arcsec per pixel. Spectropolarimetry is possible by inserting a quarter–wave plate, and/or an half–wave plate, and a beamsplitter. The polarizing beamsplitter is a modified Glan–Taylor prism (see section 2.3.1), which uses a second total reflection of the ordinary ray to make it parallel to the extraordinary one, and additional optics to make the optical path equal for the two rays (Fig. 21). The wave plates and the beamsplitter are inserted just after the slit and cover a field of about 25 arcsec.

The rotatable half–wave and quarter–wave plates are used for measuring both linear and circular polarization. Imaging polarimetry over the small field of view is also possible by replacing the grating with a mirror. An example of spectropolarimetric observations obtained with LRIS is given in Figure 22 (Cimatti et al. 1996).

7.2.2. *Polarimetric capabilities of instruments for the ESO Very Large Telescope*

Of particular interest for European astronomers are the polarimetric capabilities of the instruments being built for the ESO Very Large Telescope (VLT), which is an array of four 8m telescopes under construction on Cerro Paranal in Chile. Most of the VLT

---

† In polarimetry the *absolute* error on the degree of polarization P is roughly equal to the *fractional* error on the total flux measurement of the source. Therefore the S/N ratio – i.e. the inverse of the fractional error – on the degree of polarization is much smaller (P times) than the S/N ratio on the total flux measurement, particularly for low polarization.

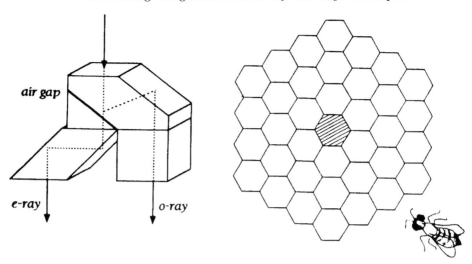

FIGURE 21. A schematic diagram of the modified Glan–Taylor beam splitter used for polarimetry in LRIS at the Keck 10m telescope (courtesy of R. Goodrich). The bee shown on the right, being able to see polarization, particularly enjoys the Keck telescope, which is the best polarimeter around.

instruments will have polarimetric capabilities in order to exploit the gains of large telescopes for polarimetry (see Table 1).

FORS, a focal reducer imager/spectrograph, is very similar to EFOSC (see section 5.1 and 6.1). It is designed to work at the Cassegrain focus between 330 and 1100 nm with spectroscopic resolution up to 2000, and will have both imaging– and spectro-polarimetry modes. These are provided by rotatable retarders and a Wollaston prism to be used in combination with filters or grisms and focal plane masks or slitlets. Given the large size of the optical beam, the retarder plates are built as mosaics of 3x3 plates. It is anticipated that the degree of linear polarization will be measured with an accuracy of 1% in one hour down to U, B, V and R magnitudes of 22–23 in imaging and down to V=17.3 in spectroscopy with 2.5Å resolution.

ISAAC, the IR imager/spectrograph for the Nasmyth focus, will work between 1 and 5μm with spectroscopic resolutions in the range 300–10,000. It will do imaging polarimetry using a fixed analyzer in one of the filter wheels, to be used in combination with filters and with rotation of the whole instrument. Although the original design of ISAAC foresees to use wire grid analyzers, the possibility of replacing them with a Wollaston prism is being considered.

Similarly, imaging polarimetry will be possible with CONICA, the high spatial resolution, near–IR (1–5μm) camera. It is designed to work at the Nasmith focus in combination with adaptive optics (AO) but the effects of the AO optical train on polarization measurements have yet to be carefully assessed. CONICA will have both wire–grid analyzers and Wollaston prisms to be used in combination with focal plane masks.

UVES is an echelle spectrograph for the Nasmyth with a spectroscopic resolution of 40,000. The possibility of doing spectropolarimetry with a polarization analyzer in the pre-slit optical train has been investigated, but is not in the present plan because of possible difficulties with the polarization induced by M3, the image slicer, the image derotator and the spectrograph. Nevertheless these problems are not insoluble and polarimetry with UVES would be useful, for example, to study the line polarization structure in AGN and emission nebulae, as well as Zeeman splitting.

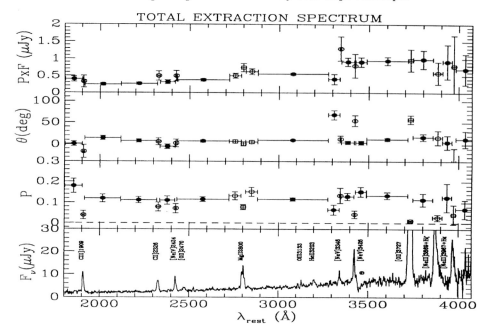

FIGURE 22. The results of spectropolarimetry of 3C 324 a radio galaxy at z=1.2, with $m_R = 21.7$, obtained with LRIS at the Keck 10m telescope from 4 exposures of 40 minutes at different position angle of the half–wave plate (courtesy of A. Cimatti).

The possibility of doing both imaging– and spectro–polarimetry with VISIR, the 8–24μm Cassegrain imager/spectrograph, has been studied and is feasible with a rotatable retarder and a wire–grid analyzer. However, the polarization components are not in the present design, although space has been reserved for them for later implementation. For distant radio galaxies the radiation emitted by stars, dust and non thermal nuclear processes in the near IR (K–band) is shifted into the VISIR range. The polarimetric capabilities of MIIS could be used to study interstellar polarization by transmission through aligned dust and scattering of nuclear radiation by electrons, whose cross–section does not depend on wavelength.

8. Polarimetry of AGN

Although this School is about instrumentation, and we all enjoy to deal with optics, detectors, electronics and other pieces of hardware, we should never forget that, in the astronomical world, instruments are not the goal, but are only the essential tools to reach the real goal, which is the understanding of the Universe. It is only by keeping this goal clear in mind that we can build the best instruments. Therefore I shall devote this last section to some astronomy that has been done with polarimetry, and I hope you will forgive me for selecting this application in a field where I have some knowledge and experience: research on Active Galactic Nuclei (AGN, see Blandford, Netzer & Woltjer, 1990, for a recent review), in particular, on Unified Models for AGN (Antonucci, 1993).

The paradigm of Unified Models is that AGN emit anisotropically: therefore their appearance depends on the viewing direction. On a single object we are restricted to a unique viewing direction and we do not immediately know how it would look like from another one. Therefore Barthel (1994) said that "Unification is a kind of religion: one

TABLE 1. Polarimetric capabilities of VLT instruments.

Instr.	λ range (μm)	$\Delta\lambda/\lambda$	Mode	Polarizers	Limiting mag. (σ_P=1%)
FORS	0.33–1.1		imag.	Woll., retard.	22–23 (U,B,V,R)
"	"	$\leq$2000	spectr.	" "	16–17 (1.3Å/pix)
ISAAC	1–5		imag.	fixed analyzer	K~18.5
CONICA	1–5		imag.	Woll., wire-grid	K=21
UVES*	0.33–1.1	90000	spectr.	?	
VISIR*	8–24		im+sp	wire-grid, retar.	

* Polarimetry is possible but not foreseen in the present design.

has to believe in things one cannot see". However polarimetry can help us to bring AGN Unification back to the realm of Science. The reason is that non–direct viewing directions to an AGN are offered through scattering, and scattered light, since it is strongly polarized, can be found and separated from other types of radiation by polarimetry (see section 3.2). Through scattering we can see anisotropic radiation even if it is not directed to us, as for example the beam of light emitted by a torch or a lighthouse, which is visible at night because it is scattered by particles in the air. The use of polarimetry to demonstrate and study anisotropic emission in AGN has been pioneered by Miller & Antonucci (1983), particularly for Seyfert galaxies. In the following I will concentrate on the use of polarimetry to understand the unification of powerful radio loud AGN.

8.1. *Alignment and polarization in distant radio galaxies*

Powerful radio loud AGN ($P_{radio} > 10^{25}WHz^{-1}$) are FRII radio galaxies (RG) and radio quasars (RQ). Their importance resides in the fact that they are the most luminous objects in the Universe and that they are hosted by the most distant galaxies (redshift up to about 4). Therefore they give information on the most energetic phenomena on one hand, and on the formation and early evolution of galaxies on the other. However, to obtain this information one has first to separate the active component from the non–active stellar one.

Early hopes that in RG, whose extended light is not swamped by the bright nucleus as in quasars, the active component would be negligible in the optical, allowing us to obtain information on the stellar populations directly from the spectra (e.g. Lilly & Longair, 1984), were frustrated by the discovery of the so called "alignment effect": the morphological structure of RG with $z \geq 0.7$, as observed in the optical range, is elongated and aligned with the radio axis. This effect is true both for the continuum, resulting in an excess in the UV rest–frame, and for the emission lines; moreover, in particular for the continuum, it does not occur at lower redshift (e.g. McCarthy et al., 1987). The link between the morphology of the extended structure and the nuclear activity, which is implied by the alignment effect, gives reasons to suspect that RG might have a non–stellar component, and to doubt that the stellar populations can be inferred directly from the spectra. Nevertheless the first explanation of the alignment effect attributed it to young stars whose formation is induced by the radio jet (e.g. De Young, 1989), and purely stellar models were then still used to explain the spectra of distant RG (Chambers & Charlot, 1990).

An alternative explanation of the alignment effect was suggested by observations of a highly ionized cloud along the radio axis of a nearby powerful RG, PKS2152-69 at z=0.0282: the continuum radiation associated with the cloud is very blue ($f_\nu \propto \nu^3$) and is strongly linearly polarized in the B band ($P_B = 12 \pm 3\%$) with the **E** vector perpendicular to the line joining the cloud to the nucleus (di Serego Alighieri et al. 1988). These observations could only be explained if the continuum in the cloud is nuclear radiation which is scattered by small dust particles at the cloud. Moreover the nuclear radiation as seen by the cloud must be much stronger than what we see directly, implying anisotropy. Tadhunter, Fosbury & di Serego Alighieri (1989) suggested that this mechanism of scattering of anisotropic nuclear radiation could occur also in distant RG: if the UV nuclear radiation were emitted into two opposite cones centred around the radio axis, dust scattering would make it visible also for those objects (RG) for which our direct line of sight is outside the cones, and would produce elongated and aligned structures, shifted in the optical by the redshift. Indeed PKS2152-69, which in the V and R band has a normal elliptical morphology with the major axis roughly perpendicular to the radio axis, becomes aligned to the radio axis in the U band, where the cloud is very prominent with respect to the stars.

The crucial test for the scattering explanation of the alignment effect was clearly to measure the optical polarization of distant radio galaxies, which however are very faint (V>20 for z>0.7). The measurements of the cloud in PKS2152-69 (B=20.6) made with the imaging polarimetry mode of EFOSC (see section 5.1) had demonstrated that polarimetry is possible on very faint objects and prompted the discovery that two distant RG (3C 277.2 at z=0.766 and 3C 368 at z=1.132) are strongly linearly polarized in the rest–frame UV, with the **E** vector perpendicular to the optical/radio axis, as expected by the scattering model (di Serego Alighieri et al. 1989). These observations unquestionably show that purely stellar models are inadequate for distant radio galaxies.

An important confirmation of these results was obtained for 3C 368 by Scarrott, Rolph & Tadhunter (1990), who also showed that the polarization was spatially extended for this RG, making it look like a very large reflection nebula. Further imaging polarimetry observations, summarized by Cimatti et al. (1993), confirmed that all powerful RG with $z > 0.7$ (about 10 have been observed so far) are perpendicularly polarized in the rest–frame UV ($P > 5\%$). At lower redshift the polarization observed in the V band is lower and not always perpendicular. It is clear that alignment and perpendicular polarization are associated phenomena in distant RG, since they both involve the rest–frame UV, in particular the extended parts, and they have the same redshift dependence. The most obvious explanation is that they are both due to scattering of anisotropic nuclear radiation, in agreement with the Unified Model. However, in order to have a safer and better understanding of the Unified Model of powerful radio loud AGN some questions have to be answered, which are discussed in the next section.

8.2. *Towards a better understanding of Unified Models of radio loud AGN*

In the previous section we have seen how imaging polarimetry of distant RG has demonstrated the presence of a non–stellar component in their rest–frame UV light, which is probably due to scattering of nuclear radiation hidden from direct view and emitted preferentially along the radio axis. Here we examine how polarimetry can improve our understanding of how a radio loud AGN looks like from different directions.

8.2.1. *What is the nuclear spectrum?*

Imaging polarimetry gives only very crude spectral information: therefore it does not tell much about the spectrum of the anisotropic nuclear light before it is scattered. This radiation could be dominated by a non thermal continuum, which is emitted anisotropically because the emitting particles are moving relativistically along the radio jet (relativistic beaming), as observed in blazars, or, like in quasars, could also have broad permitted emission lines, which are seen only in certain directions because of obscuration by optically thick material. The answer to this question is crucial to decide whether Unification is between RG and blazars or between RG and RQ and then to put constraints on the solid angle of the anisotropic radiation from the statistics of the "parent populations" and from the physical mechanism producing the anisotropy.

The key to this problem is to observe broad permitted lines in the polarized spectrum of distant RG, and must be tackled with spectropolarimetry. Broad polarized MgII2800 emission has been observed in a few RG with $z \sim 0.8$ using 4m class telescopes (di Serego Alighieri et al. 1994, 1996). However these observations are at the limit of what can be done with 4m telescopes, and are now made with a much larger accuracy with the 10m Keck telescope (Dey et al. 1996, Cimatti et al. 1996). They demonstrate that distant RG harbour an hidden quasar in their nuclei. If one can make some assumptions on the spectral shape of the intrinsic polarization of the scattered component, it is possible to obtain the spectrum of the hidden nucleus from the polarized light spectrum, i.e. the product of the degree of polarization times the total light spectrum (see section 6 and Figure 22).

8.2.2. *Where is the hidden nuclear source?*

Since in RG the nuclear source is probably hidden from direct view, it is often difficult to locate the nucleus on optical images. Also in this case polarimetry offers the solution: in case of scattering, if the scattered source is point like and if it emits unpolarized light, the pattern of **E** vectors in a polarization map defines the position of the source. In fact the source must be in the plane perpendicular to the **E** vector at every location where scattered light is detected. It is therefore sufficient to measure the direction of the **E** vector in two positions to locate the central source. If one has measurements at more than two positions, as is normally the case, one can use this redundancy to assess the accuracy in the determination of the source position and the likelihood that the central source is indeed point like and intrinsically unpolarized. Examples of how this technique might be applied to distant RG are given by Scarrott et al. (1990) and by Cohen et al. (1996).

8.2.3. *What is the scattering medium?*

Scattering around RG can be due to electrons or to dust particles. Polarimetry helps in the determination of the scattering medium in several ways. First, the early imaging polarimetry observations of 3C 368 by di Serego Alighieri et al. (1989) were done both in the V and in the R band. The observed polarized flux density is:

$$f_\nu^{pol}(2650\text{Å}) \sim 2 f_\nu^{pol}(3200\text{Å}),$$

where the observed wavelengths are given in the rest frame. If scattering is due to electrons, which have a wavelength independent cross section (see section 3.2.1), then the incident spectrum must have the same slope as the scattered one:

$$f_\nu^{inc} \propto \nu^{2.5},$$

which is much bluer than ever observed directly in RQ in this wavelength range. There-fore it is more likely that scattering is due to dust, which makes the incident radiation bluer.

Second, if scattering is due to electrons, the thermal motion of the electrons would pro-duce a smearing of the input spectrum (see section 3.2.1). Since the scattered MgII2800 emission line observed in distant radio galaxies has the same width as the lines observed directly in quasars, scattering cannot be produced by electrons hotter than about 10^5K (di Serego Alighieri et al., 1994), as would be those in the X–ray halos around cD galaxies in clusters.

Third, the scattering efficiency per unit mass is much higher for dust particles, than for electrons in a gas. Therefore dust scattering dominates in an interstellar medium which has a gas–to–dust ratio similar to that which we find in our Galaxy. For example, in order to explain the scattered light observed in distant RG at very large distances from the nucleus, the required dust mass to do the scattering is $M_{dust} \sim 3 \times 10^8 M_\odot$, which is a plausible amount of gas for a galaxy, while the gas mass for electron scattering is $M_{gas} \sim 2 \times 10^{12} M_\odot$, which is larger than the *total* mass of most galaxies (di Serego Alighieri et al., 1994).

However, when scattering occurs close to the nucleus, where the dust is destroyed by the strong radiation field and gas densities are high, then electron scattering can dominate, as in some Seyfert 2 galaxies. In most situations the interstellar medium is a mixture of gas and dust and both contribute to scattering. A discussion of dust scattering in radio galaxies is given by Manzini & di Serego Alighieri (1996).

8.2.4. *How much dilution?*

We have seen in Section 6 that the determination of the amount of dilution requires the knowledge of the intrinsic polarization of the polarized component, which is rarely available. However changes in the observed polarization across spectral features are a clear sign of the presence of dilution. For example the drop in the degree of the observed polarization, which occurs in some RG at 4000Å, shows the presence of diluting radiation with a strong 4000Å break, like the light from an evolved stellar population (di Serego Alighieri et al., 1994).

Furthermore, if the observed spectrum is the sum of a polarized and of an unpolarized component ($I_{tot} = I_p + I_u$, see Fig. 14), and if the polarized component has a spectral line which is not present in the unpolarized one, then the line is seen also in the polarization spectrum ($P_{obs}(\lambda)$). In addition, if the intrinsic polarization of the polarized component is constant across the line, then the relative amount of dilution can be estimated from the ratio of the line equivalent width in the total and in the polarized spectra:

$$\frac{I_u}{I_{tot}} = 1 - \frac{EW_{tot}}{EW_{pol}}.$$

An attempt to constrain the amount of young stars in distant RG using the polarization of the broad MgII2800 line has been made by di Serego Alighieri et al. (1994). However more accurate data are necessary to solve this problem.

8.2.5. *Are narrow lines isotropically emitted?*

The conventional wisdom is that narrow forbidden lines in AGN, since they are emitted in low density gas, must come from regions far from the nucleus, and therefore must be isotropic, since they cannot be obscured at those distances. If this is true, then the luminosity of the narrow lines must be the same in RG as in RQ, for Unification to hold. However Jackson & Browne (1990) have found that the [OIII]5007 line luminosity is a factor of 5–10 higher in RQ than in RG. The conclusion is that either RG and RQ cannot be unified, or the [OIII]5007 line is emitted anisotropically, at least partially. This second possibility, which is not so unplausible since the critical density of O^{++} is high enough for it to be partially within the obscuring torus, can be tested with spectropolarimetry, since an anisotropic line would be partially polarized in RG. With Andrea Cimatti, Bob Fosbury and Ronald Hes we have observed in spectropolarimetry 6 RG and 3 RQ. The preliminary results are that the [OIII]5007 line is polarized in 5 out of 6 RG, indicating anisotropic emission, while it is not polarized in the 3 RQ (di Serego Alighieri, 1996).

8.2.6. *Why does the polarization depend on redshift?*

We have seen in section 8.1 that the perpendicular polarization observed in the V–band in RG depends strongly on redshift in the sense that it sets in only for $z \leq 0.7$, similarly to the alignment effect. This could be simply due to the the fact that at higher redshift the measurements are made in the rest-frame UV and that the observed polarization is stronger in the UV, because of dilution by red stars. In fact the redshift threshold occurs where the V-band starts sampling the near UV below the 4000Å break of the diluting stellar light. In this case low redshift radio galaxies should also be aligned and polarized in the UV. This is indeed true in some cases, like for 3C 195 (z=0.11, Cimatti & di Serego Alighieri, 1995), but is not such a universal property as at higher redshift. Therefore there could be some evolution in the dust properties or polarization could depend on radio power (nearby objects are biased to lower power).

8.3. *A test of the Einstein Equivalence Principle*

I would like to end this course by giving you an example of how polarimetry of celestial sources can be used to tests the foundations of physics. In fact, quite surprisingly, the polarization measurements of distant RG described in the previous sections provide a test of the Einstein equivalence principle (EEP), on which all metric theories of gravitation are based. The equivalence principle comes in three forms: 1. The weak EP (WEP) stating the equivalence of a gravitational field and of a uniformly accelerated frame, as far as the motion of freely falling bodies is concerned; it implies that the gravitational mass is equal to the inertial mass. 2. The EEP extends the equivalence to all experiments involving non-gravitational forces. 3. The strong EP (SEP) extends the EEP also to gravitational experiments.

The WEP is tested with accuracy of about 10^{-12} by Eötvös–type experiments (Braginsky & Panov, 1972). On the other hand the EEP is tested by gravitational redshift experiments to an accuracy of only about 10^{-4}. Schiff (1960) has conjectured that any consistent Lorentz–invariant theory of gravity which obeys the WEP would necessarily obey the EEP. If true, this conjecture would imply that the EEP is automatically tested to the same precision as the WEP, thereby greatly increasing our experimental confidence in all metric theories of gravity. However Ni (1977) has found a unique counter–example to Schiff's conjecture: a pseudoscalar field Φ that couples to electromagnetism in a Lagrangian of the form:

$$L = -\frac{1}{16\pi}\sqrt{g}\Phi e^{\mu\nu\rho\sigma}F_{\mu\nu}F_{\rho\sigma} \tag{8.1}$$

leading to violation of the EEP, while obeying the WEP.

Carroll & Field (1991) have shown that, if a coupling such as (8.1) were significant in a cosmological context, then the plane of polarization of light coming from very distant objects would be rotated during its journey across a considerable fraction of the size of the Universe. If we could show that such a rotation is not observed, we could then put limits on the parameter Φ and conclude that the EEP is not violated in this unique fashion.

Carroll, Field & Jackiw (1990) have checked that this rotation is smaller than $10°$ for a sample of radio sources with redshifts between 0.4 and 1.5, using the fact that the distribution of position angle differences between the radio polarization and the radio axis peaks around $90°$ with a smaller peak around $0°$, without sign of a redshift dependence. However this is a statistical test, which relies on the correction of the observed polarization angle for Faraday rotation, and is not based on a physical understanding of the phenomenon.

On the other hand Cimatti et al. (1994) have shown that the perpendicularity between the optical polarization angle and the optical/radio axis holds within $10°$ for every single radio galaxy with a polarization measurement, and with redshift larger than 0.5 and up to 2.6. This perpendicularity is precisely predicted by the scattering model and does not relay on the correction for Faraday rotation, which is negligible at optical wavelengths. Therefore these polarization measurements provide a very stringent test of the EEP.

Acknowledgements

Many people have contributed to these lessons and to the success of the course. In particular I would like to thank José Miguel Rodríguez Espinosa for the organization of the School and for giving me an opportunity to teach on a topic that I like, Begoña López and Nieves Villoslada for the kind support during the School, my colleague teachers Jacques Beckers, Michael Irwin, David Gray, Barbara Jones, Ian McLean, Richard Puetter, and Keith Taylor for pleasant and instructive conversations, Andrea Cimatti, Bob Fosbury, Egidio Landi Degli Innocenti and Jeremy Walsh for many useful discussions and for their comments on the manuscript, and Simone Esposito for lending me a couple of polarizers.

REFERENCES

ANTONUCCI, R. 1993, *Ann. Rev. Astron. Astrophys.* **31**, 473.

BARTHEL, P.D. 1994, in *The Physics of Active Galaxies*, ed. by G.V. Bicknell et al., ASP Conference Series, Vol. 54, p. 175.

BLANDFORD, R.D., NETZER, H., & WOLTJER, L. 1990 *Active Galactic Nuclei*, Springer Verlag.

BRAGINSKY, V.B., & PANOV, V.I. 1972 *Sov. Phys. JETP.* **34**, 463.

CARROLL, S.M. & FIELD, G.B. 1991 *Phis. Rev. D* **43**, 3789.

CARROLL, S.M., FIELD, G.B., & JACKIW, R. 1990 *Phis. Rev. D* **41**, 1231.

CHAMBERS, K.C., & CHARLOT, S. 1990 *Ap. J. Lett.* **348**, L1.

CIMATTI, A., DI SEREGO ALIGHIERI, S., FOSBURY, R.A.E., SALVATI, M., & TAYLOR, D. 1993 *MNRAS* **264**, 421.

CIMATTI, A., DI SEREGO ALIGHIERI, S., FIELD, G.B., & FOSBURY, R.A.E. 1994 *Ap. J.* **422**, 562.

CIMATTI, A., & DI SEREGO ALIGHIERI, S. 1995 *MNRAS* **273**, 7p.

CIMATTI, A., DEY, A., VAN BREUGEL, W., ANTONUCCI, R., & SPINRAD, H. 1996 *Ap. J.*, in press.

CLARKE, D., & STEWART, B.G. 1986 *Vistas in Astronomy* **29**, 27.

COHEN, M.H., TRAN, H.D., OGLE, P.M., & GOODRICH, R.W. 1996, in *Extragalactic Radio Sources*, ed. by Fanti et al., Kluwer Academic Publ., in press.

DE YOUNG, D.S. 1989 *Ap. J. Lett.* **342**, L59.

DEY, A., CIMATTI, A., VAN BREUGEL, W., ANTONUCCI, R., & SPINRAD, H. 1996 *Ap. J.*, in press.

DI SEREGO ALIGHIERI, S. 1996, in *Extragalactic Radio Sources*, ed. by Fanti et al., Kluwer Academic Publ., in press.

DI SEREGO ALIGHIERI, S., BINETTE, L., COURVOISIER, T.J.-L., FOSBURY, R.A.E., & TADHUNTER, C.N. 1988 *Nature* **334**, 591.

DI SEREGO ALIGHIERI, S., FOSBURY, R.A.E., QUINN, P.J., & TADHUNTER, C.N. 1989 *Nature* **341**, 307.

DI SEREGO ALIGHIERI, S., CIMATTI, A., & FOSBURY, R.A.E. 1994 *Ap. J.* **431**, 123.

DI SEREGO ALIGHIERI, S., CIMATTI, A., FOSBURY, R.A.E., & PEREZ-FOURNON, I. 1996 *MNRAS*, in press.

DIRAC. P.A.M. 1958 *The Principles of Quantum Mechanics*, Clarendon Press.

FOSBURY, R.A.E., CIMATTI, A., & DI SEREGO ALIGHIERI, S. 1993 *The Messenger* **47**, 11.

JACKSON, N., & BROWNE, I.W.A. 1990 *Nature* **343**, 43.

KITCHIN, C.R. 1984 *Astrophysical Techniques*, Adam Hilger.

KLIGER, D.S., LEWIS, J.W., & RANDALL, C.E. 1990 *Polarized Light in Optics and Spectroscopy*, Academic Press.

LANDI DEGL'INNOCENTI, E. 1992, in *Solar Observations: Techniques and Interpretation*, ed. by F. Sanchez et al., Cambridge University Press, p. 71.

LILLY, S.J., & LONGAIR, M. 1984 *MNRAS* **211**, 833.

MANZINI, A., & DI SEREGO ALIGHIERI, S. 1996 *A & A*, in press.

MCCARTHY, P.J. 1993 *ARAA* **31**, 639.

MCCARTHY, P.J., VAN BREUGEL, W.J.M., SPINRAD, H., & DJORGOWSKI, S. 1987 *Ap. J. Lett.* **321**, L29.

MELNICK, J., DEKKER, H., & D'ODORICO, S. 1989 EFOSC (ESO Faint Object Spectrograph and Camera), ESO Operating Manual No. 4.

MILLER, J.S., & ANTONUCCI, R.R.J. 1983 *Ap. J. Lett.* **271**, L7.

MINNAERT, M. 1954 *Light and Colour in the Open Air*, Dover.

NI, W.T. 1977 *Phys. Rev. Lett.* **38**, 301.

OKE, J.B., COHEN, J.G., CARR, M., CROMER, J., DINGIZIAN, A., HARRIS, F.H., LUCINIO, R., SCHAAL, W., EPPS, H., & MILLER, J. 1995 *P.A.S.P.* **107**, 375.

RYBICKI, G.B., & LIGHTMAN, A.P. 1979 *Radiative Processes in Astrophysics*, John Wiley and Sons.

SCARROTT, S.M., ROLPH, C.D., & TADHUNTER, C.N. 1990 *MNRAS* **243**, 5p.

SCHIFF, L.I. 1960 *Am. J. Phys.* **28**, 340.

SHURCLIFF, W.A. 1955 *J. Opt. Soc. Amer.* **45**, 399.

SERKOWSKI, K., MATHEWSON, D.S., & FORD, V.L. 1975 *Ap. J.* **196**, 261.

TADHUNTER, C.N., FOSBURY, R.A.E., & DI SEREGO ALIGHIERI, S. 1989 *BL Lac Objects: 10 years after* ed. by L. Maraschi et al., Springer, p. 79.

VAN DE HULST, H.C. 1957 *Light Scattering by Small Particles*, John Wiley & Sons, Inc.

WALSH, J.R. 1992, in *4th ESO/ST-ECF Data Analysis Workshop* ed. by P.J. Grosbøl & R.C.E. de Ruijsscher, ESO Conf. and Workshop Proc. No. 41, p. 53.

WHITE, R.L. 1979 *Ap.J.* **229**, 954.

WHITTET, D.C.B. 1992 *Dust in the Galactic Environment*, Inst. of Phys. Publ., Bristol.

WILKING, B.A., LEBOFSKY, M.J., & RIEKE, G.H. 1982 *A.J.* **87**, 695.